IOP Series in Coherent Sources, Quantum Fundamentals, and Applications

About the Editor

F J Duarte is a laser physicist based in Western New York, USA. His career has covered three continents while contributing within the academic, industrial, and defense sectors. Duarte is editor/author of 15 laser optics books and sole author of three books: *Tunable Laser Optics, Quantum Optics for Engineers*, and *Fundamentals of Quantum Entanglement*. Duarte has made original contributions in the fields of coherent imaging, directed energy, high-power tunable lasers, laser metrology, liquid and solid-state organic gain media, narrow-linewidth tunable laser oscillators, organic semiconductor coherent emission, N-slit quantum interferometry, polarization rotation, quantum entanglement, and space-to-space secure interferometric communications. He is also the author of the generalized multiple-prism grating dispersion theory and pioneered the use of Dirac's quantum notation in N-slit interferometry and classical optics. His contributions have found applications in numerous fields, including astronomical instrumentation, dispersive optics, femtosecond laser microscopy, geodesics, gravitational lensing, heat transfer, laser isotope separation, laser medicine, laser pulse compression, laser spectroscopy, mathematical transforms, nonlinear optics, polarization optics, and tunable diode-laser design. Duarte was elected Fellow of the Australian Institute of Physics in 1987 and Fellow of the Optical Society of America in 1993. He has received various recognitions, including the *Paul F Foreman Engineering Excellence Award* and the *David Richardson Medal* from the Optical Society.

Coherent Sources, Quantum Fundamentals, and Applications

Since its discovery the laser has found innumerable applications from astronomy to zoology. Subsequently, we have also become familiar with additional sources of coherent radiation such as the free electron laser, optical parametric oscillators, and coherent interferometric emitters. The aim of this book Series in Coherent Sources, Quantum Fundamentals, and Applications is to explore and explain the physics and technology of widely applied sources of coherent radiation and to match them with utilitarian and cutting-edge scientific applications. Coherent sources of interest are those that offer advantages in particular emission characteristics areas such as broad tunability, high spectral coherence, high energy, or high power. An additional area of inclusion are the coherent sources capable of high performance in the miniaturized realm. Understanding of quantum fundamentals can lead to new and better coherent sources and unimagined scientific and technological applications. Application areas of interest include the industrial, commercial, and medical sectors. Also, particular attention is given to scientific applications with a bright future such as coherent spectroscopy, astronomy, biophotonics, space communications, space interferometry, quantum entanglement, and quantum interference.

A full list of titles published in this series can be found here: https://iopscience.iop.org/bookListInfo/series-in-coherent-sources-and-applications.

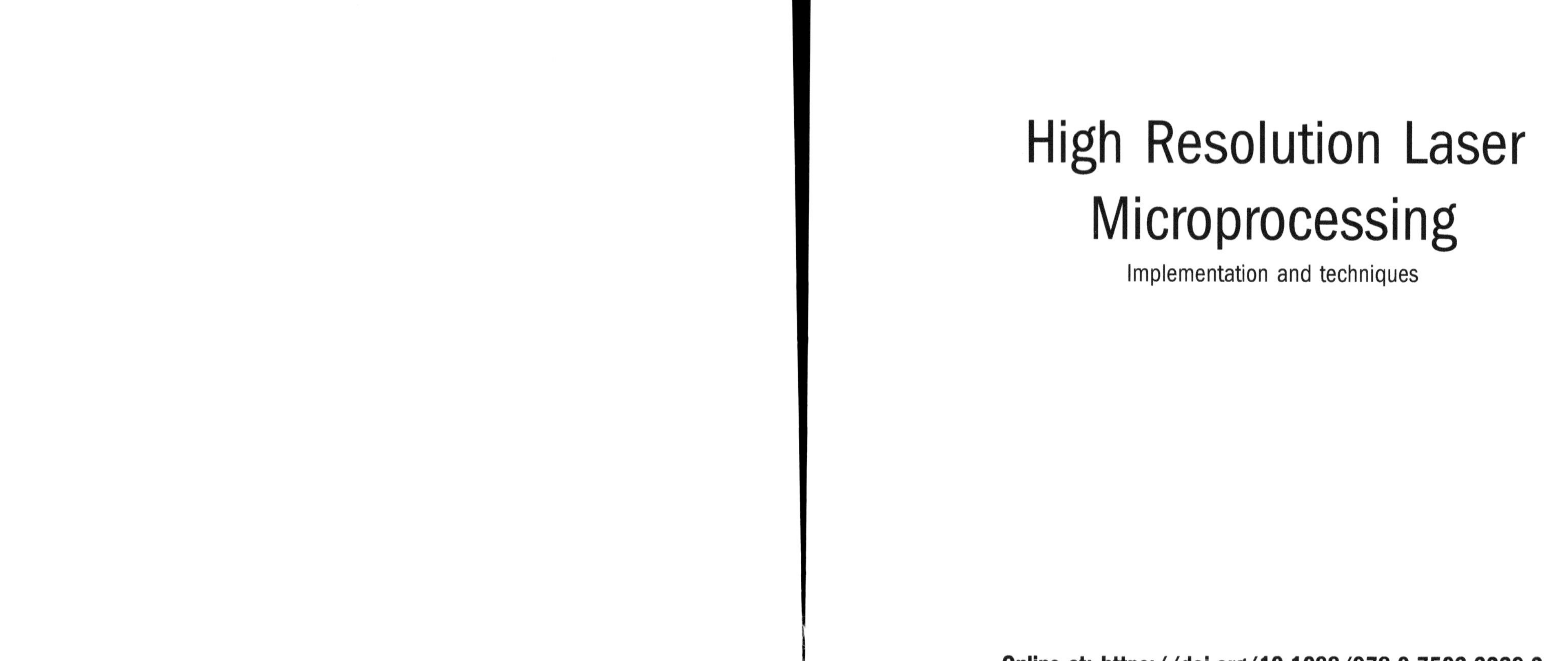

High Resolution Laser Microprocessing

Implementation and techniques

Online at: https://doi.org/10.1088/978-0-7503-3239-2

High Resolution Laser Microprocessing

Implementation and techniques

Bogdan Ştefăniţă Călin
Center for Advanced Laser Technologies, National Institute for Laser, Plasma and Radiation Physics, Magurele, Romania

Marian Zamfirescu
Center for Advanced Laser Technologies, National Institute for Laser, Plasma and Radiation Physics, Magurele, Romania

Niculae Puşcaş
Department of Physics, National University of Science and Technology Politehnica Bucharest

IOP Publishing, Bristol, UK

ISBN   978-0-7503-3239-2 (ebook)
ISBN   978-0-7503-3237-8 (print)
ISBN   978-0-7503-3240-8 (myPrint)
ISBN   978-0-7503-3238-5 (mobi)

DOI   10.1088/978-0-7503-3239-2

Version: 20251101

IOP ebooks

British Library Cataloguing-in-Publication Data: A catalogue record for this book is available from the British Library.

Published by IOP Publishing, wholly owned by The Institute of Physics, London

IOP Publishing, No.2 The Distillery, Glassfields, Avon Street, Bristol, BS2 0GR, UK

US Office: IOP Publishing, Inc., 190 North Independence Mall West, Suite 601, Philadelphia, PA 19106, USA

Contents

Preface

This book has been the subject of many hours of work, debate, writing, discussing, checking, and rewriting yet again. Information is available and easy to access almost everywhere in the world, and has been so for quite a long time. That is why it is increasingly difficult to write anything technical, be it a book, a chapter, or even an opinion article about something that is already more or less well known. There are many books on optics and lasers, well written and exhaustive on particular subjects. Their content has already been filtered through the minds of readers and authors, being improved over time.

But even so, in our daily work-related responsibilities, we still encountered situations where we wanted to discuss something specific with our colleagues, and when we wanted to reference something for further details, we sometimes hit a wall. Things that many people learn or teach by hands-on methods, things that professionals may consider obvious or too simple to write down, and things that could be more quickly understood with just a bit of insight. And for those things, we tried compiling the information contained in this book. This book is not an exhaustive document on optics or lasers, far from it. It is not a complete undergraduate manual, it is not a handbook. However, we tried our best to make it a useful document. We touched on some of the most important concepts, often times using a more practical approach. We also added a series of complete Python scripts that were used to generate more complex plots found in this text, in the hope that readers can use these scripts to either explore the notions described in them, or use them for further analysis and insight.

The book starts with a discussion on the nature of light itself. Specifically, we quickly but clearly describe the route from an electromagnetic wave to ray optics, as well as describing the Lorentz model for dielectrics, with the purpose of reaching the notion of TART—Transmission, Absorption, Reflection, Transmission. We further move into discussing laser physics, starting from the base concepts, discussing practical aspects such as beam measurement and characterization, and moving towards more contemporary methods, issues and hot topics. In the following chapter, we return to a more simple but practical approach, where we show step-by-step descriptions and guides, for example, on how to align a laser beam. In the end we discuss emergent microprocessing technologies that are rising quickly in popularity and have already been used to make significant scientific advances.

Author biograpies

Bogdan Ştefăniţă Călin

Bogdan Ştefăniţă Călin is a junior scientific researcher at the Center for Advanced Laser Technologies from the National Institute for Laser, Plasma and Radiation Physics in Măgurele, Bucharest, Romania. He received an engineering degree in physical engineering, a master's degree in lasers and particle accelerators engineering, and a PhD in Physics, from University Politehnica of Bucharest, Romania. His current field is high resolution laser processing, micro-scale 3D printing and their applications in optics, particle acceleration and tissue engineering.

Marian Zamfirescu

Marian Zamfirescu is a senior scientific researcher at the Center for Advanced Laser Technologies from the National Institute for Laser, Plasma and Radiation Physics in Măgurele, Bucharest, Romania. He received an engineering degree in physical engineering and a master's degree in optics and nonconventional techniques using lasers and plasma, from University of Bucharest, Romania, and a PhD in physics from Blaise-Pascal University from Clermont-Ferrand, France. His scientific research activity is currently dealing with laser spectroscopy and laser micro- and nano-structuring using ultrafast pulsed lasers.

Niculae Puşcaş

Niculae Puşcaş is an emeritus professor at the Department of Physics – University Politehnica of Bucharest, Romania. He received a bachelor's degree in physics from University Babeş – Bolyai of Cluj-Napoca, Romania, and a PhD in physics from University of Bucharest. His teaching and research activities include courses for laser physics and engineering, integrated optics and photonics, optical materials, and 37 research projects.

IOP Publishing

High Resolution Laser Microprocessing
Implementation and techniques
Bogdan Ştefăniţă Călin, Marian Zamfirescu and Niculae Puşcaş

Chapter 1

Optical physics for laser materials processing

One of the best ways to develop a strong intuition regarding optical systems and devices is to clearly study and analyze the propagation of light at different scales. As such, this chapter provides a theoretical background to support a better understanding of the topics discussed further. This chapter contains fundamental theoretical derivation, all while providing arguments supporting its importance and the circumstances of its utility. The scope of this book is more inclined to developing practical skills in the laboratory, and therefore the theoretical derivations are kept to the fundamentals. Their purpose is to act as either a proper introduction, where all steps of derivations are shown and explained, or as a reminder for those who already have a good background in physics of light. We are, however, providing references towards a more detailed approach, whenever necessary.

The first subsection provides an introductory electromagnetic optics background. It follows a different progression than often encountered in other texts, relating to the topic of light propagation and its mathematical description. Although it is a fundamental approach, we consider this to be of significant importance when talking about high-resolution laser processing. It's not so much about retaining information, such as rules, formulas, values and other similar things, but more about understanding a mechanism and deriving required information when needed, rather than just recalling it. In the case of laser processing in particular, we consider that understanding the propagation of light through a system and its interaction with matter along the way (lenses, mirrors, samples, etc) helps the experimentalist develop their intuition and design better experimental setups.

1.1 Electromagnetic theory of light propagation

One usual method of describing light propagation, through a classical approach, is using Maxwell's equations. It's probably the most encountered method, we might add, and for good reason. This set of equations can be used to derive, most importantly, the electromagnetic wave propagation, the material response to such waves, and the

doi:10.1088/978-0-7503-3239-2ch1

connections between the two. We start by describing wave propagation. Using appropriate approximations, we can obtain several equations that are applicable under specific conditions: the Helmholtz equation, the paraxial approximation, and the eikonal equation. We will later discuss why those approximations are useful, where and when to use them. However, let's first recap the mathematical description. According to [1], Maxwell's equations in differential form in free space are as follows:

$$\nabla \cdot \mathbf{E} = \frac{\rho}{\varepsilon_0} \tag{1.1a}$$

$$\nabla \cdot \mathbf{B} = 0 \tag{1.1b}$$

$$\nabla \times \mathbf{E} = -\frac{\partial \mathbf{B}}{\partial t} \tag{1.1c}$$

$$\nabla \times \frac{\mathbf{B}}{\mu_0} = \varepsilon_0 \frac{\partial \mathbf{E}}{\partial t} + \mathbf{J} \tag{1.1d}$$

where $\mathbf{E}$ is the electric field, $\mathbf{B}$ is the magnetic field, $\mathbf{J}$ is the current density, ρ is the charge density, and ε_0 and μ_0 are the vacuum electric permittivity and magnetic permeability, respectively. In order to derive the wave equation from equation (1.1), we can take the curl of equation (1.1c):

$$\nabla \times (\nabla \times \mathbf{E}) = -\frac{\partial}{\partial t}(\nabla \times \mathbf{B}) \tag{1.2}$$

Next we use equation (1.1d) to replace $\nabla \times \mathbf{B}$. Using the vector identity:

$$\begin{aligned} \nabla \times (\nabla \times \mathbf{E}) &= \nabla (\nabla \cdot \mathbf{E}) - \nabla^2 \mathbf{E} \\ &\text{and} \\ \nabla^2 \mathbf{E} &= \nabla \cdot (\nabla \mathbf{E}) \end{aligned} \tag{1.3}$$

to expand the left-hand side term, therefore obtaining:

$$\nabla(\nabla \cdot \mathbf{E}) - (\nabla \cdot \nabla)\mathbf{E} = -\frac{\partial}{\partial t}\left(\mu_0 \varepsilon_0 \frac{\partial \mathbf{E}}{\partial t} + \mu_0 \mathbf{J}\right) \tag{1.4}$$

$$\nabla^2 \mathbf{E} - \frac{1}{c^2}\frac{\partial^2 \mathbf{E}}{\partial t^2} = \mu_0 \frac{\partial \mathbf{J}}{\partial t} + \frac{\nabla \rho}{\varepsilon_0} \tag{1.5}$$

which can be further simplified if we consider propagation through vacuum, i.e., no free charges, $\rho = 0$, $\mathbf{J} = 0$, resulting in:

$$\nabla^2 \mathbf{E} - \frac{1}{c^2}\frac{\partial^2 \mathbf{E}}{\partial t^2} = 0 \tag{1.6}$$

where $c = \frac{1}{\sqrt{\mu_0 \varepsilon_0}}$ is the speed of light in vacuum. This is the simplest form of the wave equation as we consider propagation through vacuum (homogeneous form).

Important phenomena for laser–matter interaction arise from introducing material parameters through the constitutive relations. Before we get there, however, let us first discuss different forms of the wave equation and their applicability.

The wave equation is one of the most utilized equations for practical applications. It can describe wave propagation under any conditions, provided there are definable boundary conditions. It can also be used to derive other equations that describe time-dependent electromagnetic phenomena. However, while it offers a considerable amount of information, it can become difficult to solve numerically, especially for more complex problems and over a large space. Depending on the application, we might not need the time-dependence information. For such situations, we can use the Helmholtz equation.

In order to derive the Helmholtz equation, we consider a plane wave solution as follows:

$$u(\mathbf{r}, t) = A\mathrm{e}^{\mathrm{i}(\omega t - \mathbf{kr} + \phi_0)}$$
$$\Leftrightarrow u(\mathbf{r}, t) = A\mathrm{e}^{-\mathrm{i}\mathbf{kr}}\mathrm{e}^{\mathrm{i}(\omega t + \phi_0)} \tag{1.7}$$
$$\Leftrightarrow u(\mathbf{r}, t) = A(\mathbf{r})T(t)$$

where $\mathbf{r}$ is the position vector, t is time, ω is the angular frequency, $\mathbf{k}$ is the wavevector, and ϕ_0 is the initial phase. We insert this solution into equation (1.5), and obtain:

$$\nabla^2[A(\mathbf{r})T(t)] = \frac{1}{c^2}\frac{\partial^2}{\partial t^2}[A(\mathbf{r})T(t)]$$
$$\Leftrightarrow \nabla^2 A(\mathbf{r})T(t) = \frac{1}{c^2}A(\mathbf{r})\frac{\partial^2 T(t)}{\partial t^2} \tag{1.8}$$
$$\Leftrightarrow \frac{\nabla^2 A(\mathbf{r})}{A(\mathbf{r})} = \frac{1}{c^2 T(t)}\frac{\partial^2 T(t)}{\partial t^2}$$

Since both sides of equation (1.7) are independent, the equation has solutions only if both sides are equal to a constant, chosen as $-k^2$ for mathematical reasons (in this form, k turns out to be the wave vector). We, therefore, obtain:

$$\frac{\nabla^2 A(\mathbf{r})}{A(\mathbf{r})} = -k^2 \tag{1.9}$$
$$\nabla^2 A(\mathbf{r}) + k^2 A(\mathbf{r}) = 0$$

where $k = \frac{2\pi}{\lambda}$ is the angular wavenumber. A similar process can be followed to obtain the equation that contains only the time dependence. Equation (1.9) is known as the homogeneous Helmholtz equation. Notice that equation (1.9) is time-independent, which is the primary reason of its applications, since it reduces the complexity of a wave problem significantly. It's worth noting that we've used the term 'wave problem', i.e., any wave, be it an electromagnetic, acoustic or even matter wave (atemporal Schrödinger's equation), albeit under certain conditions. In other words, most situations where we deal with steady-state oscillations can be analyzed using the Helmholtz equation.

For example, let's look at Schrödinger's equation. With a few simple substitutions we can transform this equation to be mathematically identical to the Helmholtz equation:

$$\frac{-\hbar^2}{2m} \nabla^2 \Psi + U\Psi = E\Psi \qquad \left| \cdot \left(-\frac{2m}{\hbar^2}\right)\right.$$

$$\Leftrightarrow \nabla^2 \Psi - \frac{2mU}{\hbar^2}\Psi = -\frac{2mE}{\hbar^2}\Psi$$

$$\Leftrightarrow \nabla^2 \Psi + k^2\Psi = F\Psi \tag{1.10}$$

where $k = \frac{\sqrt{2mE}}{\hbar}$, $F = \frac{2mU}{\hbar^2}$, $\hbar$ is the reduced Planck's constant, m is the mass of the particle, U is the potential, E is the kinetic energy, and Ψ is the **wave** function. The equation itself is the same, as it should be, since it describes the propagation of a wave, albeit a wave of matter instead of electromagnetic waves.

1.2 Paraxial approximation

When working with laser radiation, especially beam delivery systems, we use **Gaussian optics**. This type of optics describes the propagation of optical waves in **paraxial** (or small-angle) approximation, i.e., waves where rays only make small angles with respect to the optical axis. Due to these small angles, we can eliminate some terms of the equation since their influence on the results is negligible, as we'll see below. We can use this approximation to more easily describe complex optical systems that contain smooth surfaces (i.e., mirrors and lenses).

To obtain the paraxial form, we first consider a plane wave propagating in the Z direction:

$$A(\mathbf{r}) = u(\mathbf{r})e^{ikz} \tag{1.11}$$

For the following section we will note $u(\mathbf{r}) = \mathbf{u}$. Inserting equation (1.11) into equation (1.9) will yield the following:

$$\nabla^2(\mathbf{u}e^{ikz}) + k^2\mathbf{u}e^{ikz} = 0 \tag{1.12a}$$

$$\left(\frac{\partial^2}{\partial x^2} + \frac{\partial^2}{\partial y^2} + \frac{\partial^2}{\partial z^2}\right)\mathbf{u}e^{ikz} + k^2\mathbf{u}e^{ikz} = 0 \tag{1.12b}$$

$$\frac{\partial^2\mathbf{u}}{\partial x^2}e^{ikz} + \frac{\partial^2\mathbf{u}}{\partial y^2}e^{ikz} + \frac{\partial}{\partial z}\left[\frac{\partial}{\partial z}(\mathbf{u}e^{ikz})\right] + k^2\mathbf{u}e^{ikz} = 0 \tag{1.12c}$$

$$\frac{\partial^2\mathbf{u}}{\partial x^2}e^{ikz} + \frac{\partial^2\mathbf{u}}{\partial y^2}e^{ikz} + \frac{\partial}{\partial z}\left(\frac{\partial\mathbf{u}}{\partial z}e^{ikz} + ik\mathbf{u}e^{ikz}\right) + k^2\mathbf{u}e^{ikz} = 0 \tag{1.12d}$$

$$\frac{\partial^2\mathbf{u}}{\partial x^2}\cancel{e^{ikz}} + \frac{\partial^2\mathbf{u}}{\partial y^2}\cancel{e^{ikz}} + \frac{\partial^2\mathbf{u}}{\partial z^2}\cancel{e^{ikz}} + ik\frac{\partial\mathbf{u}}{\partial z}\cancel{e^{ikz}} + ik\frac{\partial\mathbf{u}}{\partial z}\cancel{e^{ikz}} - \cancel{k^2\mathbf{u}e^{ikz}} + \cancel{k^2\mathbf{u}e^{ikz}} = 0 \tag{1.12e}$$

$$\frac{\partial^2 \mathbf{u}}{\partial x^2} + \frac{\partial^2 \mathbf{u}}{\partial y^2} + \frac{\partial^2 \mathbf{u}}{\partial z^2} + 2ik\frac{\partial \mathbf{u}}{\partial z} = 0 \qquad (1.12\mathrm{f})$$

We've mentioned previously that the k parameter in equation (1.9) represents the angular wavenumber and that the paraxial approximation can be made when rays only make small angles with respect to the optical axis. As such, the paraxial approximation is valid for:

$$\left| k\frac{\partial \mathbf{u}}{\partial z} \right| \gg \left| \frac{\partial^2 \mathbf{u}}{\partial z^2} \right| \qquad (1.13)$$

which basically means that, for the direction of propagation, only first-order derivatives have a significant influence on the wave propagation. Therefore, all second order derivatives can be eliminated, which means that equation (1.12f) becomes:

$$\Delta_\perp \mathbf{u} + 2ik\frac{\partial \mathbf{u}}{\partial z} = 0 \qquad (1.14)$$

where $\Delta_\perp = \nabla_\perp^2 = \frac{\partial^2}{\partial x^2} + \frac{\partial^2}{\partial y^2}$ is the transverse Laplacian. Equation (1.13) represents the **paraxial approximation of the homogeneous Helmholtz equation**. Using this approximation it is clear that it simplified the mathematics that describe the propagation. However, it is worth reiterating that this approximation can *only* be done when rays of light propagate with small angular differences between them.

1.3 The eikonal equation—material polarization approach

Another form of the wave equation is the eikonal equation. It is used to bridge (or rather support) the connection between wave optics and ray optics. Before we can derive this equation, we first have to discuss about *material* polarization and refractive index (not to be confused with the polarization of light). In equation (1.1d) we have introduced the current density, J, which we considered to be zero in order to derive the homogeneous wave equation. We know from electricity that the current density, $\mathbf{J}$ is defined as the current, $\mathbf{I}$, per unit area, S:

$$\mathbf{J} = \frac{d\mathbf{I}}{dt} \qquad (1.15)$$

where $\mathbf{I} = \frac{dq}{dt}$, q being the electric charge. Current, and therefore current density, is not only defined by the transport of *free* charges, but by *any variation of charge density*. Therefore, current density, $\mathbf{J}$ can be decomposed into three terms as follows:

$$\mathbf{J} = \mathbf{J}_{\mathrm{free}} + \mathbf{J}_{\mathrm{m}} + \mathbf{J}_{\mathrm{p}} \qquad (1.16)$$

where $\mathbf{J}_{\mathrm{free}}$ describes currents generated by the movement of free charges, $\mathbf{J}_{\mathrm{m}}$ describes internal currents of the atoms that make up the media through which light propagates, and $\mathbf{J}_{\mathrm{p}}$ represents the current given by *material polarization*. As an oscillating electromagnetic field propagates through a material, atoms and molecules suffer a charge density displacement determined by the electric field. In the simplest case, they become dipoles while under the influence of the field. In other words, atoms behave like oscillators and elongate in the direction of the electric field. Therefore, atoms drain energy from the propagating wave, only to release it in the same form, as an electromagnetic wave, when reverting to the original position (i.e., moving charges generate fields).

In this case, we assume that the dipole oscillations have the same frequency as the stimulating electromagnetic wave, albeit with a phase difference. The phase difference appears because the material response is not instantaneous, it takes time for the charge density to move around under the influence of the electric field. The resulting generated current, $\mathbf{J}_p$, is called the *polarization current* and is related to the polarization, $\mathbf{P}$ through the following relation [1]:

$$\mathbf{J}_p = \frac{\partial \mathbf{P}}{\partial t} \tag{1.17}$$

Using a similar ansatz, the charge density can also be decomposed as follows:

$$\rho = \rho_{\text{free}} + \rho_p \tag{1.18}$$

In equation (1.5) we also have the charge density on the right-hand side. We want to solve the equation in terms of polarization, therefore we need to relate the charge density, ρ, to the polarization, $\mathbf{P}$. To do that, we can use the continuity equation. A simple way to derive this equation is to apply the div–curl identity to equation (1.1d):

$$\left. \begin{aligned} \nabla \cdot (\nabla \times \mathbf{A}) &= 0 \\ \nabla \times \mathbf{B} &= \varepsilon_0 \mu_0 \frac{\partial \mathbf{E}}{\partial t} + \mu_0 \mathbf{J} \end{aligned} \right\} \Rightarrow \nabla \cdot \left(\varepsilon_0 \mu_0 \frac{\partial \mathbf{E}}{\partial t} + \mu_0 \mathbf{J} \right) = 0$$

$$\Rightarrow \mu_0 \left(\varepsilon_0 \frac{\partial}{\partial t} \nabla \cdot \mathbf{E} + \nabla \cdot \mathbf{J} \right) = 0 \tag{1.19}$$

and since $\mu_0 \neq 0$, and considering equation (1.1a), we obtain:

$$\nabla \cdot \mathbf{J} + \frac{\partial \rho}{\partial t} = 0 \tag{1.20}$$

We make the assumption that the waves we want to analyze propagate through a non-magnetic dielectric medium, i.e., $\mathbf{J}_{\text{free}}$, $\mathbf{J}_m$, and ρ_{free} are zero. This means (1.19) becomes:

$$\nabla \cdot \mathbf{J}_p + \frac{\partial \rho_p}{\partial t} = 0 \tag{1.21}$$

If we substitute equation (1.17) in equation (1.20) we obtain:

$$\nabla \cdot \frac{\partial \mathbf{P}}{\partial t} + \frac{\partial \rho_p}{\partial t} = 0$$

$$\Rightarrow \nabla \cdot \mathbf{P} + \rho_p = 0 \tag{1.22}$$

$$\Rightarrow \rho_p = - \nabla \cdot \mathbf{P}$$

If we substitute the newly derived equations, (1.17)–(1.22), in the general wave equation, (1.5), while assuming propagation through an isotropic, homogeneous, dielectric medium, the wave equation is reduced to:

$$\Delta \mathbf{E} - \frac{1}{c^2}\frac{\partial^2 \mathbf{E}}{\partial t^2} = \mu_0 \frac{\partial^2 \mathbf{P}}{\partial t^2} \tag{1.23}$$

where we have noted $\nabla^2 = \Delta$ (Laplacian operator). In order to determine the refractive index (through the dispersion relation), we have to solve equation (1.23). We can do this by inserting plane-wave solutions of the form:

$$\begin{aligned}\mathbf{E} &= E_0 e^{i(\mathbf{k}\cdot\mathbf{r}-\omega t)}\\\mathbf{P} &= P_0 e^{i(\mathbf{k}\cdot\mathbf{r}-\omega t)}\end{aligned} \tag{1.24}$$

in the mentioned equation. However, we have two time-dependent functions, the electric field, $\mathbf{E}$, and the polarization of the material, $\mathbf{P}$. Before we can continue, we have to set an explicit relation between the two functions. We are considering a linear regime, therefore we can use [2]:

$$\mathbf{P} = \chi \varepsilon_0 \mathbf{E} \tag{1.25}$$

where χ represents *susceptibility*, a dimensionless parameter that describes the material response to an applied field, i.e., the degree of polarization in this case. It can be thought of as a proportionality constant. To advance in our discussion, let's introduce the plane waves, (1.24), in the wave equation, (1.23):

$$\left(\frac{\partial^2}{\partial x^2} + \frac{\partial^2}{\partial y^2} + \frac{\partial^2}{\partial z^2}\right)(E_0 e^{i(\mathbf{k}\cdot\mathbf{r}-\omega t)}) - \varepsilon_0\mu_0\frac{\partial^2}{\partial t^2}(E_0 e^{i(\mathbf{k}\cdot\mathbf{r}-\omega t)}) = \mu_0\frac{\partial^2}{\partial t^2}(P_0 e^{i(\mathbf{k}\cdot\mathbf{r}-\omega t)}) \tag{1.26}$$

For this derivation we will skip some steps since it is similar to equation (1.12), with the mention that we can write:

$$\begin{aligned}\mathbf{E} &= E_0 e^{i(\mathbf{k}\cdot\mathbf{r}-\omega t)}\\&= E_0 e^{i(\mathbf{k}\cdot\mathbf{r})}e^{-i\omega t}\end{aligned} \tag{1.27}$$

which makes the derivation easier. The result is:

$$-k^2 E_0 e^{i(\mathbf{k}\cdot\mathbf{r}-\omega t)} + \varepsilon_0\mu_0 E_0 e^{i(\mathbf{k}\cdot\mathbf{r}-\omega t)} = -\mu_0\omega^2 P_0 e^{i(\mathbf{k}\cdot\mathbf{r}-\omega t)} \tag{1.28}$$

To simplify equation (1.28), we use equation (1.25) to substitute P_0, after which we observe we can reduce all fields, and therefore obtain:

$$\begin{aligned}-k^2 E_0 e^{i(\mathbf{k}\cdot\mathbf{r}-\omega t)} + \varepsilon_0\mu_0 E_0 e^{i(\mathbf{k}\cdot\mathbf{r}-\omega t)} &= -\mu_0\omega^2\varepsilon_0\chi E_0 e^{i(\mathbf{k}\cdot\mathbf{r}-\omega t)}\\-k^2 + \varepsilon_0\mu_0 &= -\varepsilon_0\mu_0\chi\omega^2\\\Rightarrow k^2 &= \varepsilon_0\mu_0\omega^2(1+\chi)\\\Rightarrow k^2 &= \frac{1}{c^2}\omega^2(1+\chi)\\k &= \frac{\omega}{c}\sqrt{1-\chi}\end{aligned} \tag{1.29}$$

where we used $c = \frac{1}{\sqrt{\mu_0\varepsilon_0}}$. **The term $\sqrt{1+\chi}$ represents the refractive index.**

We usually encounter the definition of the refractive index as being $n = \frac{c}{v_{\mathrm{m}}}$, where v_{m} is the speed of light inside the medium. That is due to a simple rationale:

$$k = \frac{2\pi}{\lambda} = \frac{\omega}{c}\sqrt{1 + \chi} = \frac{n\omega}{c}$$

$$\Rightarrow n = \frac{2\pi c}{\omega\lambda} = \frac{c}{\nu\lambda} \tag{1.30}$$

$$\Rightarrow n = \frac{c}{v_{\mathrm{m}}}$$

where k is the wavenumber, λ is the wavelength, ω is the *angular* frequency, and ν is the frequency. Defining the refractive index in terms of material susceptibility, rather than the ratio between the velocity of light in vacuum and its velocity in a material, offers a more clear interpretation of the parameter, i.e., the material's response to electromagnetic radiation. We have analyzed the case for a monochromatic plane wave, i.e., considering only one frequency, ω. In practical cases, however, materials respond differently to each frequency, therefore $\chi = \chi(\omega)$. Moreover, since $\mathbf{E}$ and $\mathbf{P}$ from equation (1.24) can oscillate with a phase difference, the susceptibility is a complex value. A complex susceptibility means the refractive index is also a complex value. The imaginary part of the refractive index accounts for losses in the material. Equation (1.30) is usually known as the dispersion relation, as $n = n(\omega)$. We will get back to the refractive index when discussing laser media and laser–matter interaction.

Until this point, we have analyzed light propagation as a wave. The paraxial approximation allows us to simplify a theoretical analysis for cases where the propagation direction closely follows the optical axis. However, even with this approximation, analysis of larger scale optical systems becomes complicated. For such systems we can use ray theory. Ray theory can be deduced from Maxwell's equations through means of the eikonal equation. As mentioned previously, the eikonal equation can be thought of as a justification of ray theory. This equation can be used to give appropriate description of light propagation through a medium of spatially varying refractive index, as it describes the *direction* a wave is propagating. It is applicable to any kind of waves, scalar or vectorial (i.e., acoustic or electromagnetic). For optics and laser processing, this equation is important not only because it can be used to describe light propagation over long distances and inhomogeneous medium, but it can also be used to deduce Fermat's principle.

There are multiple methods to derive the eikonal equation for optics, leading to different forms. While the approach we're taking is slightly more difficult than the alternatives, from a mathematical point of view, it offers a better image of the underlying thought process, making it easier to interpret the results.

We start from the wave equation considering a inhomogeneous dielectric medium, where the refractive index depends on position. This should not be confused with an anisotropic medium, where the refractive index depends on propagation *direction*.

$$\Delta\mathbf{E}(\mathbf{r},\, t) - \frac{n(\mathbf{r})}{c^2}\frac{\partial^2\mathbf{E}}{\partial t^2} = 0 \tag{1.31}$$

Since we're trying to determine the behavior of a wave propagating through an inhomogeneous medium, a plane wave solution is no longer suitable for equation (1.31), because we have to take into account the wavefront and its deformations during propagation, due to the spatially varying refractive index. In [1] the trial solution for equation (1.31) is:

$$\mathbf{E}(\mathbf{r}, t) = \mathbf{E}_0(\mathbf{r})e^{i[k_{\text{vac}}R(\mathbf{r})-\omega t]} \tag{1.32}$$

where $k_{\text{vac}} = \frac{\omega}{c} = \frac{2\pi}{\lambda_{\text{vac}}}$. The function $R(\mathbf{r})$ is scalar and depicts the wavefront mentioned previously, at a given moment. In other words, $R(\mathbf{r})$ represent phase isosurfaces of the propagating wave. Substituting equation (1.32) in equation (1.31) and performing the time derivative, we obtain:

$$\Delta(\mathbf{E}_0(\mathbf{r})e^{i[k_{\text{vac}}R(\mathbf{r})-\omega t]}) - \frac{n(\mathbf{r})}{c^2}\frac{\partial^2}{\partial t^2}(\mathbf{E}_0(\mathbf{r})e^{i[k_{\text{vac}}R(\mathbf{r})-\omega t]}) = 0$$

$$\Leftrightarrow \Delta(\mathbf{E}_0(\mathbf{r})e^{i[k_{\text{vac}}R(\mathbf{r})-\omega t]}) + \frac{n(\mathbf{r})\omega^2}{c^2}(\mathbf{E}_0(\mathbf{r})e^{i[k_{\text{vac}}R(\mathbf{r})-\omega t]}) = 0 \tag{1.33}$$

$$\Leftrightarrow \Delta(\mathbf{E}_0(\mathbf{r})e^{i[k_{\text{vac}}R(\mathbf{r})-\omega t]}) + n(\mathbf{r})k_{\text{vac}}^2(\mathbf{E}_0(\mathbf{r})e^{i[k_{\text{vac}}R(\mathbf{r})-\omega t]}) = 0$$

We can split the exponential term similarly to equation (1.27), after which we can eliminate part of it as follows:

$$\Delta(\mathbf{E}_0(\mathbf{r})e^{i[k_{\text{vac}}R(\mathbf{r})]})e^{-i\omega t} + n(\mathbf{r})k_{\text{vac}}^2\mathbf{E}_0(\mathbf{r})e^{i[k_{\text{vac}}R(\mathbf{r})]}e^{-i\omega t} = 0$$

$$\Leftrightarrow \Delta(\mathbf{E}_0(\mathbf{r})e^{i[k_{\text{vac}}R(\mathbf{r})]}) + n(\mathbf{r})k_{\text{vac}}^2\mathbf{E}_0(\mathbf{r})e^{i[k_{\text{vac}}R(\mathbf{r})]} = 0 \tag{1.34}$$

Calculating the Laplacian in equation (1.34) is more complex due to the dependencies of position. The computation of the Laplacian below is cited from [1], with the mention that the authors divided equation (1.34) to k_{vac}^2, for a subsequent approximation. The complexity arises from the fact that the wave depends on all three directions, because we are considering an arbitrary wavefront. As such, the authors of [1] start by exemplifying the Laplacian for one direction, starting with the *gradient* for the x component:

$$\nabla[E_{0x}(\mathbf{r})e^{ik_{\text{vac}}R(\mathbf{r})}] = [\nabla E_{0x}(\mathbf{r})]e^{ik_{\text{vac}}R(\mathbf{r})} + ik_{\text{vac}}E_{0x}(\mathbf{r})[\nabla R(\mathbf{r})]e^{ik_{\text{vac}}R(\mathbf{r})} \tag{1.35}$$

Using equation (1.35), we apply ∇ once more to compute the *Laplacian* for the x component:

$$\nabla \cdot \nabla [E_{0x}(\mathbf{r})e^{ik_{\text{vac}}R(\mathbf{r})}] = \{ \nabla^2 E_{0x}(\mathbf{r}) - k_{\text{vac}}^2 E_{0x}(\mathbf{r})[\nabla R(\mathbf{r})] \cdot [\nabla R(\mathbf{r})]$$
$$+ ik_{\text{vac}}E_{0x}(\mathbf{r})[\nabla^2 R(\mathbf{r})] + 2ik_{\text{vac}}[\nabla E_{0x}(\mathbf{r})] \cdot [\nabla R(\mathbf{r})]\}e^{ik_{\text{vac}}R(\mathbf{r})} \tag{1.36}$$

By applying a similar computation to each component of the wave and then summing them, we obtain:

$$\nabla^2[\mathbf{E}_0(\mathbf{r})e^{ik_{\text{vac}}R(\mathbf{r})}] = (\nabla^2 \mathbf{E}_0(\mathbf{r}) - k_{\text{vac}}^2\mathbf{E}_0(\mathbf{r})[\nabla R(\mathbf{r})] \cdot [\nabla R(\mathbf{r})] + ik_{\text{vac}}\mathbf{E}_0(\mathbf{r})[\nabla^2 R(\mathbf{r})]$$
$$+ 2ik_{\text{vac}}\{\mathbf{i}[\nabla E_{0x}(\mathbf{r})] \cdot [\nabla R(\mathbf{r})] + \mathbf{j}[\nabla E_{0y}(\mathbf{r})] \cdot [\nabla R(\mathbf{r})] \tag{1.37}$$
$$+ \mathbf{k}[\nabla E_{0z}(\mathbf{r})] \cdot [\nabla R(\mathbf{r})]\})e^{ik_{\text{vac}}R(\mathbf{r})}$$

where $\mathbf{i}$, $\mathbf{j}$, and $\mathbf{k}$ are the unit vectors for each of the Cartesian directions. Inserting equation (1.37) into equation (1.34) we obtain (equation (9.6) from [1]):

$$\left[\nabla R(\mathbf{r}) \cdot \nabla R(\mathbf{r}) - [n(\mathbf{r})]^2\right] \mathbf{E}_0(\mathbf{r}) = \frac{\nabla^2 \mathbf{E}_0(\mathbf{r})}{k_{\mathrm{vac}}^2} + \frac{i}{k_{\mathrm{vac}}}\nabla^2 R(\mathbf{r}) + \frac{2i}{k_{\mathrm{vac}}}i\nabla E_{0x}(\mathbf{r}) \cdot \nabla R(\mathbf{r})$$
$$+ \frac{2i}{k_{\mathrm{vac}}}\mathbf{j}\nabla E_{0y}(\mathbf{r}) \cdot \nabla R(\mathbf{r}) + \frac{2i}{k_{\mathrm{vac}}}\mathbf{k}\nabla E_{0z}(\mathbf{r}) \cdot \nabla R(\mathbf{r})$$

$$\Leftrightarrow \nabla R(\mathbf{r}) \cdot \nabla R(\mathbf{r}) - [n(\mathbf{r})]^2 = \frac{1}{\mathbf{E}_0(\mathbf{r})}\,\boxed{\frac{1}{k_{\mathrm{vac}}}}\left(\frac{\nabla^2 \mathbf{E}_0(\mathbf{r})}{k_{\mathrm{vac}}} + i\nabla^2 R(\mathbf{r}) + 2ii\nabla E_{0x}(\mathbf{r}) \cdot \nabla R(\mathbf{r})\right.$$
$$\left. + 2i\mathbf{j}\nabla E_{0y}(\mathbf{r}) \cdot \nabla R(\mathbf{r}) + 2i\mathbf{k}\nabla E_{0z}(\mathbf{r}) \cdot \nabla R(\mathbf{r})\right) \tag{1.38}$$

We observe that equation (1.38) can be heavily simplified, as Peatross and Ware suggest [1] (p230), by taking '*the limit of a very short wavelength. With it we lose the effects of diffraction so that only macroscopic features are relevant.*':

$$\lambda_{\mathrm{vac}} \to 0$$
$$k_{\mathrm{vac}} = \frac{2\pi}{\lambda_{\mathrm{vac}}} \Rightarrow \frac{1}{k_{\mathrm{vac}}} \to 0 \tag{1.39}$$

Therefore, everything on the right-hand side is reduced and equation (1.39) becomes:

$$\nabla R(\mathbf{r}) \cdot \nabla R(\mathbf{r}) - [n(\mathbf{r})]^2 = 0$$
$$\nabla R(\mathbf{r}) \cdot \nabla R(\mathbf{r}) = [n(\mathbf{r})]^2 \tag{1.40}$$

Eequation (1.40) represents the *eikonal equation*. It can also be written as (9.8) from [1]:

$$\nabla R(\mathbf{r}) = n(\mathbf{r})\mathbf{s}(\mathbf{r}) \tag{1.41}$$

where $\mathbf{s}(\mathbf{r})$ is a unit vector normal to the surface of the wavefront, at any given point (see Figure 1.1).

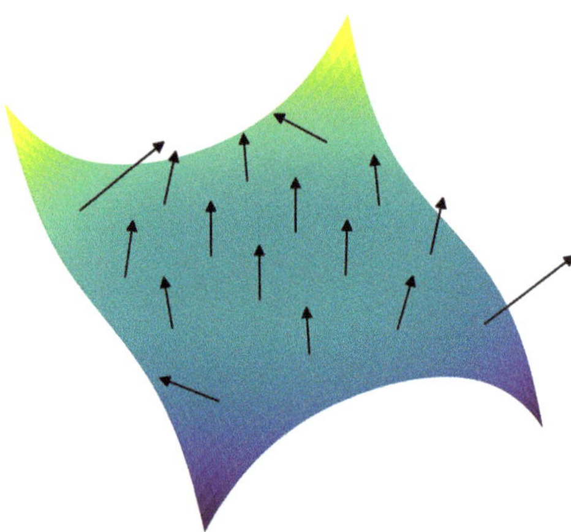

Figure 1.1. Phase isosurface (i.e., wavefront) for a wave propagating in an inhomogeneous dielectric medium and the direction of propagation, determined by $\nabla R(\mathbf{r})$, represented by the arrows.

This form of the eikonal equation makes it easier to interpret. It shows that the direction a wave is propagating is normal to the surface of the wavefront at any given point, i.e., in the direction defined by $\mathbf{s}(\mathbf{r})$. We can also observe that $\mathbf{s}(\mathbf{r})$ defines the direction and $n(\mathbf{r})$ defines the magnitude of $\nabla R(\mathbf{r})$, referencing the angle of refraction.

1.4 The eikonal equation—Fermat's principle approach

Another approach for the direction of propagation relative to the refractive index is considering Fermat's principle. It is commonly stated as the fact that *light travels the distance between two points on the path that takes the least time*. We consider that a more intuitive statement is that *light travels the **shortest optical path** between two points*. The reason is that an extremum in the time of propagation between two points corresponds to an extremum in the optical path. Fermat's principle can be derived using several approaches. Although less intuitive, we can use the eikonal equation (1.41) to derive Fermat's principle. We consider this approach offers better in-depth understanding of the principle. We also present the often encountered straightforward approach afterwards, for comparison.

To derive Fermat's principle using equation (1.41), we are interested in finding a relation between the *refractive index* and the *path* light takes between two points, knowing the propagation direction is given by $\mathbf{s}(\mathbf{r})$. We can use Stokes' theorem:

$$\oint_{\Gamma} \mathbf{u} \cdot d\Gamma = \iint_{\Sigma} \nabla \times \mathbf{u} \cdot d\Sigma \tag{1.42}$$

where Γ is a closed curve, $\mathbf{u}$ is a vector field, and Σ is an oriented open surface defined by Γ. We can apply a curl to equation (1.41) to eliminate $\nabla R(\mathbf{r})$ (curl of a gradient is always zero for twice-differentiable functions), and we're left with:

$$\nabla \times [n(\mathbf{r})\mathbf{s}(\mathbf{r})] = 0 \quad |\leftarrow \text{apply} \iint_{\Sigma} (\text{o})d\Sigma$$

$$\Rightarrow \iint_{\Sigma} \nabla \times n(\mathbf{r})\mathbf{s}(\mathbf{r}) \cdot d\Sigma = \oint_{\Gamma} n\mathbf{s}(\mathbf{r}) \cdot d\Gamma = 0 \tag{1.43}$$

Any two points, let's say A and B, can be connected through a closed loop. Therefore, equation (1.43) shows that:

$$\int_{A}^{B} n(\mathbf{r})\mathbf{s}(\mathbf{r})d\Gamma \quad \text{does not depend on the path} \tag{1.44}$$

At this point we should turn our attention to the dot product in equation (1.43), or more specifically how it is expressed mathematically (geometric definition):

$$\mathbf{a} \cdot \mathbf{b} = ab \cos \theta \tag{1.45}$$

where θ is the angle between the two vectors, **a** and **b**. A path that is parallel to **s** will result in the cosine to remain equal to 1 over the whole integration. Any other path will cause the cosine values to vary with values below 1. The result of the integral is the same, no matter the path, as points A and B do not change. As such, if we eliminate the dot factor (note: **s** is a *unit vector* that gives the direction normal to the surface of a wavefront), the result of the integral will be greater for any path, *other than the one parallel to* **s**.

Another way of reasoning is using Maupertuis' principle (special case of **the principle of least action**), or, in mathematical terms:

$$\int_A^B n(\mathbf{r})\mathbf{s}(\mathbf{r})\mathrm{d}\Gamma = \min\left(\int_A^B n(\mathbf{r})\mathrm{d}\Gamma\right) \tag{1.46}$$

The right-hand side of equation (1.46) represents the *optical path length*.

A simpler way of arriving to the same conclusion is by considering the time, τ, it takes a wave to travel between A and B, which in mathematical terms can be expressed as:

$$\tau = \int_{t_1}^{t_2} \mathrm{d}t$$
$$\tau = \int_{t_1}^{t_2} \frac{\mathrm{d}t}{\mathrm{d}r}\frac{\mathrm{d}r}{\mathrm{d}t}\mathrm{d}t \quad |\cdot\frac{c}{c}$$
$$\tau = \frac{1}{c}\int_{t_1}^{t_2} \frac{c}{v_m}\frac{\mathrm{d}r}{\mathrm{d}t}\mathrm{d}t \tag{1.47}$$
$$\tau = \frac{1}{c}\int_A^B n\ \mathrm{d}r$$

where c is vacuum speed of light, v_m is light velocity in a medium of refraction index $n = \frac{c}{v_m}$. Using a simple reasoning that velocity is distance over time, therefore, from equation (1.47) we get:

$$c = \frac{\int_A^B n\ \mathrm{d}r}{\tau} \tag{1.48}$$
$$\Rightarrow \int_A^B n\ \mathrm{d}r = \tau c = \gamma$$

is the *optical path length*. Both results suggest the same concept: *light travels between any two given points following the shortest optical path length*. The relation between the optical path length, γ, and the physical (geometrical) path length, r, is given by the refractive index:

$$r = \gamma n \tag{1.49}$$

Fermat's principle and, more directly, the optical path length is used to describe and explain many important optical phenomena, such as the working principle of lenses, phase differences when light traverses a medium with a different refractive index, interference, and others. One important application for laser processing is increasing the numerical aperture of a microscope objective using immersion oil.

1.5 Propagation of electromagnetic waves through matter. Lorentz oscillator

In the previous sections (1.1–1.4) we discussed the relation between Maxwell's and wave equations, but the analysis itself didn't consider the environment in detail, i.e., most of our discussion was done considering wave propagation through vacuum and shortly mentioning the polarization of a material once we reached the eikonal equation. Let us turn our attention to electromagnetic wave propagation through a material and focus on what happens to the material itself. One of the most important fundamental topics in optics is the refractive index. It helps us understand and predict many optical phenomena such as refraction, dispersion, and even absorption. In most explanations relating to the refractive index, regardless of the delivery (books, courses, videos, etc), it is often described in the context of geometric optics, where light refraction is used as a defining physical phenomenon. In the previous paragraphs, we instead took a different path, via using Maxwell's equations to derive and describe the behavior and properties of light. What is, then, the connection between Maxwell's equations and the refractive index? What about absorption and dispersion? What exactly is this physical property called 'refractive index'?

To achieve a basic understanding, albeit simplified, let us consider how the material responds to the presence of an electromagnetic field.

1.6 Lorentz oscillator model

The principal idea of this chapter is based on an analogy between a harmonic oscillator and the response of a material to the presence of an electromagnetic field, at the atomic scale.

There is a theoretical model, developed by Heaviside [3] that describes the fact that an electric charge that changes its position, i.e., it is moving, will generate a wave that propagates with the speed of light (see figure 1.2).

On the left side we have a stationary charge, and the electric field lines radiating from it. On the right side we observe that the movement of the charge itself generates a wave radiating outwards. Note what happens to the wave in the direction of charge movement.

We also know that there is a force applied to an electric charge when it is found in the presence of an electric field:

$$\mathbf{F} = q\mathbf{E} \tag{1.50}$$

One of the simplest ways to describe an atom is to say it is composed of a positively charged nucleus, surrounded by a cloud of negatively charged electrons. Overall,

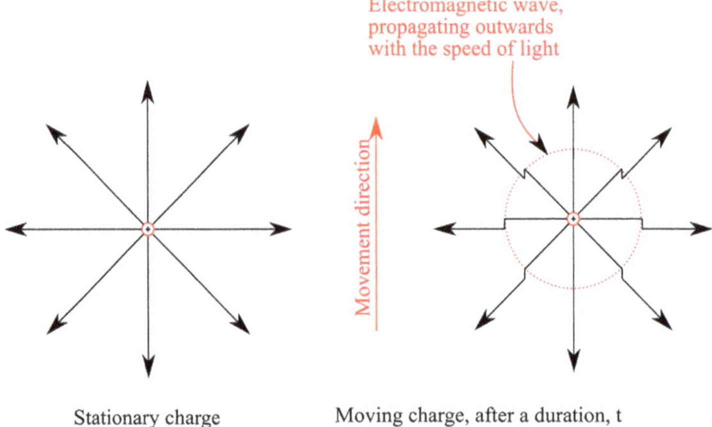

Figure 1.2. Schematic of an electromagnetic wave generated by a moving charge.

the atom is electrically neutral if there is no external influence, or, in other words: the center of the positive charge coincides with the center of the negative charge. However, that changes if the atom is found in the presence of an electric field (see figure 1.3). An external electric field that acts upon an atom will pull the positive charge towards the direction of the field lines with a force **F**, and the negative charge in the opposite direction, with the same force **F**. This in turn determines a small shift in the center of both positive and negative charges, until the attraction force between these charges is equal to the force determined by the external electric field.

When the external electric field is static, the system can reach an equilibrium state and we describe the material as being polarized (not to be confused with polarization of an electromagnetic field). However, when a time-varying electric field propagates through the material (as is the case of electromagnetic waves), the positive and negative charges will tend to follow the field lines. At this moment we can point out two main ideas:

- Each atom 'tries' to keep up with the frequency of the electric field, but the material response is not instant as the charged particles have mass and therefore inertia, i.e., there is a phase difference between the oscillating electric field and the atoms themselves.
- As previously mentioned, moving charges generate waves, therefore the atoms absorb energy from the oscillating electric field (they get polarized), but they release it back as a slightly different oscillating electric field (due to phase changes).

These re-emitted waves will overlap with the incident wave, but due to the phase difference induced by the delay between the polarization and the re-emission, the propagation velocity will 'slow down'. This reduction is characterized by the refraction index, which, as previously mentioned, is related to the electric permittivity of the material.

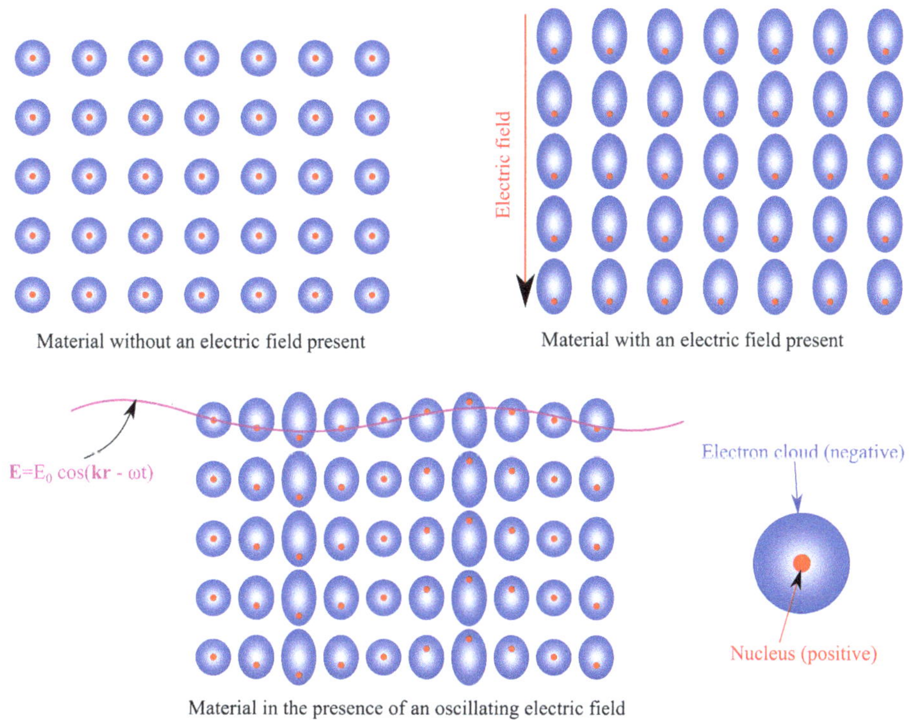

Figure 1.3. Schematic representation of the material response in the presence of an electric field.

The equation of motion for this case is similar to a forced oscillator:

$$m\frac{\partial^2 \mathbf{r}}{\partial t^2} + m\Gamma\frac{\partial \mathbf{r}}{\partial t} + m\omega_0^2 \mathbf{r} = -q\mathbf{E} \qquad (1.51)$$

where m is the mass of the electron cloud, $\mathbf{r}$ is the position vector, t is time, Γ is the attenuation coefficient (friction), ω_0 is the resonance frequency, q is the electric charge and $\mathbf{E}$ is the electric field. In the particular case of an atom, the nucleus is considered stationary due to the considerable mass difference in comparison to electrons, i.e., we only consider the oscillations of the electron cloud.

Equation (1.51) can be solved using different approaches. One approach is to use the Fourier transform. Therefore, if we apply the Fourier transform to equation (1.51) we obtain:

$$m(-i\omega)^2\mathbf{r}(\omega) + m\Gamma(-i\omega)\mathbf{r}(\omega) + m\omega_0^2\mathbf{r}(\omega) = -q\mathbf{E}(\omega) \qquad (1.52)$$

which, when simplified (use $\mathbf{r}(\omega)$ as a common denominator), we can rewrite as:

$$\left(-m\omega^2 - i\omega m\Gamma + m\omega_0^2\right)\mathbf{r}(\omega) = -q\mathbf{E}(\omega) \qquad (1.53)$$

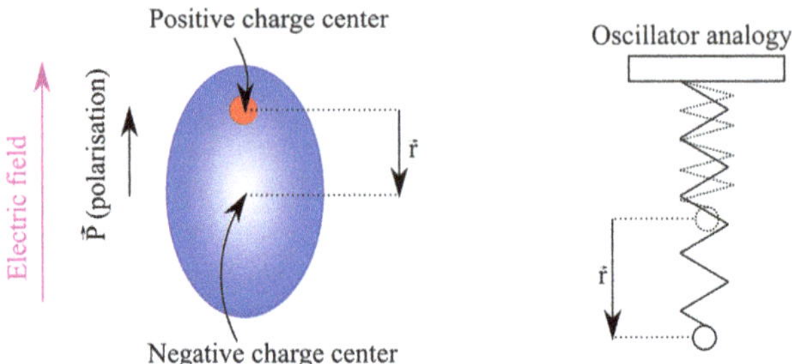

Figure 1.4. Schematic representation of the response of an atom in the presence of an electric field.

The next step is to solve equation (1.52) for $\mathbf{r}(\omega)$. The function $\mathbf{r}(\omega)$ represents the distance between the center of positive and negative charges (see figure 1.4).

The solution of equation (1.53) is:

$$\mathbf{r}(\omega) = -\frac{q}{m}\frac{\mathbf{E}(\omega)}{\omega_0^2 - \omega^2 - i\omega\Gamma} \tag{1.54}$$

We have reached a point where it is helpful to introduce a physical quantity that can be used to describe the phenomenon discussed above, i.e., the electric charge and the separation itself. We therefore define the electric dipole moment as:

$$\boldsymbol{\mu}_{\mathrm{e}}(\omega) = -q\mathbf{r}(\omega) \tag{1.55a}$$

$$\Rightarrow \boldsymbol{\mu}_{\mathrm{e}}(\omega) = \frac{q^2}{m}\frac{\mathbf{E}(\omega)}{\omega_0^2 - \omega^2 - i\omega\Gamma} \tag{1.55b}$$

$$\Rightarrow \boldsymbol{\mu}_{\mathrm{e}}(\omega) = [\alpha(\omega)]\mathbf{E}(\omega) \tag{1.55c}$$

Addendum 1 *The electric dipole moment, $\boldsymbol{\mu}_{\mathrm{e}}$, does not represent the magnetic permeability. Similarly, $\alpha(\omega)$ is not the absorption coefficient.*

In equation (1.55), the term $[\alpha(\omega)]$ represents a tensor, used generally for anisotropic materials. In this case, however, we consider the material as isotropic, to keep the analysis simple, i.e., optical properties are the same, regardless of propagation direction of the incident electromagnetic wave. This term is called Lorentz polarizability and describes how easily electric charges can be displaced from their equilibrium state.

Up to this point, we have considered what happens to a single atom. However, we can scale everything to a large amount of atoms, i.e., a material. As such, we can express material polarization as:

$$\mathbf{P}(\omega) = \frac{1}{V}\sum_{V}\boldsymbol{\mu}_{e_i}(\omega) \tag{1.56}$$

Addendum 2 *In equation (1.56) we have expressed the electric dipole moment as mediated over the whole volume of a material.*

In other words, in equation (1.56) we averaged the electric dipole moment. Each atom is polarized slightly differently, but if we average this over the entire volume, we can express material polarization as follows:

$$\mathbf{P}(\omega) = N\langle\boldsymbol{\mu}_{e}(\omega)\rangle \tag{1.57a}$$

$$\Rightarrow \mathbf{P}(\omega) = \frac{Nq^2}{m}\frac{1}{\omega_0^2 - \omega^2 - i\omega\Gamma}\mathbf{E}(\omega) \tag{1.57b}$$

where N represents the number of atoms per unit volume, and $\langle\boldsymbol{\mu}_{e}(\omega)\rangle$ is the dipole moment, averaged per unit volume.

We can (re)introduce the electric susceptibility χ_{e} (used in section 1.3) This term is used to describe how easily an electric field can polarize a certain material. Therefore, accounting for the equation for the polarizability of a material previously derived, we can deduce the following equation for the electric susceptibility:

$$\left.\begin{array}{l} \mathbf{P}(\omega) = \varepsilon_0\chi_{e}(\omega)\mathbf{E}(\omega) \\[2mm] \mathbf{P}(\omega) = \dfrac{Nq^2}{m}\dfrac{1}{\omega_0^2 - \omega^2 - i\omega\Gamma}\mathbf{E}(\omega) \end{array}\right\} \Rightarrow \chi_{e}(\omega) = \frac{N\alpha(\omega)}{\varepsilon_0} \tag{1.58a}$$

$$\Rightarrow \chi_{e}(\omega) = \left(\frac{Nq^2}{\varepsilon_0 m}\right)\frac{1}{\omega_0^2 - \omega^2 - i\omega\Gamma} \tag{1.58b}$$

Note that in equation (1.58b) we grouped all constants together. These terms are also known as 'plasma frequency':

$$\omega_{p}^2 = \frac{Nq^2}{\varepsilon_0 m} \tag{1.59}$$

We will not go into detail regarding ω_p, but it is important to remember that any material has a specific value of this term. For the moment, we will express the electric susceptibility as a function of ω_p:

$$\chi_{e}(\omega) = \frac{\omega_{p}^2}{\omega_0^2 - \omega^2 - i\omega\Gamma} \tag{1.60}$$

In other words, we can say that the electric susceptibility is a transfer function in an oscillating system. We must mention, however, that we did not consider the position

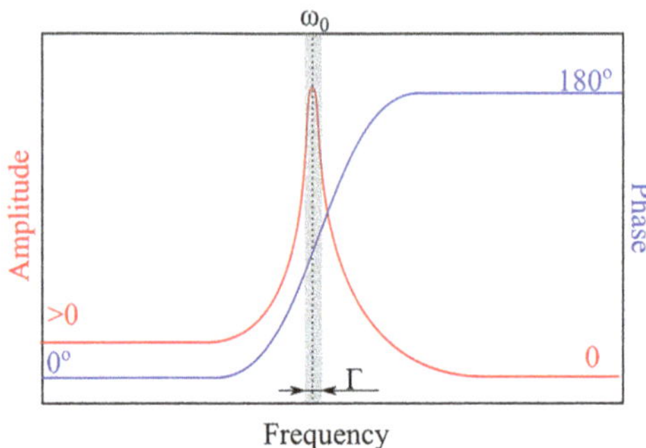

Figure 1.5. Electric susceptibility as a function of frequency (for $\omega_0 = 3$ and $\Gamma = 1$).

of the atoms. This is important as they can influence each other and therefore affect the electric dipole oscillations.

In figure 1.5 we show the electric susceptibility as a function of frequency.

Note that figure 1.5 is similar to the response of a harmonic oscillator. Things get more complicated around the resonance frequency. We can also observe that the Γ term determines the shape of the amplitude for χ_e. At low frequency, we observe that the electric susceptibility is not zero, but:

$$|\chi_e(\omega = 0)| = \left(\frac{\omega_p}{\omega_0}\right)^2 \tag{1.61}$$

The phase, however, at $\omega = 0$, is 0. At high frequencies we obtain:

$$|\chi_e(\omega = \infty)| = 0 \tag{1.62}$$

In this case, the phase difference is 180°. When $\omega = \omega_0$, i.e., at resonance, we obtain:

$$|\chi_e(\omega = \omega_0)| = \frac{\omega_p^2}{\omega_0 \Gamma} \tag{1.63}$$

The phase difference at resonance is 90°, again, similarly to a harmonic oscillator.

Addendum 3 *Phase differences, discussed above, refer to the phase difference between the oscillating electric field and the response of the material through which it propagates.*

1.7 Lorentz model for dielectrics

In the previous section we discussed a general approach to Lorentz oscillator. In this section we will discuss its consequences on relative permittivity, after which we can extract some conclusions regarding some parameters such as the refractive index and material dispersion.

Considering that:

$$\mathbf{D} = \epsilon_0 \,\underset{\text{vacuum}}{\mathbf{E}} + \underset{\text{material}}{\mathbf{P}} \tag{1.64}$$

We have previously talked about the fact that material polarization, **P**, represents the response of a material to an external electric field. In other words, it shows the energy (from the electric field) absorbed by the material via movement of the center of electric charge at the atomic level (both positive and negative).

Addendum 4 *You may correlate* **P** *to* **E** *via the potential energy of a harmonic oscillator.*

Introducing equation (1.58a) in equation (1.64), we obtain:

$$\mathbf{D} = \varepsilon_0 \mathbf{E} + \varepsilon_0 \chi_e \mathbf{E} \tag{1.65a}$$

$$\Rightarrow \mathbf{D} = \varepsilon_0 (1 + \chi_e)\mathbf{E} \tag{1.65b}$$

$$\Rightarrow \mathbf{D} = \varepsilon_0 \varepsilon_r \mathbf{E} \tag{1.65c}$$

The term ε_r is called relative permittivity and represents the ratio between the electric permittivity (absolute value) of a material and the vacuum permittivity, ε_0. The absolute permittivity is a measure of the polarizability of a material. The relative permittivity, therefore, is described by:

$$\varepsilon_r(\omega) = \frac{\varepsilon(\omega)}{\varepsilon_0} \tag{1.66}$$

Knowing what the susceptibility, χ_e, is (see equations (1.59), (1.60)), we may rewrite the relative permittivity $\varepsilon_r(\omega)$ as:

$$\varepsilon_r(\omega) = 1 + \chi_e = 1 + \frac{\omega_p^2}{\omega_0^2 - \omega^2 - i\omega\Gamma} \tag{1.67}$$

Equation (1.67) represents the dielectric function for the Lorentz model with a single resonance ('single resonance' refers to a single variable introduced in our simplified, unidimensional model, i.e., each and every atom responds individually to the incident electric field, without influencing other atoms, and not considering complex atomic systems such as molecules). We notice that the dielectric function takes on a complex form. To move forward, it is useful to separate $\varepsilon_r(\omega)$ in its real and imaginary parts. We observe that in equation (1.67) the complex part is the denominator. Therefore, we need to multiply and divide by the complex conjugate of the denominator:

$$\epsilon_r(\omega) = \epsilon_r'(\omega) + i\epsilon_r''(\omega) = 1 + \frac{\omega_p^2}{(\omega_0^2 - \omega^2) - i\omega\Gamma}\,\frac{(\omega_0^2 - \omega^2) + i\omega\Gamma}{(\omega_0^2 - \omega^2) + i\omega\Gamma} \tag{1.68a}$$

$$= 1 + \omega_p^2 \frac{\left(\omega_0^2 - \omega^2\right) + i\omega\Gamma}{\left(\omega_0^2 - \omega^2\right)^2 + \omega^2\Gamma^2} \tag{1.68b}$$

$$= 1 + \omega_p^2 \frac{\omega_0^2 - \omega^2}{\left(\omega_0^2 - \omega^2\right)^2 + \omega^2\Gamma^2} + i\omega_p^2 \frac{\omega\Gamma}{\left(\omega_0^2 - \omega^2\right)^2 + \omega^2\Gamma^2} \tag{1.68c}$$

From equation (1.68) we obtain the real and imaginary parts:

$$\varepsilon_r'(\omega) = 1 + \omega_p^2 \frac{\omega_0^2 - \omega^2}{\left(\omega_0^2 - \omega^2\right)^2 + \omega^2\Gamma^2} \tag{1.69a}$$

$$\varepsilon_r''(\omega) = \omega_p^2 \frac{\omega\Gamma}{\left(\omega_0^2 - \omega^2\right)^2 + \omega^2\Gamma^2} \tag{1.69b}$$

In figure 1.6 you can observe the real and imaginary parts of the relative permittivity, as a function of frequency. Notice that at high frequencies, ε_r asymptotically approaches 1, which indicates that the propagation of a high frequency electromagnetic wave tends to be similar to vacuum propagation.

Let us see what happens, in this context, to the refractive index. In its complex form, the refractive index can be defined as:

$$\tilde{n} = n_0 + i\kappa = \pm\sqrt{\mu_r \varepsilon_r} \tag{1.70}$$

In equation (1.70), the real part, n_0, is defined as the 'normal refractive index', and it represents the same material parameter that is most often referred to as 'refractive index', as is the case of Snell's law, for example. Notice, however, that there is also an imaginary part, κ, which is referred to as the 'extinction coefficient'. In our case, we consider the magnetic permeability $\mu_r \approx 1$, i.e., we neglect the material response to the magnetic field of the incident wave. We can, therefore, directly correlate the complex refractive index to the relative permittivity, ε_r:

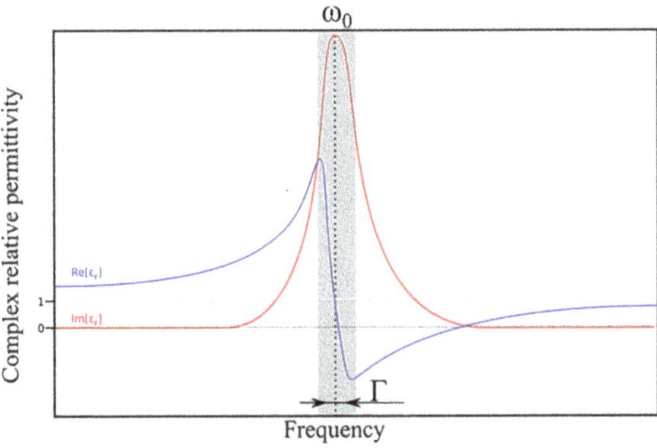

Figure 1.6. Schematic representation of the relative electric permittivity as a function of frequency.

$$(n_0 + i\kappa)^2 = \varepsilon_r' + i\varepsilon_r'' \tag{1.71a}$$

$$n_0^2 + i2n_0\kappa - \kappa^2 = \varepsilon_r' + i\varepsilon_r'' \tag{1.71b}$$

$$(n_0^2 - \kappa^2) + i(2n_0\kappa) = \varepsilon_r' + i\varepsilon_r'' \tag{1.71c}$$

$$\Rightarrow \varepsilon_r' = n_0^2 - \kappa^2 \tag{1.71d}$$

$$\Rightarrow \varepsilon_r'' = 2n_0\kappa \tag{1.71e}$$

If we know the complex refractive index, we can calculate the relative permittivity. Similarly, if we know the relative permittivity, we can calculate the refractive index. In order to calculate the refractive index using permittivity, we must solve equation (1.71e) for κ, and then replace κ in equation (1.71e), therefore obtaining:

$$\kappa = \frac{\varepsilon_r''}{2n_0} \tag{1.72a}$$

$$n_0^2 - \left(\frac{\varepsilon_r''}{2n_0}\right)^2 = \varepsilon_r' \tag{1.72b}$$

Expanding the parenthesis from equation (1.72b) and rearranging the terms, we obtain a second-degree equation for n_0^2:

$$n_0^4 - \varepsilon_r'n_0^2 - \frac{\varepsilon_r''^2}{4} = 0 \tag{1.73}$$

Solving equation (1.73) we obtain:

$$n_0^2 = \frac{\varepsilon_r' + \sqrt{\varepsilon_r'^2 + \varepsilon_r''^2}}{2} \tag{1.74a}$$

$$\Rightarrow n_0 = \sqrt{\frac{|\varepsilon_r| + \varepsilon_r'}{2}} \tag{1.74b}$$

In a similar manner, we can obtain an expression for the extinction coefficient, κ:

$$\kappa = \sqrt{\frac{|\varepsilon_r| - \varepsilon_r'}{2}} \tag{1.75}$$

Addendum 5 *Notice the forms of n_0 si κ as a function of the complex relative permittivity. What do you notice?*

Although not rigorously developed in this section, the theoretical model discussed above does allow us to describe the attenuation of a wave inside a dielectric. Therefore, let us further discuss the attenuation coefficient, often noted as α. When an electromagnetic wave propagates through an isotropic dielectric material, we can mathematically describe the phenomenon as:

$$E(z) = E_0 e^{ik_0 \tilde{n} z} = E_0 e^{ik_0(n_0 + i\kappa)z} = E_0 e^{-k_0 \kappa z} e^{ik_0 n_0 z} \tag{1.76}$$

Addendum 6 *The wavevector, k_0, has the index 0 so as not to be confused with the imaginary part of the refractive index, κ. Notice the negative sign, ' $-$ ', after expanding the exponential.*

Generally, a wave can be intuitively described using the attenuation constant, α, and the phase, β, as:

$$E(z) = E_0 \; \underset{\text{attenuation}}{\underline{e^{-\alpha z}}} \; e^{i\beta z} \tag{1.77}$$

Comparing equations (1.76) and (1.77), we can deduce that:

$$\alpha = k_0 \kappa \tag{1.78a}$$

$$\beta = k_0 n_0 \tag{1.78b}$$

At this point, the attenuation coefficient was introduced rather intuitively, as opposed to using a rigorous mathematical deduction. However, the reasoning for this approach is to connect the propagation of an electromagnetic wave to the response of a material to that oscillating wave, a response which shows specific characteristics. These characteristics are inter-dependent, expressed through terms such as 'permittivity', 'refractive index', 'attenuation', and others. We can graphically represent the refractive index (real and imaginary parts, separately), as functions of frequency (see figure 1.7).

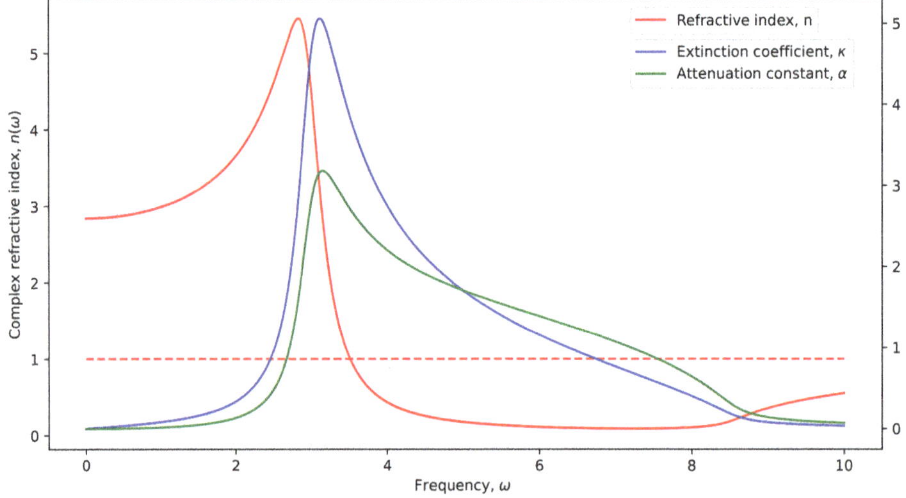

Figure 1.7. Refractive index, extinction coefficient and attenuation coefficient, as functions of frequency.

What can we notice in figure 1.7? First of all, we notice that, similar to relative permittivity, ε_r, and electric susceptibility, χ_e, the interaction between the incident wave and the dielectric material becomes highly complex around the resonance frequency. We notice that somewhere after the resonance frequency, the refractive index can reach negative values, however, as frequency rises, the refractive index gets asymptotically closer to 1, i.e., the wave will behave as in vacuum (once again similar to ε_r and χ_e). We do notice, on the other hand, that the extinction coefficient, κ, goes to 0, as well as the attenuation, but this happens for both low and high frequencies.

Without going into further details, we now reach a point where we can conclude on some fundamental aspects regarding the behavior of dielectrics in the presence of an electromagnetic wave. In figure 1.8 we have shown a TART plot—Transmission, Absorption, Reflection, Transmission. In the upper plot we can observe the

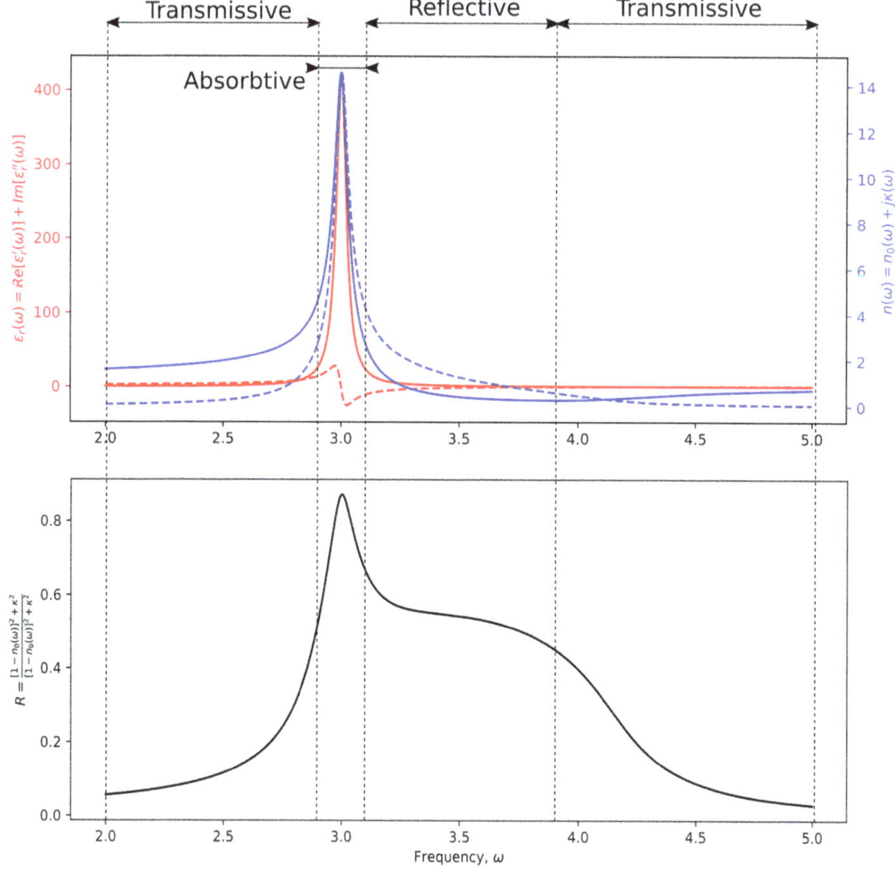

Figure 1.8. General TART plot for dielectrics, i.e., Transmission, Absorption, Reflection, Transmission. Upper plot shows the refractive index and electric permittivity (real and imaginary parts), lower plots shows reflectance.

refractive index and electric permittivity (real and imaginary parts). In the lower plot, we have the reflectivity as a function of frequency.

A dielectric material shows high transmission at frequencies below resonance, where the attenuation coefficient, κ, is small. As the frequency of the incident electromagnetic wave increases, the material response is different. Around resonance, a dielectric shows high absorption. When frequency increases further from resonance, the material shows high reflectivity, which remains as such until the frequency of the incident wave approaches the plasma frequency, ω_p although the reflectivity variation is not as fast as the case of absorption. After ω_p, high transmission is again shown.

References

[1] Ware M and Peatross J 2015 *Physics of Light and Optics* (Lulu.com)
[2] Feynman R, Leighton R and Sands M 2011 *The Feynman Lectures on Physics* **Vol II** New Millennium edition (New York: Basic Books)
[3] Heaviside O 1894 Electromagnetic waves, the propagation of potential, and the electromagnetic effects of moving charge *Electrical Papers* **Vol 2** (Cambridge: Cambridge Univ. Press) pp 490–9

IOP Publishing

High Resolution Laser Microprocessing
Implementation and techniques
Bogdan Ştefăniţă Călin, Marian Zamfirescu and Niculae Puşcaş

Chapter 2

Principles of laser physics

This section contains information regarding the generation of laser light, and its properties. Although the information is presented using a step-by-step bottom-up approach, we will present arguments for why each element is important when learning about and working with lasers.

Let's first discuss lasers and laser radiation, in order to try and paint the bigger picture. Laser radiation is electromagnetic radiation, similar to light. It is light, in fact, but has some particular properties. It is usually found that laser radiation should have three main characteristics: it should be monochromatic, directional, and coherent. As we will see further on, this affirmation, while not wrong, is not entirely correct.

Firstly, there is no purely monochromatic light, as that would imply it stretches infinitely in space. Based on Fourier theory, any signal can be described using an infinite sum of sinusoidal functions. But to have a purely monochromatic signal, i.e., one pure frequency, would mean it should stretch to infinity. Otherwise, it would not be a sinusoidal function, but rather a wavepacket (sum of sinusoidal functions). This sum, in turn, determines a broadening in the frequency domain. To understand this relation easier, we can refer to Heisenberg's uncertainty principle:

$$\Delta E \cdot \Delta t \geqslant \frac{\hbar}{2} \tag{2.1}$$

where ΔE is the energy band, Δt is the time band, and $\hbar$ is the reduced Planck's constant. We are talking about light, and we know the energy of a photon is $E_f = \hbar \omega$. By a simple substitution we can obtain:

$$\Delta \omega \cdot \Delta t \geqslant \frac{1}{2} \tag{2.2}$$

A perfectly monochromatic wave would mean $\Delta \omega = 0$, which in turn means $\Delta t \to \infty$, i.e., the wave would have to propagate forever, or in other words, into

doi:10.1088/978-0-7503-3239-2ch2   2-1

an infinite space. Therefore, what we mean by 'monochromaticity' when it comes to laser beams actually refers to a very narrow spectrum. The width is determined by various broadening mechanisms, which will be described in the following chapter.

However, it goes further than that! You may already know that many lasers have short and ultrashort pulses. What does that mean for the spectrum? If we look at the same uncertainty principle above, it would mean that the shorter a pulse is, the smaller Δt is. In a similar manner, the minimum $\Delta \omega$ increases. As we will see further in the text, this becomes important for ultrashort laser systems and their applications.

The second particular property of laser light is denoted as directionality, i.e., laser light propagates as a narrow collimated beam. This has to do with how most resonant cavities and amplifiers are designed and it is not necessarily an intrinsic property of laser light. Most laser systems deliver a well-collimated beam with a Gaussian or top-hat transverse intensity profile. However, although useful, it is not a necessary requirement of laser radiation. In many situations, laser radiation has an increased divergence right after generation and/or amplification, and it is collimated before the output window of the system. This directionality is less significant as the laser system is smaller. More precisely, the smaller the laser beam is transversally, the higher its divergence. For example, non-collimated laser diodes have a typical divergence of the order of tens of degrees, while He–Ne lasers have a divergence of the order of milliradians.

Why is it considered an important and rather specific property of laser radiation then? Let us compare it with a conventional light source, say, an incandescent light bulb. If we consider the filament to be small enough to be considered a point source, then the radiation it produces is propagating uniformly in all directions, and the intensity obeys the inverse square law, i.e., the intensity is inversely proportional to the square of the distance from the source. Light from a laser system, in most cases, is delivered as a quasi-parallel beam, and its intensity remains roughly the same over a long distance. Moreover, because it is quasi-parallel, it can be more easily tightly focused.

The third property of laser light is coherence. This is the most definitive property and its source lies in the generation mechanism of laser light. More specifically, it is the stimulated emission process that generates coherence in laser radiation. Coherence in lasers is often encountered as spatial and temporal. These terms are used to describe conditions where different rays of the laser beam are propagating in phase. However, it is more easily understood if we analyze spatial coherence as transverse coherence, and temporal coherence as longitudinal coherence.

2.1 Introductory atomic physics for laser transitions

At the core of laser systems and their applications lie atomic transitions and light–matter interaction. To better understand how laser radiation interacts with the active medium, what a gain bandwidth is, and other phenomena taking place in order to obtain a laser output, let us first look at the basic physics of an atom. More precisely, what the quantum numbers are and where they come from. That is because all light generation and light–matter interaction revolve around these numbers.

There were several theoretical models of an atom over the years, such as the famous plum-pudding model and the planetary model, each trying a different approach to explaining the observed phenomena involving atoms. There are, however, two models that are often used to describe an atom, each for different reasons. One uses Bohr's postulates, and the other is a quantum mechanical approach. While the correct description of an atom is indeed the quantum mechanical one, Bohr's model is still useful when describing some fundamentals. As such, we will start with Bohr's model to derive the principal quantum number, and then move to the more elegant approach of solving Schrödinger's equation for a hydrogen-like atom.

Bohr's model of the atom is also known as the Rutherford–Bohr model. In the early 1900s, Hans Geiger and Ernest Marsden, under the supervision of Ernest Rutherford, conducted an experiment where they used α particles from a radon-222 isotope source to irradiate a thin gold foil, in order to verify Thomson's plum-pudding model. If Thomson's model was to be validated, the α particles should have passed through the thin foil without much deviation from the initial path. What they observed, however, was that some of the particles were strongly deviated, and even scattered backwards. This suggested that the atoms had a positively charged nucleus, with significant mass, that was confined in a very small volume, in great contrast to what the plum-pudding model was suggesting. From this information arose the planetary model (also known as the Rutherford model), where negatively charged electrons were orbiting a positively charged nucleus, the latter being confined in a significantly small volume compared to the whole atom. But this model was flawed for several reasons. The first reason was that, in this model, orbiting electrons are accelerated (due to centripetal force), and according to classical electrodynamics, any accelerated point charge will emit electromagnetic energy. As such, the electron would lose energy as it orbits, and fall in on the nucleus in ≈ 16 ps. Moreover, as the electron spirals down towards the nucleus, it increases the speed and completes a revolution faster, therefore the frequency of the emitted electromagnetic radiation would smoothly increase, which would result in a continuous spectrum. However, that was completely different from the experimental observations, as it was seen that atoms emit specific discrete wavelengths.

Even so, experimental observations were compelling and a planetary system-like model seemed to fit at the time, albeit with something missing. Therefore, to overcome these deficiencies, Bohr's model starts with three postulates, or rather three imposed constraints:

1. There are stationary orbits where the electron can move without radiating power. These orbits are found at a distance where the angular momentum is an integer number of Planck's constant h (from de Broglie's relationships). In mathematical terms, this means:

$$mv_n r_n = nh \qquad (2.3)$$

where m is the electron mass, v_n is its velocity, r is the distance from the nucleus where it orbits, and n is an integer.

2. The electron can switch orbit only by absorbing or emitting radiation, given by:

$$E_i - E_j = h\nu_{ij} \tag{2.4}$$

where E_i and E_j are the energies of the stationary orbits between which the electron transits.

3. The interaction between the electron and the nucleus is electrostatic. In mathematical terms, it means the centrifugal and electrostatic forces are equal:

$$\frac{mv_n^2}{r_n} = \frac{Ze^2}{4\pi\varepsilon_0 r_n^2} \tag{2.5}$$

where Z is the charge number, e is the elementary charge, and ε_0 is the vacuum permittivity. Note that the right-hand side of equation (2.5) is positive, as we consider the influence of the (positive) nucleus over the electron.

From equation (2.3), we can obtain:

$$v_n = \frac{n\hbar}{mr_n} \tag{2.6}$$

and replacing v_n in equation (2.5) we obtain:

$$
m\frac{n^2\hbar^2}{m^2 r_n^2} = r_n\frac{Ze^2}{4\pi\varepsilon_0 r_n^2}
$$
$$
\Rightarrow r_n = \frac{4\pi\varepsilon_0 n^2\hbar^2}{Ze^2} \tag{2.7}
$$
$$
\Rightarrow r_n = r_1 \cdot n^2
$$

where r_1 is the radius of the first Bohr orbit, $r_1 \approx 0.5$ Å.

Using equations (2.6) and (2.7) we can obtain v_n as:

$$
v_n = \frac{n\hbar}{mr_n} = \frac{\hbar}{mnr_n}
$$
$$
\Rightarrow v_n = \frac{v_1}{n} \tag{2.8}
$$

where v_1 is the velocity of the electron found on the first Bohr orbit, and $v_1 \approx 2.2 \times 10^6$ m s^{-1}.

Now that we have an expression for the velocity, as well as for the potential energy (equation (2.5)), we can obtain the total energy as:

$$E_n \quad = \quad \frac{mv_n^2}{2} - \frac{Ze^2}{4\pi\varepsilon_0 r_n}$$

$$= \quad \frac{1}{2} \cdot \frac{Ze^2}{4\pi\varepsilon_0 r_n} - \frac{Ze^2}{4\pi\varepsilon_0 r_n} \tag{2.9}$$

$$\Rightarrow E_n = \frac{-Ze^2}{8\pi\varepsilon_0 r_n}$$

Using equation (2.7) in equation (2.9), we obtain:

$$E_n = -\frac{(Ze^2)^2 m}{2(4\pi\varepsilon_0)^2 n^2 \hbar^2} \tag{2.10}$$

$$\Rightarrow E_n = \frac{E_1}{n^2}$$

where E_1 is the energy of the electron found in Bohr's first orbit, and $E_1 \approx -13.6\,\text{eV}$. The negative value of E_1 means that the electron is in a bound state (needs 13.6 eV to escape the influence of the nucleus).

Let us look at how energy levels are stacked around a nucleus (see figure 2.1).

The n parameter represents the **principal quantum number**. We can also observe that the energy levels get closer and closer together as n increases.

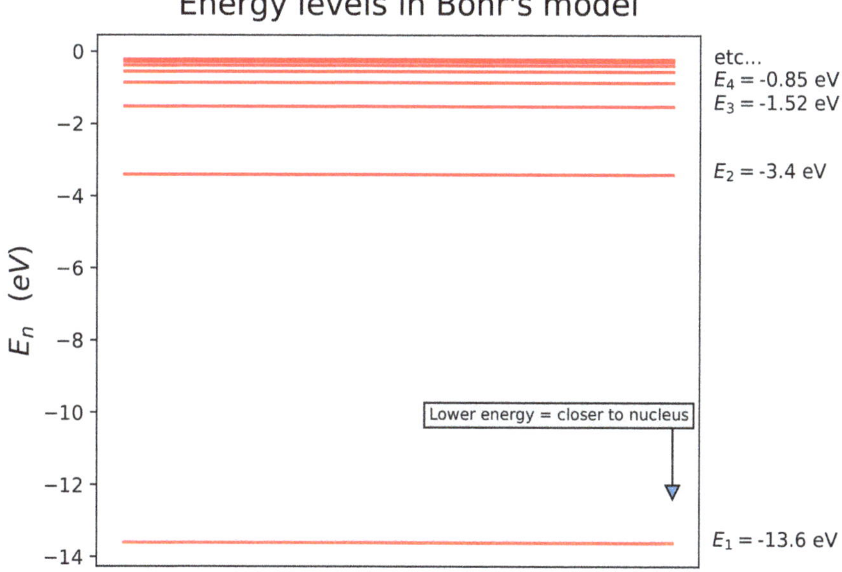

Figure 2.1. Energy levels according to Bohr's atomic model.

Bohr's model has some inconsistencies. While often referred to as 'wrong', it is rather an incomplete model. It can only be applied to single electron systems. It is most often used as a gateway model to quantum mechanical description. It can predict the energy levels for the hydrogen atom, and the difference between them. It cannot explain, however, the existence of the other quantum numbers and the dynamics described by them. Bohr's model has a problem with the very first orbit, if we take Heisenberg's uncertainty principle, i.e., $\Delta p \Delta x \geqslant \frac{\hbar}{2}$. Moreover, it was experimentally demonstrated to be wrong (or rather incomplete) when the fine structure was measured. Even so, the model offers accurate values for the energy levels.

Visualization is important in this case. In Bohr's model, the length of an orbit is equal to an **integer of de Broglie wavelengths**. That number happens to be n (see figure 2.2).

There were other questions that remained unanswered by Bohr's model. The best approach to describe the structure of an atom and its dynamics is to take the quantum mechanical approach. Solving Schrödinger's equation for an atom is not an easy task, but it is important to at least outline the issue. Let us consider the hydrogen atom, for it is the simplest to analyze.

Did you notice anything odd about Bohr's model so far? It is flat, i.e., two-dimensional. For a quantum mechanical analysis, we have to consider a fully three-dimensional system. The position of the electron is described by three parameters: position r, azimuth angle ϕ and elevation angle θ (see figure 2.3). These are spherical coordinates and we use them in this case due to the central potential of the nucleus (proton) in which the electron is moving. Moreover, as we will see further, we will need to use the variable separation method, a feat that cannot be accomplished using Cartesian coordinates, for the case of the central potential. Schrödinger's equation in this case (Cartesian coordinates) is defined as:

$$H = -\frac{\hbar^2}{2\mu}\Delta\Psi(r,\,\theta,\,\phi) + U(r)\Psi(r,\,\theta,\,\phi) = E\Psi(r,\,\theta,\,\phi) \qquad (2.11)$$

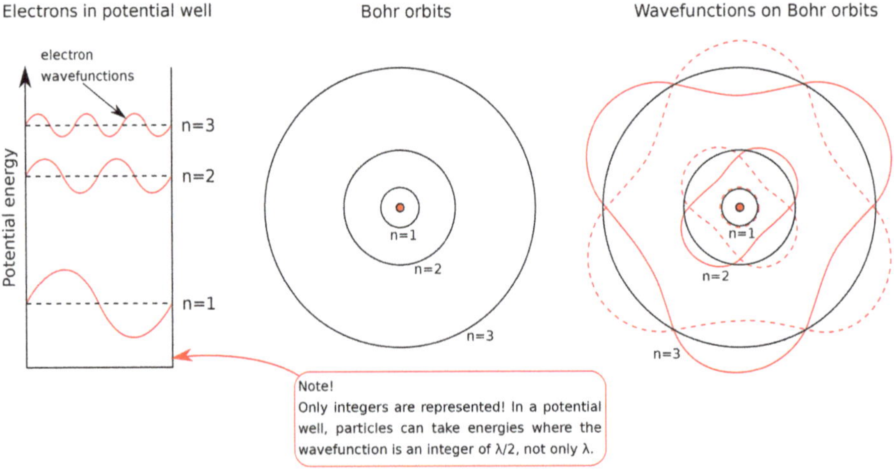

Figure 2.2. Schematic representation of the Bohr model of the atom.

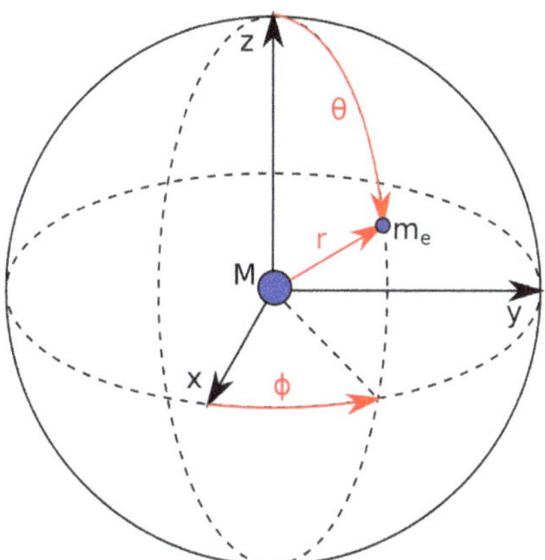

Figure 2.3. Positioning parameters for the electron in a hydrogen atom.

where $\Psi(r, \theta, \phi)$ is the wavefunction of the electron, $\mu = \frac{m_e M}{m_e + M}$ is the reduced mass, and $U(r) = \frac{-e^2}{4\pi\varepsilon_0 r}$ is the electrostatic potential.

Since the position of the electron is determined by r, θ, and ϕ, we need to switch to spherical coordinates for the Δ operator found in equation (2.11). Using the conversion formulas:

$$
\begin{aligned}
x &= r \sin\theta \cos\phi \\
y &= r \sin\theta \sin\phi \\
z &= r \cos\theta
\end{aligned}
\tag{2.12}
$$

and knowing that, for a differentiable function, A:

$$
\frac{\mathrm{d}A}{\mathrm{d}x} = \frac{\mathrm{d}A}{\mathrm{d}r} \cdot \frac{\mathrm{d}r}{x}
\tag{2.13}
$$

we obtain Schrödinger's equation, in spherical coordinates, as:

$$
\frac{-\hbar^2}{2\mu}\left[\frac{1}{r^2\sin\theta}\sin\theta\frac{\partial}{\partial r}\left(r^2\frac{\partial\Psi}{\partial r}\right) + \frac{\partial}{\partial\theta}\left(\sin\theta\frac{\partial\Psi}{\partial\theta}\right) + \frac{1}{\sin\theta}\frac{\partial^2\Psi}{\partial\phi}\right] - \frac{e^2}{4\pi\varepsilon_0 r}\Psi
$$
$$
= E\Psi(r, \theta, \phi)
\tag{2.14}
$$

Moving E to the left side, and multiplying equation (2.14) with $-\frac{2\mu}{\hbar^2}$, we obtain:

$$
\frac{1}{r^2\sin\theta}\sin\theta\frac{\partial}{\partial r}\left(r^2\frac{\partial\Psi}{\partial r}\right) + \frac{\partial}{\partial\theta}\left(\sin\theta\frac{\partial\Psi}{\partial\theta}\right) + \frac{1}{\sin\theta}\frac{\partial^2\Psi}{\partial\phi} + \frac{2\mu}{\hbar^2}\left(\frac{e^2}{4\pi\varepsilon_0 r} + E\right)\Psi = 0
\tag{2.15}
$$

We can see that some of the terms depend on the radial part, r, and the others on the angular parts, θ and ϕ. As such, we can employ the variable separation method, and consider a solution of the form:

$$
\Psi(r, \theta, \phi) = R(r)Y(\theta, \phi)
\tag{2.16}
$$

where $R(r)$ are radial functions, and $Y(\theta, \phi)$ are spherical harmonics.

If we insert equation (2.16) into equation (2.15), do the derivatives of Ψ and rearrange the terms, we obtain:

$$\frac{1}{R}\left(r^2\frac{dR}{dr}\right) + \frac{2\mu r^2}{\hbar^2}\left(E + \frac{e^2}{4\pi\varepsilon_0 r}\right) + \frac{1}{Y\sin\theta}\frac{\partial}{\partial\theta}\left(\sin\theta\frac{\partial Y}{\partial\theta}\right) + \frac{1}{Y\sin^2\theta}\frac{\partial^2 Y}{\partial\phi^2} = 0 \quad (2.17)$$

We observe that the angular part of equation (2.17) can further be split into two parts, applying the variable separation again. Let us then consider that:

$$Y(\theta, \phi) = P(\theta) \cdot F(\phi) \quad (2.18)$$

If we insert equation (2.18) into equation (2.17) and, again, do the required derivatives and rearrangements, we obtain:

$$\underbrace{\frac{1}{R}\left(r^2\frac{dR}{dr}\right) + \frac{2\mu r^2}{\hbar^2}\left(E + \frac{e^2}{4\pi\epsilon_0 r}\right)}_{\text{radial}} + \underbrace{\frac{\sin\theta}{P}\frac{d}{d\theta}\left(\sin\theta\frac{dP}{d\theta}\right) + A\sin^2\theta}_{\text{elevation}} + \underbrace{\frac{1}{\phi}\frac{d^2F}{d\phi^2}}_{\text{azimuth}} = 0$$

$$(2.19)$$

We can observe that there are three equations that we must solve. We can write $\Psi(r, \theta, \Phi)$ as:

$$\Psi(r, \theta, \phi) = R(r)P(\theta)F(\phi) \quad (2.20)$$

Each of these three parts of Ψ gives us a quantum number:

1. $R(r)$—gives the **principal quantum number**, $n = 1, 2, 3, \ldots$
   Each number corresponds to a specific orbit, and are also noted as K, L, M, N, …
2. $P(\theta)$—gives the **orbital quantum number**, $l = 0, 1, 2, \ldots, n - 1$
   These values are also often noted as s, p, d, f, ….
3. $F(\phi)$—gives the **magnetic quantum number**, $m_l = -l, -l + 1, \ldots, l$
   This gives us the spatial orientation of the angular momentum, and therefore affects the energy of an electron only in the presence of external magnetic field.

There is a **fourth quantum number**, $m_s = \pm\frac{1}{2}$, which represents the spin of the electron (rotation about its own axis).

These quantum numbers define the energy levels of an atom, under certain conditions. There are atoms, however, that have electrons with the same n and l values, but different m_l and electron spin (m_s). The energy level difference between these electrons and another electron, say closer to the nucleus, is the same. These are called degenerate energy levels, and we will see how this influences the transitions and light generation inside a material in chapter 2.2. In other words, a level is said to be degenerate if it can 'hold' at least two electrons, albeit with different quantum numbers. The magnetic quantum number, m_l, splits an energy level in the fine structure of an atom, when the atom is in the presence of an external magnetic field.

This removes the degeneracy and creates the so-called hyperfine structure of the atom.

The unnormalized general solutions for each equation (radial, polar, and azimuth), i.e., each part of equation (2.20), are as follows:

$$R(r) \propto r^l e^{-\frac{r}{na_0}} L_{n+1}^{2l+1}\left(\frac{2r}{na_0}\right)$$

$$P(\theta) \propto P_l^{m_l}(\cos\theta)$$

$$F(\phi) \propto e^{im_l\phi}$$

$$(2.21)$$

where n, l, m_l are the quantum numbers listed above, r is the radial distance, $a_0 = r_1$ is the first Bohr radius, $L_{n+1}^{2l+1}(\frac{2r}{na_0})$ are associated Laguerre polynomials, and $P_l^{m_l}(\cos\theta)$ are the associated Legendre polynomials. A detailed description of solving Schrödinger's equation for the hydrogen atom can be found in Arthur Beiser's *Concepts of Modern Physics*, chapter 6 [1]. As an example, Table 2.1 shows normalized wavefunctions for the hydrogen atom for principal quantum numbers of $n = 1$ and $n = 2$.

Radial wavefunction, probability density, and probability distribution are presented in figure 2.4. For all n where $l = 0$, the radial probability density will be maximum over the center of the system, i.e., the nucleus. This, however, does **not** mean the electron is most likely found inside the center of the nucleus. It is a common misconception. The probability density is, indeed, proportional to Ψ^2. However, the probability of finding an electron in a certain volume is found by **integrating the probability density function**. In other words, the probability of finding an electron in an infinitesimal volume element, dV, is $\Psi^2 dV$. The element, dV, in spherical coordinates, is:

$$dV = r^2 dr \sin\theta d\theta d\phi$$

$$(2.22)$$

For the radial function, we have no dependency of the angular parts of the integration, $d\theta$ and $d\phi$, i.e., we integrate over all angles. Considering the separable variables as well, this results in:

Table 2.1. Normalized wavefunctions for the hydrogen atom for $n = 1$ and $n = 2$ [1].

n	l	m_l	$R(r)$	$P(\theta)$	$F(\phi)$	$\Psi(r,\theta,\phi)$
1	0	0	$\dfrac{2}{a_0^{3/2}} e^{-\frac{r}{a_0}}$	$\dfrac{1}{\sqrt{2}}$	$\dfrac{1}{\sqrt{2\pi}}$	$\dfrac{1}{\sqrt{\pi}\, a_0^{3/2}} e^{-\frac{r}{a_0}}$
2	0	0	$\dfrac{1}{2\sqrt{2}\, a_0^{3/2}}\left(2 - \dfrac{r}{a_0}\right) e^{-\frac{r}{2a_0}}$	$\dfrac{1}{\sqrt{2}}$	$\dfrac{1}{\sqrt{2\pi}}$	$\dfrac{1}{4\sqrt{2\pi}\, a_0^{3/2}}\left(2 - \dfrac{r}{a_0}\right) e^{\frac{r}{2a_0}}$
2	1	0	$\dfrac{1}{2\sqrt{6}\, a_0^{3/2}}\dfrac{r}{a_0} e^{-\frac{r}{2a_0}}$	$\dfrac{\sqrt{6}}{2}\cos\theta$	$\dfrac{1}{\sqrt{\sqrt{2\pi}}}$	$\dfrac{1}{4\sqrt{2\pi}} a_0^{3/2}\dfrac{r}{a_0} e^{-\frac{r}{2a_0}}\cos\theta$
2	1	± 1	$\dfrac{1}{2\sqrt{6}\, a_0^{3/2}}\dfrac{r}{a_0} e^{-\frac{r}{2a_0}}$	$\dfrac{\sqrt{3}}{2}\sin\theta$	$\dfrac{1}{\sqrt{2\pi}} e^{\pm i\phi}$	$\dfrac{1}{8\sqrt{\pi}\, a_0^{3/2}}\dfrac{r}{a_0} e^{-\frac{r}{2a_0}}\sin\theta e^{\pm i\phi}$

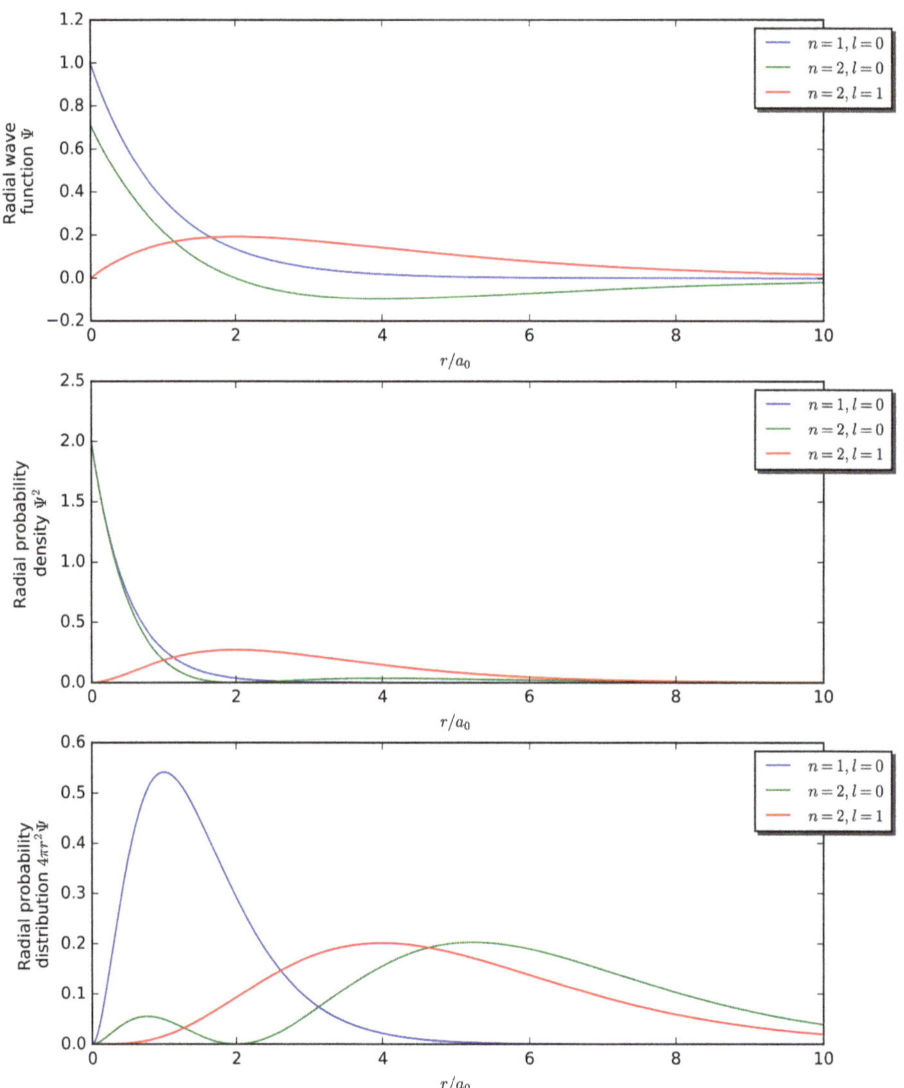

Figure 2.4. Radial wavefunction, probability density, and probability distribution for different quantum numbers.

$$\Omega = \int_0^{2\pi} \mathrm{d}\phi \int_0^{\pi} \sin\theta\,\mathrm{d}\theta$$
$$= \int_0^{2\pi} \mathrm{d}\phi (-\cos\theta)\Big|_0^{\pi}$$
$$= 2\int_0^{2\pi} \mathrm{d}\phi$$
$$= 4\pi \tag{2.23}$$

which in turn, results that:

$$dV = 4\pi r^2 dr$$
$$\Rightarrow \Psi^2 dV = \Psi^2 4\pi r^2 dr$$

(2.24)

Integrating equation (2.24) results in the probability distribution, i.e., the probability of finding the electron in the space surrounding the nucleus (see figure 2.4). The radius at which we have the highest probability of localizing an electron fits to Bohr's model for n.

For all n, where $l = 0$ we can see solutions do not present an angular dependence, i.e., they have spherical symmetry. When $l \neq 0$ and $m_l = 0$, $\Psi = \Psi(r, \theta)$, the probability distribution is symmetrical with respect to the xy plane (see figure 2.5). The probability distribution of finding an electron in the yz plane that goes through the center of the nucleus for cases presented in figure 2.4 are presented in figure 2.5:

So far we have analyzed several possible situations that one electron can be in. When it comes to lasers and active media, there are many electrons involved in every atom or molecule. How, then, are multiple electrons occupying energy levels? We can first refer to Pauli's exclusion principle, which tells us that identical fermions cannot occupy the same quantum state in a system, i.e., electrons cannot be found on levels described by the same quantum numbers. Moreover, we can also refer to Hund's first rule. The first rule says that, for a given electronic configuration, the lowest energy states are those that maximize the total spin quantum number (i.e., parallel spins). Next, we can refer to Madelung's rule and, implicitly, the aufbau principle (basically meaning 'construction principle'). The aufbau principle is that electrons occupy the lowest energy levels first, which, along Hund's first rule, indicates that subshells are occupied with electrons that have parallel spins first (also known as the 'bus seat rule'). Madelung's rule tries to establish the order in which electrons occupy energy levels in an atomic quantum system. By 'try', we mean that Madelung's rule is not entirely applicable to atoms with 20 or more electrons, as interactions between the nucleus and electrons become highly complex and exceptions to this rule arise. As an example, the electronic configuration for argon is written as:

$$1s^2 2s^2 2p^6 3s^2 3p^6$$

(2.25)

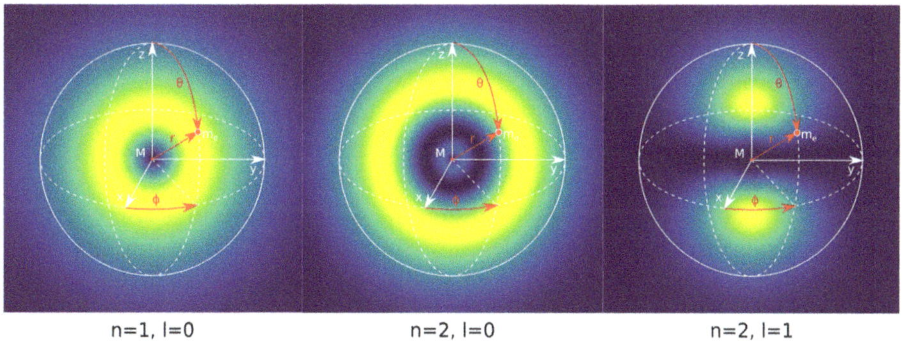

Figure 2.5. Probability distribution of an electron in the YZ plane for different values of n and l.

Oftentimes, when it comes to laser media, the spectroscopic notation is used:

$$N^{2s+1}L_j \tag{2.26}$$

where N is the principal quantum number (most often omitted), s is the total spin quantum number, L is the orbital quantum number (in the 's, p, d, $\cdots$' notation), and j is the total angular momentum quantum number.

2.2 Principles of laser physics

Laser physics refers to the phenomena underlying laser generation. As such, we will begin with an analysis of light–matter interaction. A laser system is comprised of three main components: an active optical gain medium, an energy pumping source and a feedback system (optical resonator). The gain medium is a material that amplified the beam through stimulated emission. It is a material with precise control over its properties, such as purity, size, active species concentration and shape. It can be a solid, liquid, gas, and even a plasma. The pump source is often a flash lamp or diodes, but it can also be a gas discharge or electrical current source.

The gain medium is not comprised only of atoms that allow stimulated emission, therefore generating laser radiation. In fact, gain media has at least two species: one is referred to as 'active species', and the other as the 'lattice'. The active species is the material or chemical element that, as the name implies, actively participates in the stimulated emission processes, i.e., generates laser light. The lattice is the material that acts as a support for the active species, often being the intermediary for energy transfer from the pump to the active species, and regulating the physical processes (absorption, emission, thermal conductivity, etc). This is the reason lasers are referred to as, for example: Nd:YAG, He–Ne, Ti:sapphire, etc. In the first case, Nd:YAG, neodymium is the active species which generates the light, and the yttrium–aluminium garnet (YAG) is the lattice ('garnet' refers to the crystal structure). In the second case, He–Ne, neon is the active species, but directly pumping this gas is inefficient, and therefore helium is used. Similarly, in the Ti: sapphire case, Ti gains the energy and generates the laser radiation, the sapphire (monocrystalline Al_2O_3) is the lattice that holds everything in place and has great thermal conductivity. In some cases, it is not a unique compatibility between the active species and the lattice. For example, neodymium can also be embedded in YVO_4 crystals, for improved thermal conductivity. Similarly, YAG can be used as a lattice for ytterbium to act as the active species.

Modern laser systems, especially ultrashort pulse sources, are comprised of a laser oscillator and a laser amplifier. The main difference in the mechanisms is that the amplifier does not have a feedback system, i.e., optical oscillator.

In order to analyze the phenomena inside the gain medium of a laser, let us first consider the transition between atomic energy levels, at thermal equilibrium. We use the term 'population' to refer to the number of atoms per unit volume that are found on a specific energy level. Let us consider the population density N_i, which represents the number of atoms per unit volume found on the energy level E_i, under thermal equilibrium at temperature T. N_i is distributed according to Boltzmann statistics:

$$N_i = N \frac{g_i \exp\left(-\dfrac{E_i}{k_\mathrm{B} T}\right)}{\sum_i g_i \exp\left(-\dfrac{E_i}{k_\mathrm{B} T}\right)} \tag{2.27}$$

where g_i represents the degeneracies, and $N = \sum_i N_i$. The term 'degeneracy' refers to those energy levels that correspond to different quantum states, i.e., different quantum numbers for the same energy level.

For the issue at hand, let us consider that light–matter interaction is done through processes of either absorption or emission of photons. Emission processes can be spontaneous or stimulated. When an atom absorbs the energy of a photon, an electron can transit to a higher energy level, if the frequency of the photon is appropriate. This electron has a certain average lifetime on that energy level, after which it falls back to the inferior energy level, and emits a photon similar to the one that was absorbed. However, emission can also be stimulated. It means that when a photon interacts with an electron that is already on a superior energy level, there is a non-zero probability that the photon determines the electron to fall to the lower energy level. This results in two photons that are in phase and propagate in the same direction. These processes are schematically presented in figure 2.6.

In order to study absorption and emission processes, let us first consider an electromagnetic wave of frequency ω and radiant energy spectral density $s_\omega(\omega, T)$, which interacts with a medium that contains atoms with two available energy levels, E_1 and E_2. We also consider g_1 and g_2 degeneracies for E_1 and E_2, respectively.

Each transition has a certain probability of happening. Stimulated transitions, absorption, and stimulated emission, i.e., those that happen under the influence of the external electromagnetic field of density $s_\omega(\omega, T)$, also depend on the intensity of the mentioned field. Time variation of the populations of each energy level, in mathematical terms, can be written as:

$$\left(\frac{\mathrm{d}N_1}{\mathrm{d}t}\right) = -B_{12} N_1 s_\omega(\omega, T) \tag{2.28}$$

where B_{12} is Einstein's absorption coefficient. The negative sign tells us that the population N_1 drops as photons are absorbed (fewer atoms have electrons on E_1

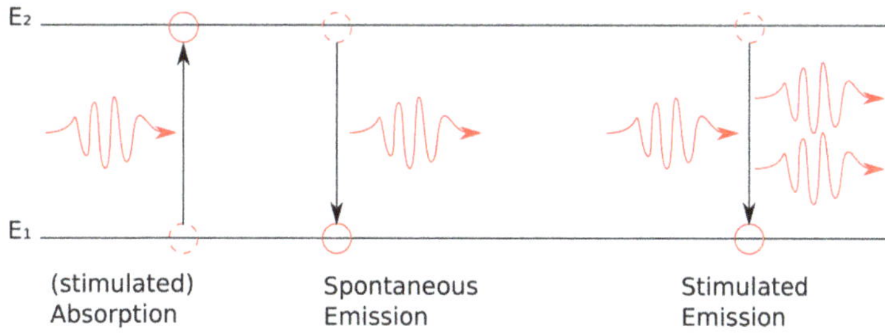

Figure 2.6. Light–matter interaction for a medium with two energy levels.

level). Notice the presence of the spectral density, $s_\omega(\omega, T)$. This means the absorption probability per time unit is $B_{12}s_\omega(\omega, T)$.

As mentioned before, an electron found on a higher energy level, in our case $E_2 > E_1$, has an average lifetime, after which it drops to the lower energy level, emitting a photon in the process. The energy of the photon is $\hbar\omega = E_2 - E_1$. This represents the spontaneous emission process. This process directly affects the population N_2:

$$\left(\frac{\mathrm{d}N_2}{\mathrm{d}t}\right)_{\text{spontaneous}} = -A_{21}N_2 \tag{2.29}$$

where A_{21} is the spontaneous emission probability. Notice that in this case, the transition probability is independent of the incident electromagnetic wave, as it is a process that happens naturally, without the need for external perturbation. If the transition is, however, determined by the incident radiation, then equation (2.29) becomes:

$$\left(\frac{\mathrm{d}N_2}{\mathrm{d}t}\right)_{\text{stimulated}} = -B_{21}N_2 s_\omega(\omega, T) \tag{2.30}$$

where B_{21} is Einstein's coefficient for stimulated emission.

We have mentioned we consider the analysis at thermodynamic equilibrium. This means absorption processes must be equal to emission processes:

$$B_{12}N_1 s_\omega(\omega, T) = A_{21}N_2 + B_{21}N_2 s_\omega(\omega, T) \tag{2.31}$$

It seems that at thermodynamic equilibrium, these processes are quite dependent on one another. How do we go about determining these coefficients? For the spontaneous emission, we can measure the lifetime under certain conditions. For stimulated processes, we can start by looking to the radiant energy spectral density, $s_\omega(\omega, T)$. From equation (2.31), we can determine that:

$$s_\omega(\omega, T) = \frac{A_{21}}{B_{12}\dfrac{N_1}{N_2} - B_{21}} \tag{2.32}$$

Let us look back at equation (2.27). We know that when N_1 population 'loses' an atom, N_2 gains one. If we consider the temperature of the medium, T, sufficiently high for Boltzmann statistics to be applicable, then the ratio between populations is:

$$\frac{N_2}{N_1} = \frac{g_2 \exp\left\{\left[-\dfrac{E_2 - E_0}{k_{\mathrm{B}}T}\right]\right\}}{g_1 \exp\left\{\left[-\dfrac{E_1 - E_0}{k_{\mathrm{B}}T}\right]\right\}} = \frac{g_2}{g_1}\exp\left\{\left[-\dfrac{E_2 - E_1}{k_{\mathrm{B}}T}\right]\right\} = \frac{g_2}{g_1}\exp\left\{\left[-\dfrac{\hbar\omega}{k_{\mathrm{B}}T}\right]\right\} \tag{2.33}$$

Introducing equation (2.33) into equation (2.32) we obtain:

$$s_\omega(\omega, T) = \frac{A_{21}}{B_{12}\dfrac{g_2}{g_1}\exp\left\{\left[\dfrac{\hbar\omega}{k_B T}\right]\right\} - B_{21}} \tag{2.34}$$

Note that the sign at the exponential function changed because in equation (2.34) we are using N_1/N_2, while in equation (2.33), the ratio is inverted. We are talking about a medium where atoms or molecules are interacting with an external field, absorbing and emitting photons. However, bear in mind that the photons re-emitted by some atoms can be reabsorbed by others. All processes happen at thermodynamic equilibrium. Therefore, we can compare these processes with Stefan–Boltzmann law for blackbody radiation:

$$s_\omega(\omega, T) - aT^4 \tag{2.35}$$

we can see that, as $T \to \infty$:

$$\lim_{T \to \infty} s_\omega \to \infty \tag{2.36}$$

If we compare equation (2.36) to equation (2.34), we see that when $T \to \infty$, then the exponential goes to 1, and, because $s_\omega \to \infty$, then the denominator in equation (2.34) must go to zero, i.e.,

$$B_{12}\frac{g_1}{g_2} - B_{21} = 0$$

$$B_{12}\frac{g_1}{g_2} = B_{21} \tag{2.37}$$

$$g_1 B_{12} = g_2 B_{21}$$

Using equations (2.37) and (2.34), we can now write the energy density as:

$$s_\omega(\omega, T) = \frac{g_2 A_{21}}{g_1 B_{12}} \cdot \frac{1}{\exp\left\{\left(\dfrac{\hbar\omega}{k_B T}\right) - 1\right\}} \tag{2.38}$$

When $\frac{\hbar\omega}{k_B T} \ll 1$, equation (2.38) must become Rayleigh–Jeans formula for blackbody radiation, i.e.,

$$\frac{\omega^2}{\pi^2 k^3}k_B T = \frac{g_2 A_{21}}{g_1 B_{12}} \cdot \frac{1}{1 + \dfrac{\hbar\omega}{k_B T} + \cdots - 1} \tag{2.39}$$

from which we can further obtain:

$$B_{12} = A_{21} \cdot \frac{g_2}{g_1} \cdot \frac{\pi^2 c^3}{\hbar\omega^3} \tag{2.40}$$

Note that in equation (2.39) we used the power series development of the exponential function, $\exp x = \sum_{k=0}^{\infty} \frac{x^k}{k!}$, and considered only the first term, since $\frac{\hbar\omega}{k_{\mathrm{B}}T} \ll 1$, therefore eliminating the exponential function itself and simplifying the equation.

Just before equation (2.29) we mentioned that for spontaneous emission, we can measure the lifetime. The coefficient A_{21} describes this spontaneous emission. Let us integrate equation (2.29), while considering that the initial N_2 population, i.e., for $t = 0$, is $N_2\,|_{t=0} = N_{20}$:

$$\frac{\mathrm{d}N_2}{\mathrm{d}t} = -A_{21}N_2$$
$$\Rightarrow \frac{1}{N_2}\mathrm{d}N_2 = -A_{21}\mathrm{d}t \left| \int ()\right. \tag{2.41}$$

We therefore obtain:

$$N_2 = N_{20}\exp\{(-A_{21}t)\} = N_{20}\exp\left\{\left(-\frac{t}{\tau}\right)\right\} \tag{2.42}$$

where the coefficient $A_{21} = \frac{1}{\tau}$, and τ is the lifetime we have mentioned beforehand. Replacing A_{21} in equation (2.40) we obtain:

$$B_{12} = \frac{1}{\tau} \cdot \frac{g_2}{g_1} \cdot \frac{\pi^2 c^3}{\hbar\omega^3} \tag{2.43}$$

Using equations (2.43) and (2.37) we can determine both Einstein's coefficients for stimulated transitions (absorption and emission) by experimentally measuring the lifetime, τ.

2.3 Population inversion and threshold

In order to obtain laser radiation, we need stimulated emission with a positive feedback. We have seen in the previous section that Einstein's coefficients for stimulated transitions are related through equation (2.37). As mentioned previously, the terms g_i represent degeneracies for the energy levels involved. If the levels are identical from a quantum point of view, then $B_{12} = B_{21}$, i.e., the transition probability is identical for both absorption and stimulated emission. One condition to obtain positive feedback, and laser radiation implicitly, is to obtain the so-called 'population inversion'. This means to have a greater number of atoms on the higher energy level, i.e., $N_2 > N_1$. In our analysis, we considered optical pumping using an external electromagnetic field of density $s_\omega(\omega, T)$. To understand the population inversion, we need to analyze the connection between light propagation through the active medium and the populations, N_1 and N_2. Let us analyze the propagation of this field through the active medium first.

Light absorption propagating through a medium is described using Lambert–Beer law:

$$\frac{\mathrm{d}I(x)}{\mathrm{d}x} = -\alpha I(x) \tag{2.44}$$

where $I(x)$ is the intensity of light, x is the propagation direction/distance, and α is the attenuation coefficient. Previously, we have used the radiant energy spectral density, $s_\omega(\omega, T)$. The intensity can be written in terms of the energy density as follows:

$$I = \frac{c}{n} s_\omega(\omega, T) \tag{2.45}$$

where c is the vacuum velocity of light, n is the refractive index, and c/n represents the velocity of light in the medium. Moreover, the energy density in the volume unit depends on the energy levels as follows:

$$s_\omega = (N_1 - N_2)\hbar\omega \tag{2.46}$$

To find an expression for the attenuation coefficient, we need to correlate $\frac{dI}{dx}$ with the energy density. In equation (2.44), the term $dx = c \cdot dt$. Replacing equation (2.46) in equation (2.45), and applying the time derivative, we obtain:

$$
\begin{aligned}
\frac{dI}{dt} &= \frac{c}{n}\frac{ds_\omega}{dt} \\
&= \frac{c}{n}\left(\frac{dN_1}{dt} - \frac{dN_2}{dt}\right)\hbar\omega \\
&= -\frac{c\hbar\omega}{n}(B_{12}N_1 s_\omega - B_{21}N_2 s_\omega) \\
&= -\frac{c\hbar\omega}{n}(B_{12}N_1 - B_{21}N_2)s_\omega \\
&= -\frac{c\hbar\omega}{n}\left(B_{12}N_1 - B_{12}\frac{g_1}{g_2}N_2\right)s_\omega
\end{aligned}
\tag{2.47}
$$

Note we did not introduce the spontaneous emission and A_{21}. The reason is that we are analyzing light–matter interactions along a specific direction, x in this case. However, spontaneous emission directs photons in a random direction, and the contribution to the x direction is negligible here. We can replace B_{12} with equation (2.43) and obtain:

$$
\begin{aligned}
\frac{dI}{dt} &= -\frac{c\hbar\omega}{n} \cdot \frac{1}{\tau} \cdot \frac{g_2}{g_1} \cdot \frac{\pi^2 c^3}{\hbar\omega^3}\left(N_1 - \frac{g_1}{g_2}N_2\right)s_\omega \\
&= -\frac{g_2 \pi^2 c^4}{g_1 \tau n \omega^2}\left(N_1 - \frac{g_1}{g_2}N_2\right)s_\omega \\
&= -\frac{g_2 \pi^2 c^3}{g_1 \tau n \omega^2}\left(N_1 - \frac{g_1}{g_2}N_2\right)s_\omega \\
&= -\frac{g_2 \pi^2 c^3}{g_1 \tau n \omega^2}\left(N_1 - \frac{g_1}{g_2}N_2\right)s_\omega
\end{aligned}
\tag{2.48}
$$

Using equations (2.48) and (2.45), we can obtain the attenuation coefficient, α:

$$\alpha = \frac{g_2}{g_1 \tau}\left(\frac{\pi c}{\omega}\right)^2\left(N_1 - \frac{g_1}{g_2}N_2\right) \tag{2.49}$$

The sign of the attenuation coefficient in equation (2.49) is determined by the relation between the populations N_1 and N_2, because all the other terms are positive. To simplify the issue, let us consider the energy levels are not degenerate, i.e., $g_1 = g_2$. As such, if $N_1 > N_2$, then $\alpha > 0$ and the light propagating through the medium is absorbed. If, however, $N_1 < N_2$, then $\alpha < 0$ and the light propagating through the medium is **amplified**. The term 'population inversion' is used to describe the state of a medium where $N_1 < N_2$, i.e., there are more atoms on the higher energy level than on the lower one.

Under thermodynamic equilibrium, atoms are distributed according to Boltzmann's equation (see equation (2.27)), and population inversion cannot be achieved. To make it possible, we need to pump sufficient energy into the system to switch it to a non-equilibrium state. A material found in such a non-equilibrium state is referred to as '**active medium**'.

For the analysis so far, we did not consider spontaneous emission, nor any other losses. The most common laser oscillator is comprised of an active medium, a pump source and a feedback system. It is schematically represented in figure 2.7. This schematic strongly resembles a general solid state oscillator, but it is worth noting that it is applicable to other oscillators as well. For example, the active medium may be a mixture of gasses, as is the case of He–Ne laser. The mirrors may also be something equivalent as well. For example, in the case of laser diodes, two of the faces are polished to act as a reflective surface, therefore removing the need for separate mirrors.

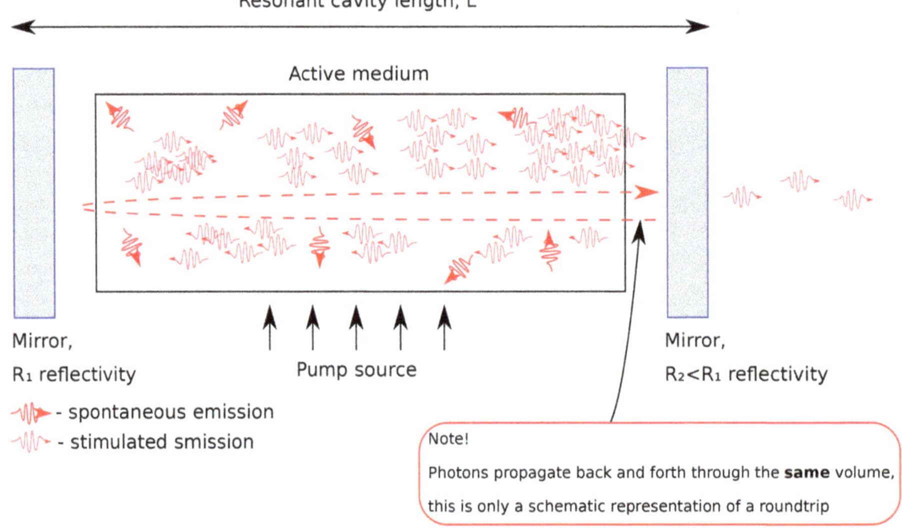

Figure 2.7. Schematic representation of a laser oscillator.

Another good example is fiber laser oscillators: the active medium is the core of the fiber itself, and mirrors are periodic structures at each end (Bragg mirrors).

Population inversion is a **necessary** condition to obtain laser radiation. But it is not **sufficient**, if we consider the spontaneous emission as well. In order to compensate the losses induced by spontaneous emission, we need to have a positive feedback system. This system is represented in figure 2.7 by the two mirrors at the sides of the active medium. For laser radiation to be achieved at the output of the oscillator, the energy we pump into the system, generally referred to as 'gain', must cover at least all losses of the system. This is the second condition for obtaining laser radiation, often encountered as **threshold gain**.

Most common losses in an optical oscillator are represented by the absorption, scattering, spontaneous emission, absorption at the mirrors and the laser output itself. In the previous analysis, we used the attenuation coefficient, α, to determine both the absorption and the gain conditions with respect to the populations, N_1 and N_2. However, let us consider α describes all losses in the system, and let us define another coefficient, β, which represents the small-signal gain of the system. Our objective is to determine what is the threshold of the gain, β_{th}, in order to obtain laser radiation at the output.

Let us begin by following the propagation of a ray of light in the resonant cavity, by following its path through a roundtrip in the system. For this analysis we will use the schematic presented in figure 2.7. As photons generated inside the active medium, they will propagate through the material and interact with its atoms. We will consider photons that are propagating along the axis of the resonant cavity, as other directions will not benefit from the feedback provided by the mirrors. Propagation through the active media is affected by both absorption and amplification:

$$I = I_0 \exp\{[(\beta - \alpha)L]\} \tag{2.50}$$

where I is the intensity at the exit of the active medium, L is the distance of propagation, α is the attenuation coefficient, β is the gain coefficient, and I_0 is the initial intensity light. The ray will then interact with the left mirror, of R_1 reflectivity. Reflectivity is usually expressed in terms of the percentage of energy it reflects. Therefore, R_1 is multiplied with the right-hand side of equation (2.50). After the ray is reflected, it goes through the active medium again and interacts with the right mirror of reflectivity R_2. Therefore, light intensity at the end of a roundtrip is described by:

$$I = I_0 R_1 R_2 \exp\{[2(\beta - \alpha)L]\} \tag{2.51}$$

Light is amplified when $I > I_0$, and the threshold for the gain, β_{th} is:

$$R_1 R_2 \exp\{[2(\beta - \alpha)L]\} = 1$$

$$\exp\{[2(\beta - \alpha)L]\} = \frac{1}{R_1 R_2} \bigg| \ln()$$

$$\Rightarrow 2(\beta - \alpha)L = \ln \frac{1}{R_1 R_2} \tag{2.52}$$

$$\Rightarrow \beta_{th} = \alpha + \frac{1}{2L} \ln \frac{1}{R_1 R_2}$$

We can further replace α using equation (2.49):

$$\beta_{\text{th}} = \frac{g_2}{g_1 \tau}\left(\frac{\pi c}{\omega}\right)^2\left(N_1 - \frac{g_1}{g_2}N_2\right) + \frac{1}{2L}\ln\frac{1}{R_1 R_2} \tag{2.53}$$

From equation (2.53) we can observe that the gain threshold, β_{th} must compensate for all losses in the resonant cavity, and depends not only on the light–matter interaction inside the cavity, described by α, but also on its design.

2.4 Optical resonators

Before we describe the resonant cavity used to generate the positive feedback necessary for achieving laser radiation, we must first approach the broadening mechanisms involved in light–matter interaction inside the cavity itself. This is, however, a topic best discussed in the context of laser–matter interaction (see section 2.7) and therefore the following paragraphs take the role of an introduction. At the beginning of the chapter, we have mentioned there is no purely monochromatic light, and argued using Fourier analysis and the uncertainty principle. Going further into this analysis is important, for it correlates with the design of a resonant cavity.

As mentioned at the beginning of the chapter, a purely monochromatic wave is spread to infinity in space. In the case of emitting atoms, the emitted radiation is not infinite, but rather has a certain duration. Therefore, it can be described as a wavepacket, i.e., a sum of plane waves. If we apply the Fourier transform to the intensity of the wave over time, we can get the frequency spectrum, and its spread, implicitly.

There is a simpler approach that is helpful in understanding where the spectral width of emission comes from. Our analysis was realized considering two energy levels. The quantum states of a two-level atom can be described by two eigenstates, Ψ_1 and Ψ_2 [4]:

$$\Psi_1 = \Phi(r)e^{\frac{iE_1 t}{\hbar}}$$
$$\Psi_2 = \Phi(r)e^{\frac{iE_2 t}{\hbar}} \tag{2.54}$$

where Φ_i is the amplitude, E_i is the energy and t is the time. A transition, be it absorption or emission, is a linear combination of the two states:

$$\Psi = \alpha\Psi_1 + \beta\Psi_2$$
$$\Rightarrow |\Psi|^2 = |\alpha\Psi_1|^2 + |\beta\Psi_2|^2 + \alpha\beta\Psi_1\Psi_2^* e^{\frac{i(E_1 - E_2)t}{\hbar}} \tag{2.55}$$

The term $\alpha\beta\Psi_1\Psi_2^* e^{\frac{i(E_1 - E_2)t}{\hbar}}$ is a time-oscillating electron density with frequency ω_{12}. Therefore, we can rewrite the exponential function in equation (2.55) as:

$$e^{\frac{i(E_1 - E_2)t}{\hbar}} = e^{-i\omega_{12}t}$$
$$\Rightarrow E_1 - E_2 = \hbar\omega_{12} \tag{2.56}$$

Let us look again to Heisenberg's uncertainty principle, which states:

$$\Delta E \cdot \Delta t \approx \hbar \tag{2.57}$$

where ΔE is the energy uncertainty and Δt is the radiative lifetime (not to be confused with the excitation lifetime). Note that:

$$E = \hbar\omega \Rightarrow \Delta E = \hbar\Delta\omega \tag{2.58}$$

In this case, Δt represents the natural lifetime of each state. We may now describe the natural lifetime as the statistical parameter related to the time it takes for the population of the excited state to decay to $1/e$ of its initial value. Using equations (2.57) and (2.58), we can deduce that the broadening of the spectrum affects each energy level that participates in a transition, absorbing or emitting a photon.

But what about the electrons found on stationary levels, in an unperturbed atom, do they have a definitive energy value? Indeed, each bound electron has precise energy levels. However, atoms in an active medium are not isolated. As we bring two identical atoms closer together, their orbitals overlap. We can refer to Pauli's exclusion principle, which dictates that identical fermions cannot occupy the same quantum state in a given quantum system. In other words, electrons cannot have the same quantum numbers in an atom. Therefore, if close enough, the orbitals of the two atoms start to overlap, and so they split into two different, but very close, energy levels. In an active medium, atoms are most often found in a crystal lattice. Let us look at one of the most commonly encountered active media, Nd:YAG. If we only look at the YAG crystal, we can observe that one molecule alone has 20 atoms [6]. 'YAG' refers to yttrium–aluminium–garnet, i.e., $Y_3Al_5O_{12}$. However, the YAG crystal itself has a primitive cell that contains a certain configuration of such molecules. The cell contains eight molecules. That means there are 160 atoms tightly packed in a volume of the order of ≈ 1 nm. Moreover, this crystal is doped with Nd^{+3} ions. These ions produce a slight displacement in the configuration of the YAG primitive cell, which in turn slightly modifies the values of the energy levels for neighboring molecules. The result is that there is a great number of energy levels that are very close together, which can be essentially treated as an energy continuum (or band) instead of discrete levels.

Broadening mechanisms can be split into two types: homogeneous and inhomogeneous. Homogeneous broadening mechanisms refer to those that affect all atoms in the same way. In this category we have lifetime and collision broadening. So far we have talked about lifetime broadening. A collision with another atom, while the atom is oscillating (emitting photons) can disrupt the phase of the emitted wave. The resulting wave can be considered as being comprised of a series of uninterrupted waves. The number of these uninterrupted waves decays exponentially, within a time period, τ, which is the mean time between collisions. The spectrum has a Lorentz profile, similar to lifetime broadening. At atmospheric pressure, $\tau \approx 10^{-10}$ s [4]. A more detailed analysis on these matters is offered in chapter 2.7.

In order to obtain a positive feedback and compensate for losses generated by light–matter interaction, the active media is placed inside an optical resonant cavity, also known as a laser cavity, or simply optical resonators. In commonly encountered cases, the resonator is an open structure comprised of two mirrors with high reflectivities, facing one another. Optical resonators that are used in the design of a laser system are usually significantly larger than the wavelength of the amplified radiation.

The optical cavity is termed 'passive' in the absence of an active medium, and 'active' in the presence of one. The electromagnetic radiation results from spontaneous or stimulated deexcitation processes of the atoms and/or molecules in the active medium, which acts as a source of energy.

2.4.1 Oscillation modes

An optical resonator can support both longitudinal and transverse oscillation modes. These represent stable configurations of the electromagnetic field inside the resonator (not to be confused with light intensity). It can be thought of as stationary waves imposed by the resonator dimensions.

To specify a mode configuration we use the $T\,EM_{rsq}$ notation. The index q refers to longitudinal modes, i.e., number of nodes of the stationary waves formed inside the resonator. The indices r and s refer to transverse modes, which refer to the nodes of the stationary waves, formed at one of the mirrors, along two directions perpendicular to one another and to the longitudinal axis of the resonator (see figure 2.8).

Let us consider a resonator of length L and height D. The various frequencies that correspond to each oscillation mode are determined by the condition:

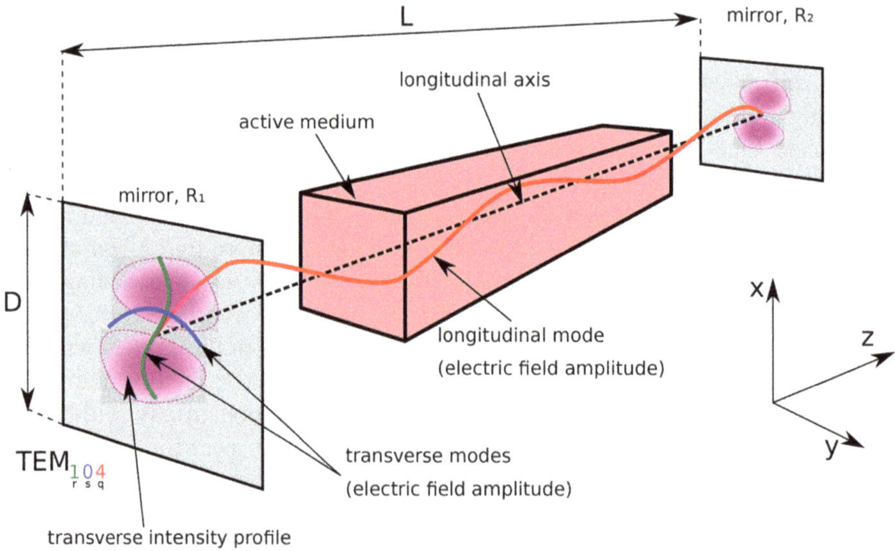

Figure 2.8. Schematic representation of longitudinal and transverse oscillation modes for an optical resonator.

$$L = q\left(\frac{\lambda_q}{2}\right) = q\left(\frac{\nu_q c}{2}\right) \tag{2.59}$$

from which we can get:

$$\nu_q = \left(\frac{qc}{2L}\right) \tag{2.60}$$

where c is the speed of light, q is the order of the oscillation mode, L is the resonator length, and λ_q is the electromagnetic radiation wavelength corresponding to the oscillation mode q.

In the visible range of electromagnetic radiation, the number of oscillation modes for stationary waves, formed along the longitudinal axis of a resonator, is of the order 10^4–10^7. Several longitudinal oscillation modes correspond to a certain transverse mode. It is worth mentioning that longitudinal modes differ from one another by the oscillation frequency, while transverse modes differ by both the oscillation frequency and the transverse field distribution.

Following stimulated emission processes, the active medium provides energy to those oscillation modes, ω_c, for which the gain is above the threshold, i.e., the spectrum, centered on Ω frequency, has a gain that exceeds the losses. The laser starts oscillating over these modes, where competitive processes take place (see figure 2.9).

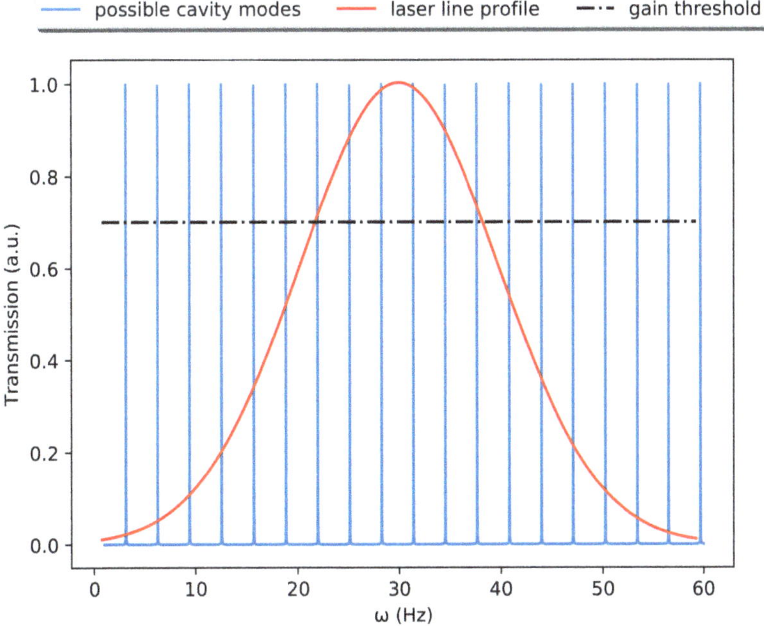

Figure 2.9. Schematic representation of laser line, oscillation modes and losses for an optical resonator.

The only amplified oscillation modes are those that are found above the threshold value and under the laser line profile. The transmittance of the resonator, T, is given by:

$$T = \frac{(1 - R_1)(1 - R_2)}{(1 - \sqrt{R_1 R_2})^2 + 4R_1 R_2 \sqrt{\sin\left(\frac{\omega L}{c}\right)}}$$

(2.61)

where R_1 and R_2 are mirror reflectivities. At resonance, i.e., $R_1 = R_2$, $T = 1$, the transmitted power is maximum.

In the case of laser resonators, the reflectivity of each mirror is high, so that the incident power tends to be (almost) equal to the emergent power, as a result from multiple reflection processes.

2.4.2 Laser resonator characteristics

The selectivity of a resonator can be described in terms of the quality factor, Q, which is generally defined as:

$$Q = \omega_0 \frac{\text{resonance stored energy}}{\text{average power drain}} = \frac{\omega_0}{\Delta\omega_{1/2}}$$

(2.62)

where ω_0 is the resonance frequency, and $\Delta\omega_{1/2}$ is the full width half maximum of a spectral line (see figure 2.10). Using equation (2.61), we can obtain:

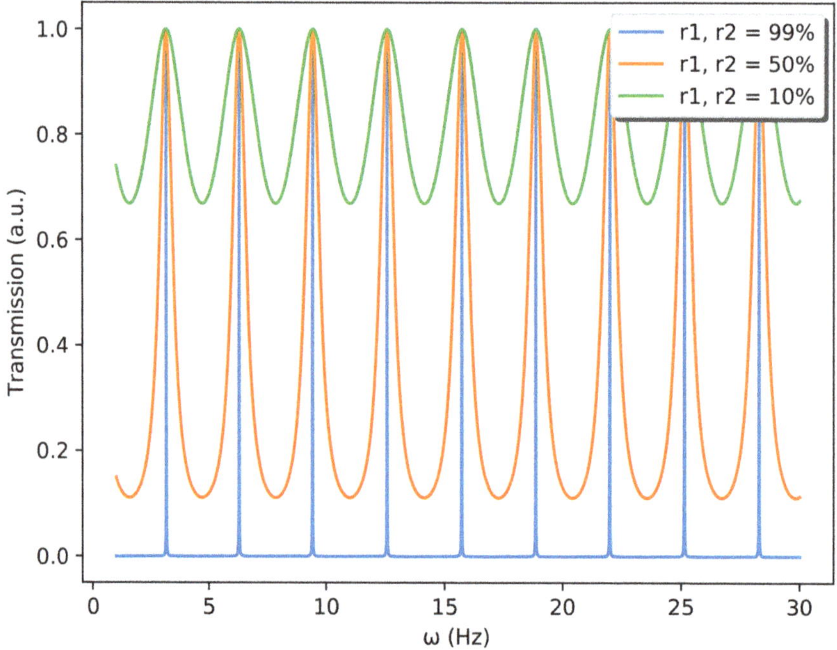

Figure 2.10. Transmission (equation (2.61)) as a function of ω, for different mirror reflectivities.

$$\Delta\omega_{1/2} = \frac{c}{2L} \frac{1 - \sqrt{R_1 R_2}}{\pi (R_1 R_2)^{\frac{1}{4}}} \tag{2.63}$$

where $\frac{c}{2L}$ is the frequency spacing in the spectrum. Therefore, we obtain the following expression for the quality factor:

$$Q = \frac{2\pi L}{\lambda_0} \frac{(R_1 R_2)^{\frac{1}{4}}}{1 - \sqrt{R_1 R_2}} \tag{2.64}$$

The finesse of the resonator is defined as:

$$F = \frac{c}{2L} \frac{1}{\Delta\omega_{1/2}} \tag{2.65}$$

In the optical range of the electromagnetic radiation spectrum, the value of $\Delta\omega_{1/2} \approx 10^2$ kHz, the quality factor $Q \approx 10^8$–10^9, and the finesse $F \approx 10^2$.

The average lifetime of the photons inside the resonator, noted as τ_c, can be calculated considering that for a roundtrip through the resonator, the photon number, n, drops to $nR_1 R_2$ (some photons are absorbed by the mirrors, since they don't reflect 100%), and:

$$\frac{dn}{dt} = -\frac{c}{2L}(1 - R_1 R_2)n \tag{2.66}$$

If we integrate equation (2.66), we will obtain the following expression for $n = n(t)$:

$$n(t) = n_0 \exp\left(-\frac{t}{\tau_c}\right) \tag{2.67}$$

where n_0 is the initial number of photons, and

$$\tau_c = \frac{2L}{c(1 - R_1 R_2)} \tag{2.68}$$

is the lifetime of the photons inside the resonator. Taking equation (2.62) into account, we can deduce the time variation of the stored energy in the resonator:

$$E(t) = E_0 \exp\left(-\frac{\omega_0 t}{Q}\right) \tag{2.69}$$

Moreover, using equations (2.64) and (2.68), we can obtain the relation between the lifetime of the photons inside the resonator, and its quality factor:

$$\tau_c = \frac{Q}{\omega_0} \quad \text{or} \quad \Delta\omega_{1/2} \cdot \tau_c = 1 \tag{2.70}$$

2.4.3 Mode distribution in an optical resonator

In our previous analysis we considered a simple resonator that has two planar mirrors at both ends of an active medium, parallel to each other. In order to calculate the mode distribution of a field inside the resonator, let us now consider a passive optical cavity (no active medium) that is comprised of two square mirrors, of height $2a$, with a spherical surface of curvature R. They are placed at a distance of $L \gg \lambda$, where λ is the wavelength of the waves propagating back and forth inside the resonator (see figure 2.11).

Analyzing wave propagation in such an optical resonator, by means of solving Maxwell's equation with the appropriate boundary conditions, quickly becomes complicated due to the high number of oscillation modes. That is why Fox and Li [5] and Boyd and Gordon [2] developed a theoretical model based on Huygens' principles and the Kirchhoff–Fresnel integral, using the scalar model of diffraction. Using this theory, we can calculate the optical field distribution inside and outside the resonator, as well as the quality factor and oscillation modes.

Let us consider that the wave inside the resonator is linearly polarized along the Y direction (see figure 2.11). Losses inside the cavity are small, due to high reflectivity of the mirrors, and therefore the wave undergoes multiple reflections inside. The field, $E(x, y)$, at the point $P(x, y)$ on the first mirror, $M1$, (left side in figure 2.11) can be evaluated using the Kirchhoff–Fresnel integral by summing the contributions of all sources at points $P'(x'y')$ found on the second mirror, $M2$, (right side in figure 2.11) as such:

$$E(x, y) = \int_{S'} \frac{ik(1 + \cos\theta')}{4\pi\rho} e^{-ik\rho} E'(x', y') \mathrm{d}S' \tag{2.71}$$

where S' is the surface of the second mirror, $M2$, ρ is the position vector from P to P', and θ' is the angle between ρ and the normal vector to the surface of $M2$.

In a similar manner, we can evaluate the field at P' found on $M2$ as:

$$E'(x', y') = \int_{S} \frac{ik(1 + \cos\theta)}{4\pi\rho} e^{-ik\rho} E(x, y) \mathrm{d}S \tag{2.72}$$

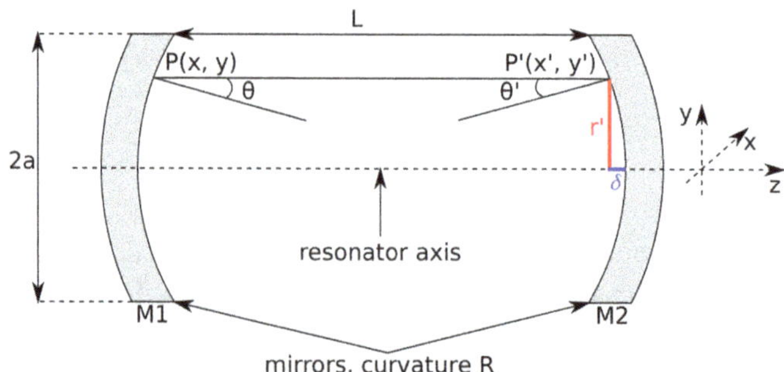

Figure 2.11. Schematic representation of a passive optical resonator with spherical mirrors.

Let us consider that, in the stationary regime, both $E(x, y)$ and $E'(x', y')$ represent the field of the same oscillation mode, produced between the two mirrors as a result of multiple, successive reflections. This means both equations must be coupled and verify the eigenvalues and eigenfunctions of the form:

$$\sigma E(x, y) = \int_{S'} \frac{ik(1 + \cos \theta')}{4\pi\rho} e^{-ik\rho} E'(x', y') \mathrm{d}S' \tag{2.73}$$

$$\sigma' E'(x', y') = \int_{S} \frac{ik(1 + \cos \theta)}{4\pi\rho} e^{-ik\rho} E(x, y) \mathrm{d}S \tag{2.74}$$

where σ and σ' are complex constants representing the eigenvalues. Solutions of equations (2.73) and (2.74) satisfy Maxwell's equations in Kirchhoff approximation and boundary conditions at the surface of the two mirrors.

By solving equations (2.73) and (2.74) we can obtain eigenfunctions that determine the resonator oscillation modes and corresponding eigenvalues that lead to resonance.

There are two methods we can approach to solve equations (2.73) and (2.74):
1. *numerical method*—it is an iterative method developed by Fox and Li;
2. *analythical method*—it consists of transforming Kirchhoff integral equations into second-kind Fredholm equations, for which general solutions are known.

In order to calculate the field distribution that determines the resonator's normal oscillation modes using the analytical method, we consider that they are reproduced over each mirror, according to:

$$E(x, t) = \sigma_r \sigma_s f_r(x) g_s(y) E_0 \tag{2.75}$$

where σ_r and σ_s determine the phase and amplitude changes along x and y directions produced by successive reflections, $f_r(x)$ and $g_s(y)$ determine field variations over the aperture of the mirrors, which are considered as formed of infinite stripes that are parallel to x and y directions, and E_0 is a constant amplitude.

If we take into account equation (2.71), then equation (2.75) can be written as:

$$E = \int_{S'} \frac{ik(1 + \cos \theta')}{4\pi\rho} e^{-ik\rho} E_0 f_r(x') g_s(y') \mathrm{d}x' \mathrm{d}y' \tag{2.76}$$

If we make a small angle approximation, i.e., small θ and θ', as well as consider $\rho = R$, where R is the curvature of the mirrors, the above equation (2.76) becomes:

$$\sigma_r \sigma_s f_r(x) g_s(y) = \int_{-a}^{a} \int_{-a}^{a} \frac{ik}{2\pi R} e^{-ikR} f_r(x') g_s(y') \mathrm{d}x' \mathrm{d}y' \tag{2.77}$$

and ρ can be calculated using:

$$\rho = \sqrt{(L - \delta - \delta')^2 + (x' - x)^2 + (y' - y)^2} \tag{2.78}$$

where:

$$\delta = R - \sqrt{R^2 - r^2} \approx \frac{r^2}{2R} \tag{2.79}$$

$$\delta' = R - \sqrt{R^2 - r'^2} \approx \frac{r'^2}{2R} \tag{2.80}$$

where r and r' represent the distances from points P and P' to the resonator axis, and both δ and δ' are the distances from the plane that contains r and r' (plane is perpendicular to the resonator axis), to the top of each mirror, found on the resonator axis, z (see figure 2.11, where r' and δ' are represented).

For a confocal resonator, i.e., $L = R$, equation (2.78) can be rewritten as:

$$\rho = R - \frac{xx' + yy'}{R} \tag{2.81}$$

From equations (2.77) and (2.81) we obtain the following integral equation:

$$\sigma_r \sigma_s f_r(x) g_s(y) = \int_{-a}^{a} \int_{-a}^{a} \frac{ik}{2\pi R} e^{-ikR} e^{\frac{ikxx'}{R}} e^{\frac{ikyy'}{R}} f_r(x') g_s(y') dx' dy' \tag{2.82}$$

If we introduce the reduced coordinates:

$$X = \frac{x\sqrt{d}}{a}$$
$$Y = \frac{y\sqrt{d}}{a} \tag{2.83}$$

where $d = \frac{a^2 k}{R} = \frac{2\pi}{N}$, and $N = \frac{a^2}{\lambda R}$ is Fresnel number, i.e., number of Fresnel zones seen from the center of one mirror on the surface of the other, and we also introduce the following notations:

$$F_r(X) = f_r(x)$$
$$G_s(Y) = g_s(y) \tag{2.84}$$

then the integral equation (2.82) is written as:

$$\sigma_r \sigma_s F_r(X) G_s(Y) = \frac{i}{2\pi} e^{-ikR} \int_{-\sqrt{d}}^{\sqrt{d}} \int_{-\sqrt{d}}^{\sqrt{d}} F_r(X') e^{ikXX'} G_s(Y) e^{ikYY'} dX' dY' \tag{2.85}$$

Equation (2.85) can be solved using the variable separation method for various configurations of the resonator, be it rectangular, circular, etc. As a result, we obtain two second-kind Fredholm integral equations, with a symmetrical kernel, an equation for each infinite strip mirror. In the case we analyzed above, the intensity of the field drops exponentially from the center towards the edges, and so the influence of the mirror size is negligible.

As such, in the case of rectangular mirrors, and using the notation:

$$\beta_r \beta_s = \frac{\sigma_r \sigma_s}{i e^{-ikR}} \tag{2.86}$$

Equation (2.85) can be separated into two unidimensional Fredholm equations:

$$F_r(X) = -\frac{i}{\sqrt{2\pi\beta_r}} \int_{-\sqrt{d}}^{\sqrt{d}} F_r(X') e^{ikXX'} dX' \tag{2.87}$$

$$G_s(X) = -\frac{i}{\sqrt{2\pi\beta_s}} \int_{-\sqrt{d}}^{\sqrt{d}} G_s(Y') e^{ikYY'} dY' \tag{2.88}$$

In the case of laser resonators, the Fresnel number is in the order of tens, and therefore we can neglect the influence of the finite size of the mirrors over the oscillation mode configuration. This means that we can extend the integration limits of equations (2.87) and (2.87) to infinity, both ways. The reason for this is that with infinity integration limits, the solutions for these equations can be approximated with Hermite polynomials.

To find the solutions, we start from the Hermite polynomials generating function:

$$\begin{aligned} h(y, t) &= \exp\{(-t^2 + 2y)\} \\ &= \exp\{(-t^2)\} \exp\{[-(t - y)^2]\} \\ &= \sum_{r=0}^{\infty} \frac{H_r(y)}{r!} t^r \end{aligned} \tag{2.89}$$

If we multiply equation (2.89) with $\exp(ixy - \frac{y^2}{2})$ and then integrate from $-\infty$ to ∞, we obtain:

$$\int_{-\infty}^{\infty} \exp\left(2yt - t^2 + ixy - \frac{y^2}{2}\right) dy = \int_{-\infty}^{\infty} \sum_{r=0}^{\infty} \frac{H_r(y)}{r!} t^r \exp\left(ixy - \frac{y^2}{2}\right) dy \tag{2.90}$$

We can use Fubini–Tonelli theorems to switch the integration and summation. Tonelli's theorem says that if a series of **continuous** functions, $f_n(x) \geqslant 0$, for all n and x, then:

$$\sum \int f_n(x) dx = \int \sum f_n(x) dx \tag{2.91}$$

After integrating the left-hand side of equation (2.90), taking into account the generating function described in equation (2.89), and then switching the integration and summation, equation (2.90) becomes:

$$\sqrt{2\pi} \sum_{r=0}^{\infty} \frac{(it)^r}{r!} H_r(x) \exp\left\{\left(-\frac{x^2}{2}\right)\right\} = \sum_{r=0}^{\infty} \frac{t^r}{r!} \int_{-\infty}^{\infty} H_r(y) \exp\left\{\left(ixy - \frac{y^2}{2}\right)\right\} dy \tag{2.92}$$

In equation (2.92), both the left-hand side and the right-hand side are equal for any values of t, and therefore the coefficients for the same powers of t are also equal, and therefore equation (2.92) can be rewritten as:

$$H_r(x)\exp\left\{\left(-\frac{x^2}{2}\right)\right\} = \frac{1}{i^r\sqrt{2\pi}} \int_{-\infty}^{\infty} H_r(y)\exp\left\{\left(ixy - \frac{y^2}{2}\right)\right\}dy \tag{2.93}$$

If we compare equations (2.87) and (2.93), we can obtain the solution as:

$$F_r(X) = H_r(X)\exp\left\{-\frac{X^2}{2}\right\} \tag{2.94}$$

where:

$$\beta_r = i^r \tag{2.95}$$

In a similar manner, we can use equation (2.88) to obtain:

$$G_s(Y) = H_s(Y)\exp\left\{-\frac{Y^2}{2}\right\} \tag{2.96}$$

where:

$$\beta_s = i^s \tag{2.97}$$

The modal distribution of the field along x and y directions of the resonator mirror is determined by the eigenfunctions $F_r(X)$ and $G_s(Y)$, given by equations (2.94) and (2.96). The functions in equations (2.94) and (2.96) are real, which leads to the fact that the reflective surfaces of the mirrors are constant phase surfaces.

Resonance condition is given by the eigenvalues of equation (2.82). As such, using equations (2.86), (2.95) and (2.97) we obtain the phase condition as:

$$\sigma_r\sigma_s = i^{r+s+1}e^{-\frac{ik}{R}} \tag{2.98}$$

At resonance, the phase change for a roundtrip inside the resonator is $2\pi q$, where $q \in \mathbb{Z}$ (is an integer). As such, we can obtain the resonance condition:

$$2\left|\frac{\pi}{2}(r + s + 1) - kR\right| = 2\pi q \tag{2.99}$$

The left side of equation (2.99) is multiplied by 2 in order to take into account the two reflections in a roundtrip.

Equation (2.99) can be written, equivalently, as:

$$2q + (r + s + 1) = \frac{4R}{\lambda} \tag{2.100}$$

The terms r, s and q that are obtained as a result of solving equation (2.86) represent the order of the longitudinal (r) and transverse (s, q) **oscillation modes** of the resonator (**refer to figure** 2.8 **for visualization**). Oscillation modes, commonly noted as TEM_{rsq}, are transverse electromagnetic, for the z direction component of the field inside the resonator is negligible. Therefore, they form an orthonormal set of eigenfunctions for any transverse field inside the resonator.

The eigenvalues, σ_r and σ_s determine the amplitude variations of the field, due to diffraction. We can use them to define a loss coefficient for a rountrip inside the resonator:

$$\alpha_d = 1 - (\sigma_r \sigma_s)^2 \tag{2.101}$$

We can now use the reduced coordinates (equation (2.83)) to determine the explicit form of the functions $f_r(x)$ and $g_s(y)$ (we have introduced them all the way back, at the beginning of this section, equation (2.75)):

$$f_r(x) = \text{const} \cdot H_r\left(x\sqrt{\frac{2\pi}{R\lambda}}\right)\exp\left\{-\frac{\pi x^2}{R\lambda}\right\} \tag{2.102}$$

A similar form can be written for $g_s(y)$. We can observe that these functions represent multiplications of Hermite polynomials and a Gaussian function. In other words, we can deduce that field distributions on the surface of a mirror is given by the Gaussian function, $\exp\left\{\left(-\pi\frac{x^2+y^2}{R\lambda}\right)\right\}$, multiplied with the appropriate Hermite polynomials.

Low-order Hermite polynomials can be found in table 2.2, where we made the notation $u = x\sqrt{\frac{2\pi}{R\lambda}}$.

The electric field of two neighboring intensity lobes has opposing phases, which in turn means that the number of nodes along a transverse direction indicates the number of phase changes that the electric field undergoes (see figure 2.12).

For an optical resonator with **rectangular symmetry**, i.e., such as the Fabry–Pérot resonator, the intensity distribution of each transverse mode, in **Cartesian coordinates**, determined by a Gaussian beam oscillating back and forth, is given by:

$$I_{rs} = I_0\left(\frac{w_0}{w}\right)^2 \cdot \left[H_r(u_x)e^{\left(-\frac{x^2}{w^2}\right)}\right]^2\left[H_s(u_y)e^{\left(-\frac{x^2}{w^2}\right)}\right]^2 \tag{2.103}$$

where I_0 is the initial value of the intensity, w_0 is the Gaussian beam waist (minimum diameter), w is the beam diameter (maximum diameter) (do not confuse w_0 and w

Table 2.2. Low-order Hermite polynomials.

1	$H_0(u) = 1$
2	$H_1(u) = 2u$
3	$H_2(u) = 4u^2 - 2$
4	$H_3(u) = 8u^3 - 12u$
5	$H_4(u) = 16u^4 - 48u^2 + 12$
6	$H_5(u) = 32u^5 - 160u^3 + 120u$

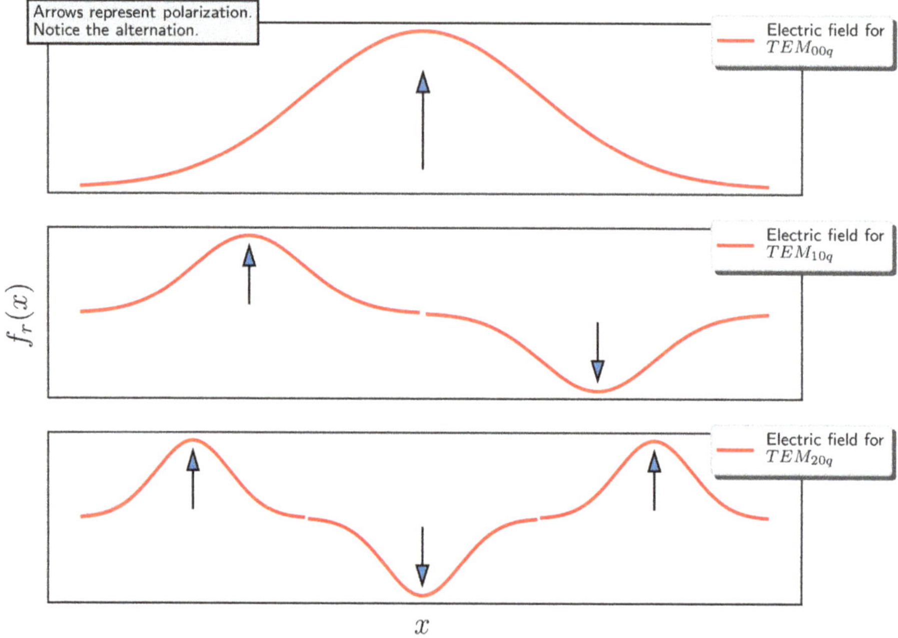

Figure 2.12. Schematic representation of $f_r(x)$, which describes modal distribution of a field on the surface of a mirror inside a resonator.

with ω_0 and ω !), x and y are the transverse coordinates, H_r and H_s are the Hermite polynomials of variables $u_x = \frac{x\sqrt{2}}{w}$ and $u_y = \frac{y\sqrt{2}}{w}$. Transverse oscillation modes of different orders, formed by a Gaussian beam in a square mirror Fabry–Pérot cavity, are presented in figure 2.13.

2.5 Optical resonator stability

2.5.1 Fabry–Pérot resonator

In order to deduce the stability condition for an optical resonator, let us first consider an optical cavity comprised of two plane mirrors (a Fabry–Pérot resonator), placed at distance L from one another (see figure 2.14, which is similar to figure 2.7, but in the absence of the active medium). The reason we choose this type of resonator is that the results can be extended to cover arbitrary mirror shapes.

The ray that is leaving point $P_1(x_1, y_1)$, found on mirror M_1, has an angle θ_{r1} with respect to the normal vector to the surface of the mirror, on the xOz plane, and ϕ_{r1} on the yOz plane. The ray arrives at point $P_2(x_2, y_2)$ found on mirror M_2 and has the angles θ_{i2} and ϕ_{i2} with the normal vector to the surface of the mirror, M_2. From this point, the ray leaves the mirror along a direction determined by angles θ_{r2} and ϕ_{r2}, with respect to the normal vector to the surface of the mirror M_2 (see figure 2.14).

Coordinates of $P_1(x_1, y_1)$ and $P_2(x_2, y_2)$ are related by:

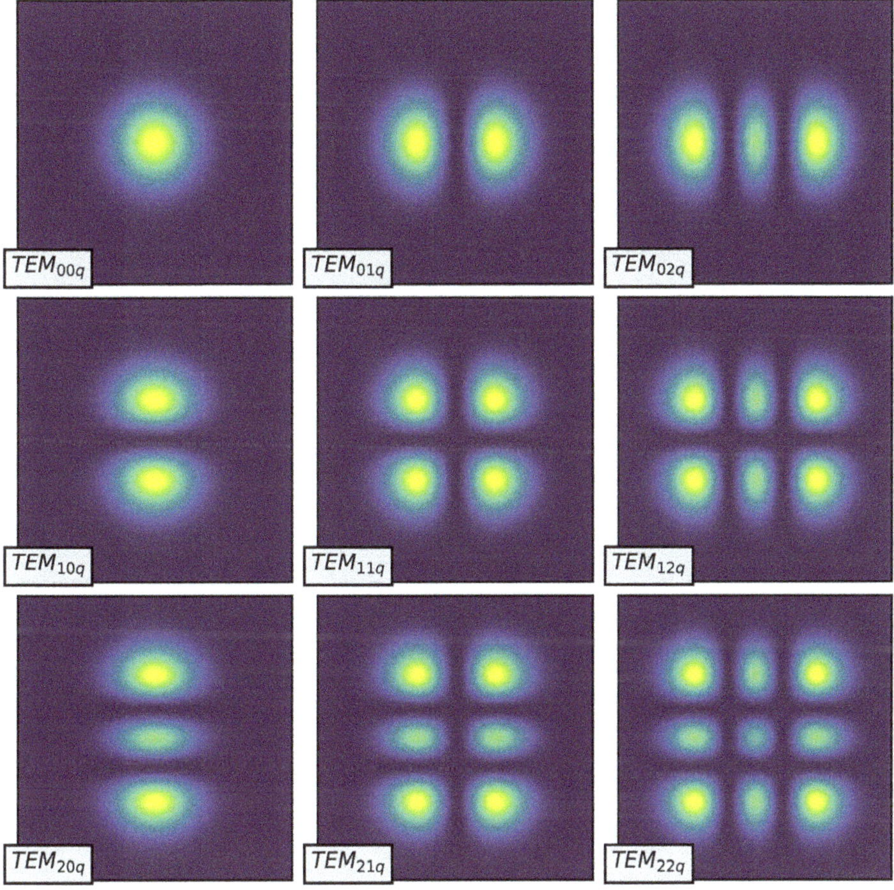

Figure 2.13. Transverse oscillation modes of different orders formed in a square mirror Fabry–Pérot cavity (see equation (2.103))

$$x_2 = x_1 + \theta L$$
$$\theta_{i2} = -\theta_{r1}$$
$$(2.104)$$

$$y_2 = y_1 + \phi L$$
$$\phi_{i2} = -\phi_{r1}$$
$$(2.105)$$

Why are equations (2.104) and (2.105) true? They are actually an approximation. If we think of an arc, A, described by an angle Ω, of a circle with the radius r_c, then its length is simply $A = \Omega \cdot r_c$. Let us now look at figure 2.14. The light rays are only schematically represented there and their reflection angles are exaggerated, but in truth, for a Fabry–Pérot resonator, rays are reflected at *very small angles within the optical cavity*. Otherwise, rays would quickly escape the cavity. The length of the resonator is much greater than the arc described by any reflection angle, θ_{ri}, on a circle of radius L (on the other mirror), i.e., $L \gg A$. This in turn means that the

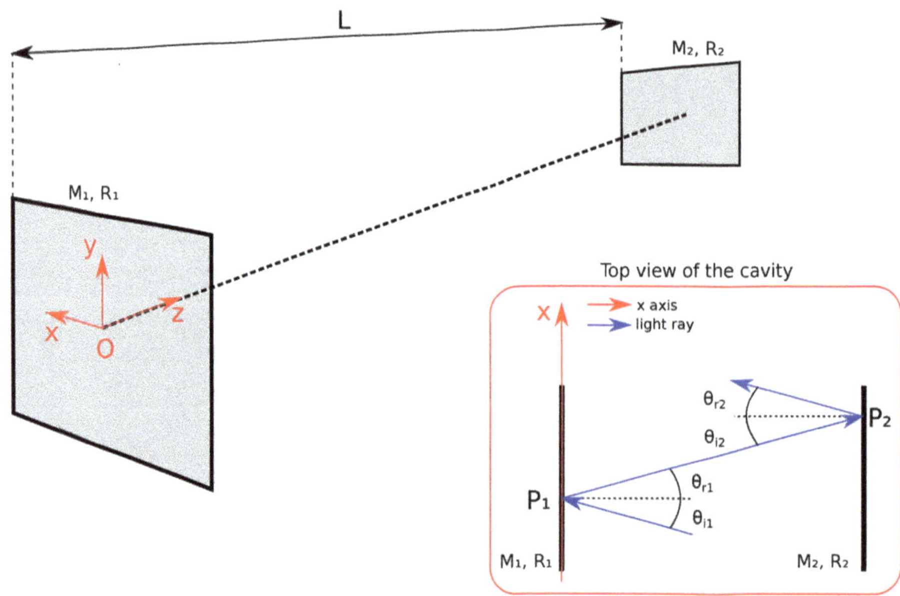

Figure 2.14. Schematic representation of a Fabry–Pérot resonator.

curvature of the arc A is negligible, and the difference in position along each axis, between two points consecutively hit by a ray of light, can be approximated with the length of the arc, A, described by angle θ_{ri} over a circle of radius L.

The roundtrip propagation of the rays between the two mirrors can be described using transfer matrices, T_x^{12} and T_y^{12}:

$$\begin{bmatrix} x_2 \\ \theta_{i2} \end{bmatrix} = T_x^{12} \begin{bmatrix} x_1 \\ \theta \end{bmatrix} \tag{2.106}$$

$$\begin{bmatrix} y_2 \\ \phi_{i2} \end{bmatrix} = T_y^{12} \begin{bmatrix} y_1 \\ \phi \end{bmatrix} \tag{2.107}$$

and, equivalently, reflection matrices, R_x and R_y:

$$\begin{bmatrix} x_2 \\ \theta_{r2} \end{bmatrix} = R_x \begin{bmatrix} x_2 \\ \theta \end{bmatrix} \tag{2.108}$$

$$\begin{bmatrix} y_2 \\ \phi_{r2} \end{bmatrix} = R_y \begin{bmatrix} y_2 \\ \phi \end{bmatrix} \tag{2.109}$$

The roundtrip of a ray of light, executed n times inside the resonator, is described by the B^n matrix, where:

$$B = RT^{12}RT^{21} \tag{2.110}$$

is the Bertolotti matrix.

Stability of a resonator can be defined by imposing the condition that, after n rountrips, parameters $|x|$, $|y|$, $|\theta|$ and $|\phi|$ must not have values greater than the initial values. Notice that all parameters are taken as absolute values, disregarding the sign, i.e., $|x|$, for example, means that the values of x parameter must remain within the interval $(-x, x)$.

We can deduce the above condition by using B^n matrix. We know that the trace and determinant of a matrix are not modified upon diagonalization, and B^n can be diagonalized using a matrix A as such:

$$B_d^n = A^{-1} B^n A = \begin{bmatrix} \lambda_1 & 0 \\ 0 & \lambda_2 \end{bmatrix} \tag{2.111}$$

The eigenvalues, λ_1 and λ_2, of B^n matrix can be obtained by solving:

$$\lambda^2 - s \cdot \lambda + p = 0 \tag{2.112}$$

where we made the notations:

$$\begin{aligned} s &= \lambda_1 + \lambda_2 \\ p &= \lambda_1 \cdot \lambda_2 \end{aligned} \tag{2.113}$$

If the condition is to be met, i.e., parameters $|x|$, $|y|$, $|\theta|$ and $|\phi|$ should not exceed the initial values, then λ_1 and λ_2 must be smaller than 1.

The reason is revealed if we look, for example, at the x parameter, in relation to B_d^n:

$$\begin{bmatrix} x_1 \\ x_2 \end{bmatrix} = \begin{bmatrix} \lambda_1 & 0 \\ 0 & \lambda_2 \end{bmatrix} \begin{bmatrix} x_1 \\ x_2 \end{bmatrix} \tag{2.114}$$

from which we can get:

$$\begin{aligned} x_1 &= \lambda_1 \cdot x_1 \\ x_2 &= \lambda_2 \cdot x_2 \end{aligned} \tag{2.115}$$

which in turn determine the following stability conditions:

$$\begin{aligned} \mathrm{Tr}[B^n] &\leqslant 2 \\ \mathrm{Det}[B^n] &\leqslant 1 \end{aligned} \tag{2.116}$$

Considering equations (2.104)–(2.109), for a resonator with square mirrors, we obtain:

$$B = \begin{bmatrix} 1 & 0 \\ 0 & -1 \end{bmatrix} \cdot \begin{bmatrix} 1 & L \\ 0 & -1 \end{bmatrix} \cdot \begin{bmatrix} 1 & 0 \\ 0 & -1 \end{bmatrix} \cdot \begin{bmatrix} 1 & L \\ 0 & -1 \end{bmatrix} = \begin{bmatrix} 1 & 2L \\ 0 & 1 \end{bmatrix} \tag{2.117}$$

and therefore:

$$B_d^n = A^{-1} B^n A = \begin{bmatrix} 1 & 2nL \\ 0 & 1 \end{bmatrix} \tag{2.118}$$

2.5.2 Concave mirror resonator

The analysis of this type of resonator is just slightly more complicated than the Fabry–Pérot resonator, for the relations between the coordinates of points consecutively hit by a ray are approximated a bit different (see figure 2.15). Let us consider a cavity with concave mirrors that have curvatures R_1 and R_2, placed at a distance L from one another. Then the relation between coordinates of points $P_1(x_1, y_1)$ and $P_2(x_2, y_2)$, that are consecutively hit by a ray, is given by:

$$x_2 = x_1 + \theta_{r1} L$$
$$\theta_{i2} = -\theta_{r1} \tag{2.119}$$

However:

$$\theta_{r2} = -2\alpha - \theta_{i2} \approx \frac{-2x_2}{R_2} - \theta_{i2} \tag{2.120}$$

The transfer and reflection matrices for the second mirror, M_2, are:

$$T_x^{12} = \begin{bmatrix} 1 & L \\ 0 & -1 \end{bmatrix}$$
$$R_x^2 = \begin{bmatrix} 1 & 0 \\ -\dfrac{2}{R_2} & -1 \end{bmatrix} \tag{2.121}$$

Using this information we can obtain the Bertolotti matrix:

$$B = \begin{bmatrix} 1 & 0 \\ -\dfrac{2}{R_1} & -1 \end{bmatrix} \cdot \begin{bmatrix} 1 & L \\ 0 & -1 \end{bmatrix} \cdot \begin{bmatrix} 1 & 0 \\ -\dfrac{2}{R_2} & -1 \end{bmatrix} \cdot \begin{bmatrix} 1 & L \\ 0 & -1 \end{bmatrix} = \begin{bmatrix} 1 & 2L \\ 0 & 1 \end{bmatrix} \tag{2.122}$$

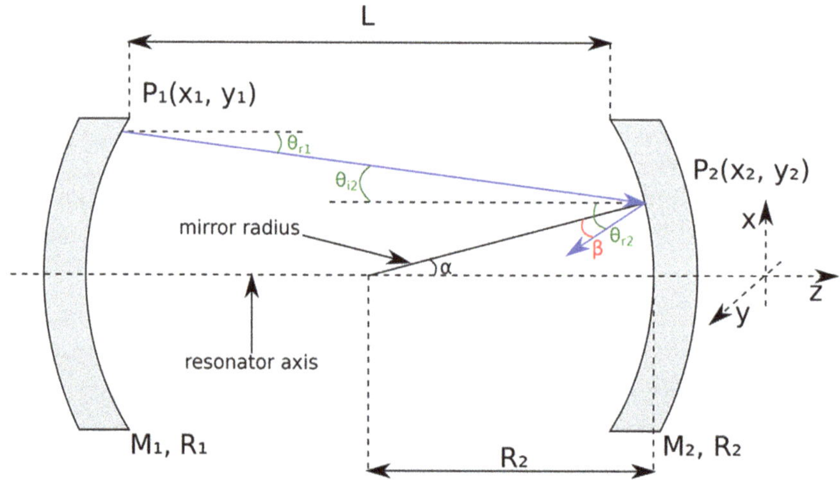

Figure 2.15. Schematic of a resonator with concave mirrors.

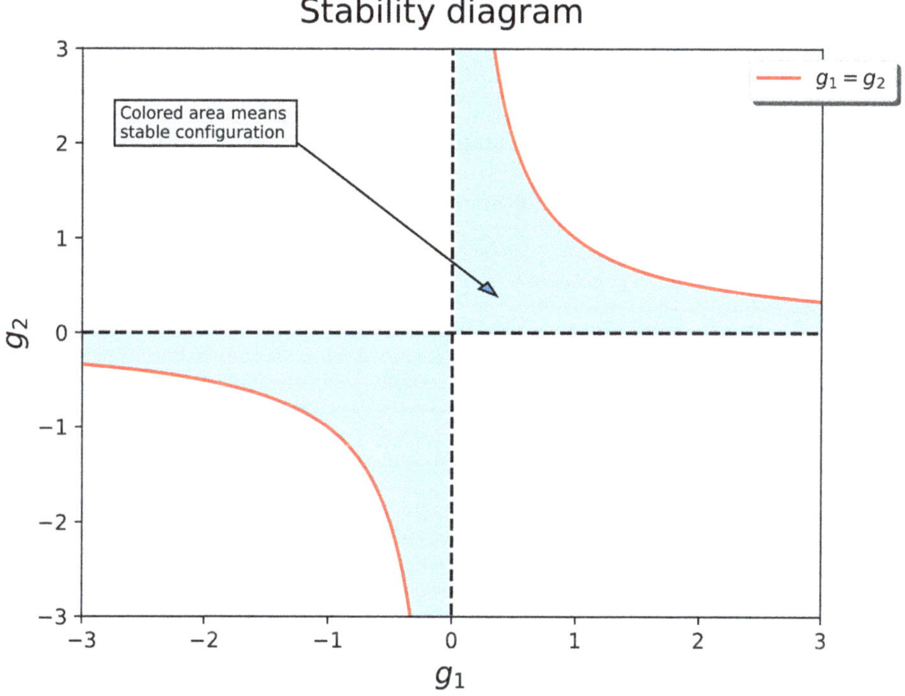

Figure 2.16. Stability diagram resulted from the stability condition (2.124).

therefore:

$$\mathrm{Tr}(B) = 1 - \frac{2L}{R_2} - \frac{4L^2}{R_1 R_2} - \frac{2L}{R_2} - \frac{4L}{R_1} + 1 \tag{2.123}$$

If we apply the stability conditions described in equation (2.116) to equation (2.123) we obtain the stability condition as:

$$\left(1 - \frac{L}{R_1}\right)\left(1 - \frac{L}{R_2}\right) \leqslant 1$$
$$\text{or :}$$
$$0 \leqslant g_1 \cdot g_2 \leqslant 1 \tag{2.124}$$

If we plot g_2 as a function of g_1, in the case of a resonator with concave mirrors of curvatures R_1 and R_2, then the area between the resulting hyperbola arcs and the axes belong to a stable resonator. This is known as the **stability diagram** (see figure 2.16).

2.5.3 Resonator types

Apart from the plane mirrors (Fabry–Pérot) and the concave mirrors resonators, there are other types. In our previous analysis, we did not approach the distance between mirrors. Optical resonators used in lasers must have an appropriate quality factor, Q, where the higher it is, the better. A higher quality factor determines a lower oscillation threshold, a better mode selectivity (i.e., better monochromaticity of the emitted radiation), an optimal mode volume that determines the size of the active medium, optimal output coupling for better efficiency, and others.

Some resonators that are often encountered are presented in table 2.3, and placed on the stability diagram in figure 2.17.

Spherical mirror resonators are stable if the reflected light is periodically focused, and unstable if light diverges following repeated reflections. Some resonators reside

Table 2.3. Common resonator types.

Number in figure 2.17	Type	Mirror radii	Distance between mirrors
1	Confocal	R	R
2	Plane-parallel	∞	Arbitrary length, L
3	Concentric	R	$2R$
4	Asymmetrical	R_1, R_2	$L < R_1 + R_2$
5	Semiconfocal	R, ∞	$R/2$
6	Semiconcentric	R, ∞	R

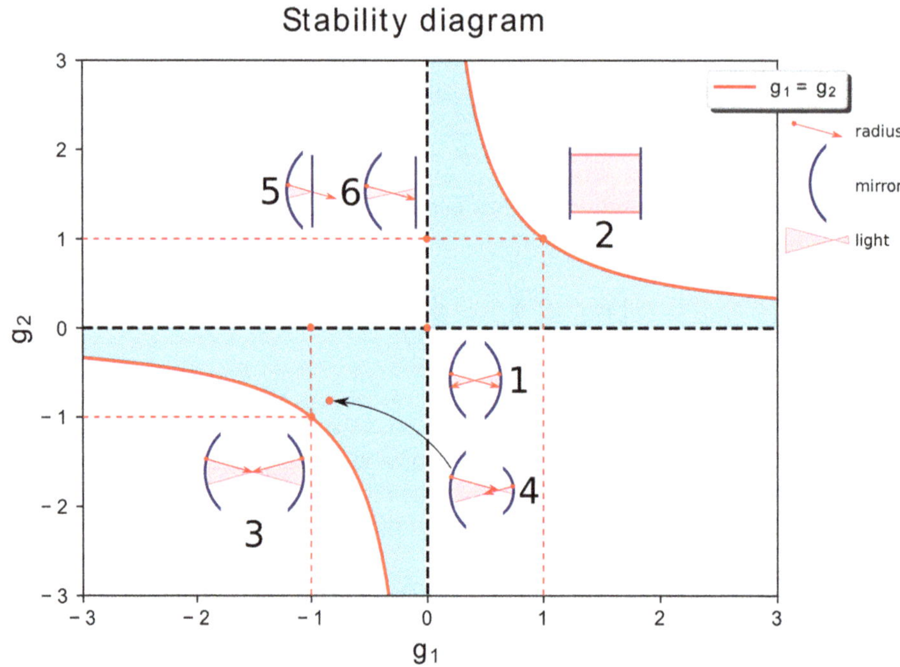

Figure 2.17. Stability diagram for resonator types listed in table 2.3.

at the boundary between stable and unstable configurations, such as the Fabry–Pérot, confocal and concentric resonators.

Plane-parallel and concentric resonators usually have higher diffraction-induced losses, and are also more prone to be affected by optical irregularities, when compared to a confocal configuration. For appropriate dimensions, the confocal resonator shows the lowest diffraction losses. Moreover, the ratio of diffraction losses of high order modes to diffraction losses of the fundamental mode, is higher when compared to other resonator types. This makes the confocal resonator favorable for single-mode laser cavities. It is also less sensitive to optical irregularities, making it easier to align the mirrors.

In order to avoid unstable configurations, the distance between mirrors, L, is chosen to be either slightly greater or smaller than the confocal length. The semiconfocal resonator can function similarly if the surface of the plane mirror is large enough. Similarly, the semiconcentric resonator can function similarly to the concentric resonator, if the plane mirror is large enough, although this makes the resonator bigger (higher volume), yet easier to adjust.

2.6 Ideal Gaussian beams

2.6.1 Gaussian beam parameters

The fundamental oscillation mode TEM_{00q} is defined by a Gaussian intensity distribution (see figure 2.13). This result is based on the Kirchhoff–Fresnel equation and can be correlated with the fact that the Fourier transforms of Gaussian functions are represented by other Gaussian functions. We can obtain an equivalent description if we search for an approximate analytical solution Helmholtz wave equation:

$$\Delta u + k^2 u = 0 \tag{2.125}$$

where $k = \frac{2\pi}{\lambda}$ is the propagation constant (modulus of the wave vector).

The solution for equation (2.125) has the form:

$$u(x, y, z) = u_0 U(x, y, z)\mathrm{e}^{-ikz} \tag{2.126}$$

where e^{-ikz} describes a plane wave propagating in the z direction, and $U(x, y, z)$ is the spacing of the wavefronts.

Let us consider a slow varying spacing with respect to the z axis (paraxial approximation, see equation (1.13)), i.e.,

$$\frac{\partial^2 U}{\partial z^2} \ll k \left| \frac{\partial U}{\partial z} \right| \tag{2.127}$$

If we replace the solution in equation (2.126) in the Helmholtz equation, equation (2.125), we obtain:

$$\frac{\partial^2 U(x, y, z)}{\partial x^2} + \frac{\partial^2 U(x, y, z)}{\partial y^2} - 2ik\frac{\partial U(x, y, z)}{\partial z} = 0 \tag{2.128}$$

Considering the radial structure of a Gaussian beam (see figure 2.13), we can consider that a particular solution of equation (2.128) has the form:

$$U(r_\perp, z) = \exp\{[-iP(z)]\}\exp\left\{\left[-ik\frac{r_\perp^2}{q(z)}\right]\right\} \qquad (2.129)$$

where $P(z)$ is the **longitudinal** parameter of the beam, $q(z)$ is the **radial** parameter of the beam, and $r_\perp^2 = x^2 + y^2$.

Inserting equation (2.129) into equation (2.128) results in:

$$-i\frac{2k}{q} - k^2\frac{r_\perp^2}{q^2} - 2k\left[\frac{dP}{dz} - k\frac{r_\perp^2}{q^2}\left(\frac{dq}{dz}\right)\right] = 0 \qquad (2.130)$$

Equation (2.130) is valid for any $r_\perp$, under the following conditions:

$$\frac{dq(z)}{dz} = 1 \qquad (2.131)$$

$$\frac{dP(z)}{dz} = -\frac{1}{q(z)} \qquad (2.132)$$

Equation (2.131) leads to the solution:

$$q(z) = q(0) + z \qquad (2.133)$$

where $q(0) = q(z = 0)$

To separate the real and imaginary parts, we introduce:

$$\frac{1}{q(z)} = \frac{1}{R(z)} - \frac{i\lambda}{\pi w^2(z)} \qquad (2.134)$$

where:

$$\frac{1}{q(0)} = \frac{i\pi w^2(z)}{\lambda} \qquad (2.135)$$

for $R(0) \to \infty$, and in $z = 0$ we have a plane wave.

Inserting equation (2.133) in equation (2.132), we obtain:

$$P(z) = i\ln\left[\frac{q(0)}{q(0) + z}\right] \qquad (2.136)$$

Consindering equations (2.134)–(2.136), the general solution for equation (2.125) becomes:

$$u = \underbrace{\left\{\frac{w_0}{w(z)}\exp\left[-\frac{r^2}{w^2(z)}\right]\right\}}_{\text{First term}}\underbrace{\exp\left\{-i\left[kz - \frac{1}{\tan\frac{z}{z_0}}\right]\right\}}_{\text{Second term}}\underbrace{\exp\left[-i\frac{kr^2}{2R(z)}\right]}_{\text{Third term}} \qquad (2.137)$$

where:

$$w_0 = \sqrt{\frac{2z_0}{k}}$$

(2.138)

is the **beam waist** and represents the value of r for which the amplitude of the wave is reduced to $1/e$ of its value on the axis.

Using a similar analogy, we define $w(z)$ as the **beam size** at distance z, which is:

$$w(z) = \sqrt{\frac{2}{kz_0}\left(z_0^2 + z^2\right)}$$
$$\Leftrightarrow w(z) = w_0\sqrt{1 + \left(\frac{\lambda z}{\pi w_0^2}\right)^2}$$

(2.139)

and the **curvature of the wavefront** of a Gaussian beam as:

$$R(z) = \frac{1}{z}\left(z_0^2 + z^2\right)$$
$$\Leftrightarrow R(z) = z\left[1 + \left(\frac{\pi w_0^2}{\lambda z}\right)^2\right]$$

(2.140)

The beam (longitudinal) intensity distribution (equation (2.137)), beam size (equation (2.139)) and wavefront curvature (equation (2.140)) of a Gaussian beam propagating in the z direction are presented in figure 2.18. The wavelength, λ, and beam waist, w_0, are equal to 632 nm.

The first term in equation (2.137) (emphasized in cyan, see equation (2.137)) describes the fact that the modal amplitude of coordinates r and z depends on the beam waist, w_0, i.e., minimum beam size. In this case, z_0 is the position where the beam size reaches $\sqrt{2}\,w_0$. This parameter is important when choosing the necessary optics for laser processing.

The second term in equation (2.137) (emphasized in orange) describes the phase variation in the propagation direction. The third term, (emphasized in olive) shows that a transverse section at an arbitrary point z does not yield an equal phase surface. The wavefront and, implicitly, equal phase surfaces, are described by $R(z)$ (see equation (2.140)), where the curvature center varies along the z-axis.

For higher distances, compared to the wavelength, from the position of the beam waist, along the z-axis, the divergence of the Gaussian beam is described by the divergence angle, θ_G:

$$\theta_G = 2\frac{dw}{dz} = \frac{2\lambda}{\pi w_0}$$

(2.141)

Field amplitude is reduced at an arbitrary point from both the r dependence, as well as the divergence.

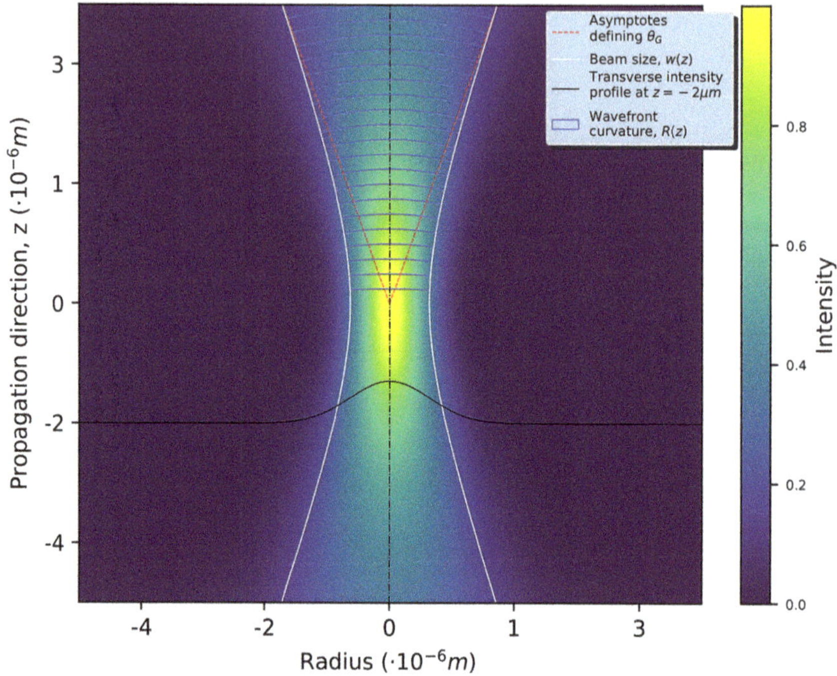

Figure 2.18. Beam size, wavefront curvature and beam intensity.

The solution of the Helmholtz equation (equation (2.125)), using the form presented in equation (2.137) defines the fundamental oscillation mode, TEM_{00}, inside a resonator (see figure 2.13).

The transverse intensity distribution can be calculated using the theories developed by Boyd, Gordon and Kogelnik [2, 3]. It drops to $1/e^2$ for $x = w_s$, where w_s is the envelope of the Gaussian beam, i.e., $w_s = w(z_{\mathrm{cut}})$, where z_{cut} is the position on the propagation axis where the **transverse** intensity distribution is considered (see figure 2.18).

2.6.2 Gaussian beams in stable resonators

The generation of oscillation modes in stable laser resonators must be accompanied by an adaptation process between the curvatures of the mirrors and the Gaussian beam. This adaptation consists of reflections following the same optical path for rays with normal incidence on the surface of the mirrors, which determine a stable self-consistent configuration of cavity modes.

In order to determine the dimensions of a stable configuration, let us consider a resonator that has two spherical mirrors with curvatures R_1 and $R_2 \neq R_1$. We must determine the position of the plane where $z = 0$ (the plane where we find the beam waist, w_0) with respect to the mirrors. This position must determine the adaptation between the mirror curvatures and the wavefront (constant phase surfaces of incident light). This problem is therefore reduced to the following equation system:

$$z_1 + z_2 = L \tag{2.142}$$

$$R(z) = -R_1 = -z\left[1 + \left(\frac{z_0}{z_1}\right)^2\right] \tag{2.143}$$

$$R(z) = R_2 = z_2\left[1 + \left(\frac{z_0}{z_2}\right)^2\right] \tag{2.144}$$

where z_1 and z_2 are the distances from the surface of each mirror, to z_0, where the beam waist is located (see figure 2.19).
Solutions for equations (2.142)–(2.144) are:

$$z_0^2 - \frac{L(R_1 - l)(R_2 - L)(R_1 + R_2 - L)}{(R_1 + R_2 - 2L)^2} \tag{2.145}$$

$$z_1 = \frac{L(R_2 - L)}{R_1 + R_2 - 2L} \tag{2.146}$$

$$z_2 = \frac{L(R_1 - L)}{R_1 + R_2 - 2L} \tag{2.147}$$

Laser mode adaptation refers to transforming Gaussian beams, described by a set of parameters that are determined by the resonator geometry, in other modes, with certain given parameters. Mode adaptation problems also arise when 'injecting'

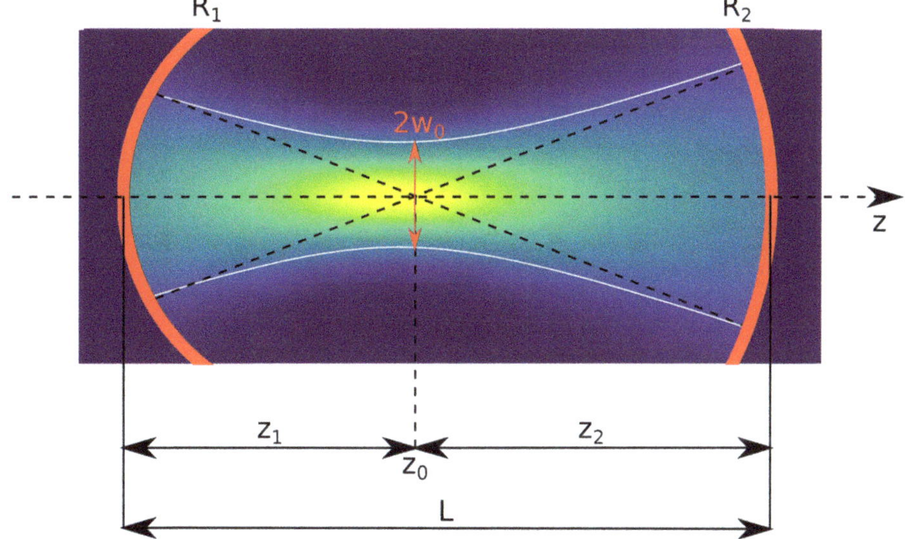

Figure 2.19. Gaussian beam in a spherical mirror asymmetric resonator.

them into other optical systems, like, for example: nonlinear crystals, scanning Fabry–Pérot interferometers, etc. In the simplest case, this adaptation can be done using a lens whose focal distance should be greater than a specific length, $f_0 = \frac{\pi w_1 w_2}{\lambda}$. This can be done by adjusting the position of the lens with respect to planes where $z = 0$, of each beams of size w_1 and w_2 (incident and outgoing from the lens).

2.7 Laser–matter interaction in quantum formalism

2.7.1 Density matrix

The state of a quantum particle can be described using a state vector, $|\Psi\rangle$. When studying quantum mechanical systems, we have pure and mixed states. A pure state can be described by a single *ket* vector, while mixed states are a combination of pure states. Mixed states imply the inability to do maximal measures of quantum states, which in turn determines incomplete information regarding these systems.

Let us consider an ensemble of N quantum systems, where $N \to \infty$, whose pure states, $|\Psi\rangle$ are not precisely known. The only possible information about a specific system is the probability, p_k, of that system to be found in a pure state, i.e., $|\Psi\rangle$, ($p_k \geqslant 0$, $\sum_k p_k = 1$). Therefore, for the whole system, we define a mixed state, which can be represented by a non-coherent superposition of pure states, which in turn means the mean value of an operator is obtained through a statistical average in the classical sense:

$$\langle \hat{Q} \rangle = \sum_k p_k \langle \hat{A} \rangle \Psi_k \tag{2.148}$$

where the probabilities, p_k, are determined through experimental measurements.

2.7.2 Density matrix formalism

The density matrix formalism was proposed by J von Neumann and allows the analysis of both pure and mixed states. To introduce the density operator we write the average of the operator $\hat{Q}$ (equation (2.148)) using the identity operator $\hat{I}$:

$$\sum_n |u_n\rangle\langle u_n| = \hat{I} \tag{2.149}$$

and therefore $\langle \hat{Q} \rangle$ becomes:

$$\begin{aligned}
\langle \hat{Q} \rangle &= \sum_n p_k \langle \Psi_n| \hat{Q} |u_k\rangle \langle u_k |\Psi_n\rangle \\
&= \sum_k \langle u_k| \hat{\rho}\hat{Q} |u_k\rangle
\end{aligned} \tag{2.150}$$

where $\hat{\rho}$ is the density operator:

$$\hat{\rho} = \sum_n p_n |\Psi_n\rangle\langle \Psi_n| \tag{2.151}$$

The average value $\langle \hat{Q} \rangle$ from equation (2.150) can be written as:

$$\langle \hat{Q} \rangle = \mathrm{Tr}(\hat{\rho}\hat{Q}) \tag{2.152}$$

Equation (2.152) represents an equivalent definition of the density operator. For a pure quantum state, the density operator degenerates into a projection operator.

The density operator has some important properties. For one, $\hat{\rho}$ is hermitian, i.e.,

$$\langle \Psi_1 | \hat{\rho} | \Psi_2 \rangle = \langle \Psi_2 | \hat{\rho} | \Psi_1 \rangle^* \tag{2.153}$$

and is normalized:

$$\mathrm{Tr}(\hat{\rho}) = 1 \tag{2.154}$$

The quantum state is pure if:

$$\mathrm{Tr}(\widehat{\rho^2}) = 1 \tag{2.155}$$

Lastly, $(\widehat{\rho^2})_{nm} = \rho_{mn}$; the density matrix for a pure state is diagonal and has one eigenvalue equal to one unit, the others being zero. The diagonal elements of the density matrix $\hat{\rho}$ are the probabilities that a particular system, from the ensemble of quantum systems, is described by the eigenstates $|u_n\rangle$, because $\hat{\rho}$ both replaces the state vectors and acts as an observable (it is hermitian, positive, with a definitive trace). The change of representation is:

$$\hat{\rho}' = \hat{U}^{-1}\hat{\rho}\hat{U} \tag{2.156}$$

where $\hat{U}$ is a unit operator. If we consider a mix of quantum states $|\Psi_n\rangle$ with probabilities p_n that describe the dynamic state of the system at a moment t_0, and which evolves in time, then at moment $t > t_0$ the system is described by the vectors $|\Psi_n\rangle_t$, with the same p_n as for t_0:

$$
\begin{aligned}
\hat{\rho}_t &= \sum_n |\Psi_n\rangle_t \, p_n \, \langle \Psi_n |_t \\
&= \sum_n \hat{T}(t, t_0) |\Psi_n\rangle_0 \, p_n \, \langle \Psi_n |_0 \, \hat{T}^+(t, t_0) \\
&= \hat{T}(t, t_0) \hat{\rho}_0 \hat{T}^+(t, t_0)
\end{aligned}
\tag{2.157}
$$

where $\hat{T}(t, t_0)$ is the unit time evolution operator.

2.7.3 Equations of motion

The unit time evolution operator is described as:

$$\hat{T}(t, t_0) = \exp\left[-\frac{\mathrm{i}\hat{H}(t, t_0)}{\hbar} \right] \tag{2.158}$$

where $\hat{H}$ is the Hamiltonian of the system.

If we consider the time evolution of operators $\hat{T}(t, t_0)$ and $\hat{T}^+(t, t_0)$, we obtain the Schrödinger equation for the density operator:

$$i\hbar\frac{d\hat{\rho}}{dt} = [\hat{H}, \hat{\rho}] \tag{2.159}$$

which formally identifies with Heisenberg's equation for operators, albeit without taking the sign into consideration.

The equation of motion for a quantum operator, $\hat{Q}$, is:

$$\frac{d\hat{Q}}{dt} = \frac{\partial\hat{Q}}{\partial t} + \frac{i}{\hbar}[\hat{H}, \hat{Q}] \tag{2.160}$$

Considering the analysis done in the previous sections (see equation (2.152)), whenever operators $\hat{Q}$ do not have an *explicit* time dependence, the equation of motion for the average value of an operator is:

$$\left\langle\frac{d\hat{Q}}{dt}\right\rangle = \text{Tr}\left(\frac{d\hat{\rho}}{dt}\hat{Q}\right) \tag{2.161}$$

If we consider equation (2.159), then the equation of motion for the elements of the density operator, in the presence of a perturbation of an atomic system, $\hat{W}(t)$, can be written as:

$$i\hbar\frac{\partial\rho_{mn}}{\partial t} = (E_m - E_n)\rho_{mn} + [\hat{W}, \hat{\rho}]_{mn} \tag{2.162}$$

Let us consider internal perturbations as well (for example between atoms and the lattice, atom–atom collisions, etc), for $\hat{W}(t)$, so that:

$$\hat{W}(t) = \widehat{W_i}(t) + \widehat{W_e}(t) \tag{2.163}$$

where $\widehat{W_i}(t)$ are internal perturbations and $\widehat{W_e}(t)$ are external perturbations.

Therefore, equation (2.162) contains relaxation terms that correspond to internal interactions. These terms modify diagonal matrix elements for ρ_{mn}. As such, in the absence of an external perturbation, equation (2.162) becomes:

$$i\hbar\frac{\partial\rho_{mn}}{\partial t} = (E_m - E_n)\rho_{mn} + \left[\widehat{W_e}, \hat{\rho}\right]_{mn} + \underbrace{\left[\widehat{W_i}, \hat{\rho}\right]_{mn}}_{\text{Relaxation term}} \tag{2.164}$$

Let us consider that for $t = 0$ the system is found in the initial state, $|u_i\rangle$. The diagonal matrix elements, ρ_{mn}, begin to exponentially decay as time increases, in the absence of an **external** perturbation, therefore the last term in equation (2.164) represents a relaxation term (in Wigner–Weisskopf approximation).

At thermal equilibrium, $\rho_{mn} = 0$ for $m \neq n$, and therefore $\hat{W_i}(t)$ modifies the non-diagonal matrix elements, ρ_{mn} according to:

$$-i\hbar\frac{\partial \rho_{mn}}{\partial t} = \hbar\omega_{mn}\rho_{mn} + [\hat{W}_{e}, \hat{\rho}]_{mn} - \frac{i\hbar}{\tau_{mn}}\rho_{mn} \tag{2.165}$$

where $\tau_{mn} = \tau_{nm}$, and $\hat{\rho}$ is hermitian. Let us consider the notation $\tau_{mn} = T_2$. This constant is the transverse relaxation time (or spin–spin relaxation time) and is correlated to the transition bandwidth, therefore being a measure of coherence time.

The diagonal elements of $\hat{\rho}$ satisfy the following equation:

$$i\hbar\frac{\partial \rho_{mn}}{\partial t} = [\hat{W}_{e}, \hat{\rho}] + i\hbar\sum_{n}(\rho_{nn}w_{nm} - \rho_{mm}w_{mn}) \tag{2.166}$$

where $\rho_{nn}w_{nm}$ and $-\rho_{mm}w_{mn}$ are transition rates that determine the increase and decrease of the probability of occupying a quantum state $|m\rangle$. At thermal equilibrium, the transition rates are equal, i.e.,

$$\rho_{mm}|_{eq}\ w_{mn} = \rho_{nn}|_{eq}\ w_{nm} \tag{2.167}$$

If we define the time constants as:

$$T_{mn} = \frac{\rho_{nn}|_{eq}}{w_{mn}} \tag{2.168}$$

and considering the notation $T_{mn} = T_1$, then equation (2.166) becomes:

$$i\hbar\frac{\partial \rho_{mn}}{\partial t} = [\hat{W}_{e}, \hat{\rho}]_{mn} + \frac{i\hbar}{T_1}\sum_{n}(\rho_{mm}|_{eq} - \rho_{mm}) \tag{2.169}$$

We considered all external perturbations are zero, i.e., $\hat{W}_{e} = 0$, therefore the system undergoes a relaxation process described by the time constant T_1. This constant is called longitudinal relaxation time (or spin–lattice relaxation). The main physical processes that determine the time constant, T_1, are spontaneous emission, lattice interactions and inelastic collisions. The other time constant, T_2, on the other hand, is determined by elastic collisions, which means $T_2 \leqslant T_1$.

If we evaluate the derivative of the average value of an operator, $\hat{Q}$, which does not *explicitly* depend on time, taking equations (2.152), (2.161), (2.162) and (2.166) into consideration, we will obtain:

$$\left\langle\frac{d\hat{Q}}{dt}\right\rangle + \frac{\langle\hat{Q}\rangle}{T_2} - \frac{\langle\hat{Q}\rangle|_{eq}}{T_1} = \frac{1}{i\hbar}\langle[\hat{Q}, \hat{H}]\rangle + \left(\frac{1}{T_2} - \frac{1}{T_1}\right)\sum_{n}\rho_{nn}Q_{nn} \tag{2.170}$$

This equation is particularly important for cases where all diagonal elements of the operator $\hat{Q}$ are zero, or the other way around, where *only* the diagonal elements are different from zero.

Let us consider an atomic or molecular system that is described by the eigenstates $|n_1\rangle$ and $|n_2\rangle$, with opposite parities, so that electric dipole transitions are allowed between these eigenstates. The Hamiltonian that determines electric dipole transitions, following the interaction between an atomic or molecular quantum system with an electric field, is given by:

$$\hat{W}_{\text{el-dip}} = -\vec{\hat{d}}\,\vec{E} = -\sum_{i=1}^{3}\vec{\hat{d}_i}\,\vec{E_i} \tag{2.171}$$

The non-diagonal form of the associated matrix for the electric dipole operator, $\hat{d}_i$ is:

$$\hat{d}_i = \begin{bmatrix} 0 & \hat{d}_i \\ \hat{d}_i^{\dagger} & 0 \end{bmatrix} \tag{2.172}$$

This is determined by the opposite parities of the non-degenerate eigenfunctions of the two-level quantum system we considered at the beginning. As such, the Hamiltonian for the electric dipole interaction, $\hat{W}_{\text{el-dip}}$, is written as:

$$\hat{W}_{\text{el-dip}} = \begin{bmatrix} 0 & -\sum_i \hat{d}_i E_i \\ -\sum_i \hat{d}_i^{\dagger} E_i & 0 \end{bmatrix} \tag{2.173}$$

For dipole transitions, the average value of the electric dipole momentum observable, can be written as:

$$\langle \hat{d}_i \rangle = \text{Tr}[\hat{\rho}, \hat{d}_i] \tag{2.174}$$

Let us consider the number of atoms, or molecules, per unit volume is:

$$N = \frac{N_0}{V} \tag{2.175}$$

The macroscopic material polarization is obtained by summing all contributions of each atomic or molecular dipole, and taking the average over all quantum systems, i.e., we can write:

$$P_i = \frac{1}{V}\sum_{k=1}^{N_0}\langle d_i \rangle^k = N\langle \tilde{d}_i \rangle \tag{2.176}$$

Averaging over all quantum systems (noted using ' $\sim$ ' symbol) is necessary because each quantum system is oriented differently. The equation of motion for the macroscopic polarization can be obtained if the equation of motion for the electric dipole momentum operator is first established. According to equation (2.172), all diagonal elements of the matrix associated with the dipole momentum operator are zero. Therefore, the time derivative of the average value for the dipole operator $\langle d \rangle$ can be obtained, considering equation (2.170), as follows:

$$\left\langle \frac{d\hat{d}_i}{dt} \right\rangle + \frac{\langle \hat{d}_i \rangle}{T_2} = \frac{1}{i\hbar}\langle [\hat{d}_i, \hat{H}] \rangle \tag{2.177}$$

where $\hat{H} = \hat{H}_0 + \hat{W}$.

The equation of motion for the electric dipole momentum observable is obtained when doing the second time derivative to equation (2.177):

$$\left\langle \frac{\mathrm{d}^2 \hat{d}_i}{\mathrm{d}t^2} \right\rangle + \left\langle \frac{\mathrm{d}\hat{d}_i}{\mathrm{d}t} \right\rangle \frac{1}{T_2} = \frac{1}{i\hbar} \left\langle \left[\frac{\mathrm{d}\hat{d}_i}{\mathrm{d}t}, \hat{H} \right] \right\rangle \tag{2.178}$$

If we replace the average of the first derivative of $\hat{d}_i$, i.e., $\left\langle \frac{\mathrm{d}\hat{d}_i}{\mathrm{d}t} \right\rangle$ (see equation (2.170)), in the right side of equation (2.178), we obtain:

$$\left\langle \frac{\mathrm{d}^2 \hat{d}_i}{\mathrm{d}t^2} \right\rangle + \frac{2}{T_2} \left\langle \frac{\mathrm{d}\hat{d}_i}{\mathrm{d}t} \right\rangle + \frac{1}{T_2^2} \langle \hat{d}_i \rangle$$
$$= -\frac{1}{\hbar^2} \left\langle \left[\left[\frac{\mathrm{d}\hat{d}}{\mathrm{d}t}, \hat{H} \right], \hat{H} \right] \right\rangle + \frac{1}{i\hbar} \left(\frac{1}{T_2} - \frac{1}{T_1} \right) \sum_i \rho_{ii} [\hat{d}_i, \hat{W}]_{ii} \tag{2.179}$$

For all important cases, the commutator $[\hat{d}_i, \hat{W}] = 0$. In order to obtain the equation of motion for $\hat{d}$, we start by looking at the commutator on the right side of equation (2.179):

$$[\hat{d}_i, \hat{H}] = [\hat{d}_i, \hat{H}_0] = \hbar\Omega \begin{bmatrix} 0 & d_i \\ -d^\dagger & 0 \end{bmatrix} \tag{2.180}$$

where $\Omega = \frac{E_2 - E_1}{\hbar}$ is the transition frequency. Then, by introducing the diagonal matrix:

$$D = \begin{bmatrix} 1 & 0 \\ 0 & -1 \end{bmatrix} \tag{2.181}$$

so that:

$$\langle \hat{D} \rangle = \mathrm{Tr}[\hat{\rho}, \hat{D}] = \rho_{11} - \rho_{22} \tag{2.182}$$

represents the difference between the probabilities to occupy the eigenstates with energies E_1 and E_2, we obtain:

$$[[\hat{d}_i, \hat{H}], \hat{H}] = \hbar^2 \Omega^2 d_i - 2\hbar\Omega d_i d_j^\dagger E_j D \tag{2.183}$$

If we consider equations (2.180)–(2.183), then the equation of motion for $\hat{d}$ operator becomes:

$$\left\langle \frac{\mathrm{d}^2 \hat{d}_i}{\mathrm{d}t^2} \right\rangle + \frac{2}{T_2} \left\langle \frac{\mathrm{d}\hat{d}_i}{\mathrm{d}t} \right\rangle + \Omega^2 \langle \hat{d}_i \rangle = \frac{2\Omega}{\hbar} (d_i d_j)(\rho_{11} - \rho_{22}) E_j^{\mathrm{local}} \tag{2.184}$$

where E_j^{local} represents the local field 'seen' by the atom or molecule, which is different from the macroscopic field, due to local polarization of the material.

In equation (2.184) we eliminated the term $\frac{\langle \hat{d}_i \rangle}{T_2^2}$ due to the condition that $\Omega^2 \gg \frac{1}{T_2^2}$.

This equation is obtained by summing and averaging equation (2.184) over all possible orientations and for every atom or molecule in the unit volume (see equation (2.176)), therefore obtaining:

$$\frac{\mathrm{d}^2 \overrightarrow{P_i}}{\mathrm{d}t^2} + \frac{2}{T_2}\frac{\mathrm{d}\overrightarrow{P_i}}{\mathrm{d}t} + \Omega^2 P_i = \frac{2\Omega}{\hbar}\left(\widetilde{d_i d_j^{\dagger}}\right)(N_1 - N_2)E_j^{\mathrm{local}} \tag{2.185}$$

where $N_1 - N_2$ is the population difference per unit volume, i.e.,

$$N_1 - N_2 = N(\rho_{11} - \rho_{22}) \tag{2.186}$$

In equation (2.186), we considered that the difference $\rho_{11} - \rho_{22}$ is roughly constant for all molecules, regardless of orientation, and that the orientation average of a product equals to the product of the two averages.

On the other hand, in an isotropic gas, comprised of anisotropic molecules, we have $(\widetilde{d_i d_j^{\dagger}}) = 0$ for $i \neq j$, because the induced electric dipole momentum must be oriented in the direction of the applied field, therefore:

$$\left|\overrightarrow{d}_{12}\right|^2 = \left|\tilde{d}_x\right|^2 + \left|\tilde{d}_y\right|^2 + \left|\tilde{d}_z\right|^2 = 3\left|\tilde{d}_x\right|^2 \tag{2.187}$$

Taking equations (2.186) and (2.187) into consideration, equation (2.185) becomes:

$$\frac{\mathrm{d}^2 \overrightarrow{P}}{\mathrm{d}t^2} + \frac{2}{T_2}\frac{\mathrm{d}\overrightarrow{P}}{\mathrm{d}t} + \Omega^2 P = \frac{2\Omega}{\hbar}\frac{\left|\overrightarrow{d}_{12}\right|^2}{3}(N_1 - N_2)\overrightarrow{E}^* \tag{2.188}$$

From equation (2.188) we can observe that material polarization acts as a harmonic oscillator with frequency Ω. It is relaxing with a time constant, T_2, in the absence of an external field, as a result of internal dephasing of individual dipoles through mutual interaction. The oscillator receives energy from the external (electric) field. The coupling between this field and the material polarization is proportional to the population difference, $N_1 - N_2$, so that when $N_1 - N_2 \to 0$, the coupling disappears, therefore resulting in a transparency of the material with respect to the field (common process for saturation phenomena).

Equation (2.186) was used to define the population different per unit volume, which represents another important observable for light–matter interaction processes. The population difference does indeed determine the field-polarization coupling strength, as previously shown, as well as the energy quantity stored inside the quantified medium.

The equation of motion for the population difference is obtained by writing equation (2.170) for the $\hat{D}$ operator, with the diagonal matrix shown in equation (2.181), therefore obtaining:

$$\left\langle\frac{\mathrm{d}\hat{D}}{\mathrm{d}t}\right\rangle + \frac{\langle\hat{D}\rangle - \langle\hat{D}\rangle^{\mathrm{eq}}}{T_1} = \frac{1}{\mathrm{i}\hbar}\langle[\hat{D}, \hat{H}]\rangle \tag{2.189}$$

or, equivalently:

$$\frac{\partial}{\partial t}(\rho_{11} - \rho_{22}) + \frac{(\rho_{11} - \rho_{22}) - (\rho_{11} - \rho_{22})^{eq}}{T_1} = \frac{1}{i\hbar}\langle[\hat{D}, \hat{H}]\rangle \tag{2.190}$$

If we consider equation (2.180), and because:

$$[\hat{D}, \hat{H}] = [\hat{D}, \hat{W}] = -2E_i \begin{bmatrix} 0 & \hat{d}_i \\ -\hat{d}_i^\dagger & 0 \end{bmatrix} \tag{2.191}$$

we can, therefore, write:

$$[\hat{D}, \hat{H}] = -\frac{2E_i[\hat{d}_i, \hat{H}]}{\hbar\Omega} \tag{2.192}$$

Imposing the condition that $\Omega T_2 \gg 1$, from equation (2.177) we obtain:

$$\left\langle \frac{d\hat{P}_i}{dt} \right\rangle = \frac{1}{i\hbar}\langle[\hat{P}_i, \hat{H}]\rangle \tag{2.193}$$

If we now look at equation (2.186), we can see that equation (2.190) becomes:

$$\frac{\partial}{\partial t}(N_1 - N_2) + \frac{(N_1 - N_2) - (N_1 - N_2)^{eq}}{T_1} = -\frac{2}{\hbar\Omega}\frac{dP_i}{dt}\vec{E}* \tag{2.194}$$

Equation (2.194) is fundamental for population difference in quantified media. We can observe that changes in the energy stored per unit volume, i.e.,

$$\rho_E = -\frac{\partial}{\partial t}\left[\frac{\hbar\Omega}{2}(N_1 - N_2)\right] \tag{2.195}$$

are determined by $\vec{P}\cdot\vec{E}$, which represents the energy given up by the field to the polarizing medium. In the absence of a field, the population difference tends towards an equilibrium value, with the longitudinal relaxation time constant T_1.

For a polarizable, isotropic, uncharged medium, field equations are obtained using Maxwell's equations, where we consider that total polarization, $\vec{P}^{tot}$, is comprised of the polarization $\vec{P}*$ determined by the transition of interest (named 'source transition'), as well as other transitions that are taking place in the medium. As such, we can write:

$$\vec{D} = \varepsilon_0\vec{E} + \vec{P}^{tot} = \varepsilon\vec{E} + \vec{P}* \tag{2.196}$$

where the effect of other transitions is considered by replacing $\varepsilon_0 \to \varepsilon$. If we introduce equation (2.196) in Maxwell's equations and do a bit of rearranging so as to obtain a wave equation (also see equation (1.23)), we can obtain the following:

$$\nabla \times (\nabla \times \vec{E}) + \frac{n}{c}\gamma\frac{\partial\vec{E}}{\partial t} + \frac{n^2}{c^2}\frac{\partial^2\vec{E}}{\partial t^2} = -\mu_0\frac{\partial^2\vec{P}*}{\partial t^2} \tag{2.197}$$

where n is the refractive index and $\gamma = \frac{\mu_0 \sigma c}{n}$ is the attenuation coefficient. A forward propagating field is described using plane waves using the following decomposition:

$$\vec{E}(\vec{r}, t) = \sum_{\alpha} \frac{\vec{E}_0^{\alpha}}{2} \exp\left[i(\omega_{\alpha} t - \vec{k}_{\alpha} \vec{r})\right] + \text{const} \tag{2.198}$$

For lasers in particular, a field enclosed in a resonant cavity is of higher importance, and therefore, for its description, Slater normal oscillation modes of the cavity are used.

If we consider an optical cavity with ideal conducting walls, a dielectric interior with permittivity ε, and boundary conditions $E_t = 0$ and $B_n = 0$ (t and n subscripts refer to transverse and normal to the walls/boundaries of the cavity), the field inside the cavity allows for the following expansions:

$$\vec{E} = -\frac{1}{\sqrt{\varepsilon}} \sum_{c} p_c(t) \vec{E}_c(\vec{r}) \tag{2.199}$$

and:

$$\vec{H} = -\frac{1}{\sqrt{\mu}} \sum_{c} \omega_c q_c(t) \vec{H}_c(\vec{r}) \tag{2.200}$$

where $\vec{E}_c(\vec{r})$ and $\vec{H}_c(\vec{r})$ are the normal oscillation modes, and the time dependence is described by coefficients $p_c(t)$ and $q_c(t)$. If we insert equations (2.199) and (2.200) in equation (2.197), and considering that oscillation modes are not orthonormal, we obtain:

$$\frac{d^2 p_c}{dt^2} + \frac{\gamma c}{n} \frac{dp_c}{dt} + \omega_c^2 p_c = \frac{1}{\sqrt{\varepsilon}} \int \frac{d^2 \vec{P}^*}{dt^2} \vec{E}_c(\vec{r}) dV \tag{2.201}$$

Let us consider that there is a single predominant oscillation mode inside the cavity, and let us introduce the radiation lifetime inside the cavity, $\tau_c = \frac{n}{\gamma c}$. This means equation (2.201) becomes:

$$\frac{d^2 \vec{E}}{dt^2} + \frac{1}{\tau_c} \frac{d\vec{E}}{dt} + \omega_c^2 \vec{E} = -\frac{1}{\varepsilon} \frac{d^2 \vec{P}}{dt^2} \tag{2.202}$$

or, for a polarization distribution analogous to the field:

$$\frac{d^2 \vec{E}}{dt^2} - \frac{1}{\tau_c} \frac{d\vec{E}}{dt} + \omega_c^2 \vec{E} = -\frac{1}{\varepsilon} \frac{d^2 \vec{P}^*}{dt^2} \tag{2.203}$$

Radiation lifetime inside optical resonators, τ_c, is determined by the reflection losses at each mirror, τ_m, the scattering and absorption losses, τ_s, and the diffraction losses, τ_d, therefore we can write:

$$\frac{1}{\tau_c} = \frac{1}{\tau_m} + \frac{1}{\tau_s} + \frac{1}{\tau_d} \tag{2.204}$$

In equations (2.188), (2.194), (2.197) and (2.203) we have to consider the fact that the local field, $\vec{E}^*$, and the source polarization, $\vec{P}^*$ are intervening as well, and therefore direct correlations between the macroscopic and microscopic must be established. It can be shown that:

$$\vec{P}^* = \frac{n^2 + 2}{3}\vec{P} = \sqrt{L}\,\vec{P} \tag{2.205}$$

$$\vec{E}^* = \frac{n^2 + 2}{3}\vec{E} = \sqrt{L}\,\vec{E} \tag{2.206}$$

where L is the Lorentz correction factor.

The expression for L, i.e., $L = (\frac{n^2+2}{3})^2$ stems from a known relation between the local and macroscopic field, i.e., $\vec{E}^* = \vec{E} + \frac{1}{3\varepsilon_0}\vec{P}^{\,\text{tot}}$, where $\vec{E}^*$ is the local field and $\vec{E}$ is the macroscopic one.

If we introduce the Lorentz correction coefficient into **equations of motion for polarization, population and electric field**, we will obtain the following set of equations of motion for electric dipole transitions:

$$\frac{d^2\vec{P}}{dt^2} + \frac{2}{T_2}\frac{d\vec{P}}{dt} + \Omega^2\vec{P} = \frac{2\Omega}{\hbar}\frac{|\vec{d}_{12}|^2}{3}(N_1 - N_2)\vec{E} \tag{2.207}$$

$$\frac{\partial}{\partial t}(N_1 - N_2) + \frac{(N_1 - N_2) - (N_1 - N_2)^{\text{eq}}}{T_1} = -\frac{2}{\hbar\Omega}\frac{d\vec{P}}{dt}\vec{E} \tag{2.208}$$

$$\nabla \times (\nabla \times \vec{E}) + \frac{n\gamma}{c}\frac{\partial\vec{E}}{\partial t} + \frac{n^2}{c^2}\frac{\partial^2\vec{E}}{\partial t^2} = -\mu_0\frac{\partial^2\vec{P}}{\partial t^2} \tag{2.209}$$

$$\frac{d^2\vec{E}}{dt^2} + \frac{1}{\tau_c}\frac{d\vec{E}}{dt} + \omega_c^2\vec{E} = -\frac{1}{\varepsilon}\vec{E}_c(\vec{r})\int\frac{d^2\vec{P}^*}{dt^2}\vec{E}_c(\vec{r})dV \tag{2.210}$$

2.8 Resonant absorption, dispersion and saturation

2.8.1 Absorption and dispersion

Absorption and dispersion processes inside a medium can be described using the electric displacement vector, $\mathbf{D} = \mathbf{D(E)}$, which can be used to define the complex susceptibility of this medium according to:

$$\mathbf{D} = \varepsilon_0(1 + \chi)\mathbf{E} + \varepsilon_0(\chi\mathbf{EE} + \chi\mathbf{EEE} + \cdots) = \varepsilon\mathbf{E} + \mathbf{P}^{\text{nl}} \tag{2.211}$$

where $\mathbf{P}^{\text{nl}}$ is the nonlinear part of material polarization.

As a first approximation, let us only consider the linear polarization, which further means that the linear susceptibility is defined by:

$$\mathbf{P} = \varepsilon_0 \chi(\omega)\mathbf{E} \tag{2.212}$$

where $\chi(\omega)$ is a complex quantity:

$$\chi(\omega) = \chi_{\text{dis}}(\omega) + i\chi_{\text{abs}}(\omega) \tag{2.213}$$

In equation (2.213), χ_{dis} describes the dispersive properties and χ_{abs} describes the absorptive properties of the medium. In order to determine the explicit frequency dependency of the linear susceptibility, we can compare equation (2.212) with the solution, $\overrightarrow{P}(\overrightarrow{E})$, given by equation (2.207), a solution which is obtained by considering the following expressions for the polarization and the electric field intensity at optical frequencies:

$$\overrightarrow{P} = \frac{1}{2}\overrightarrow{P_0}e^{i(\omega t - kx)} + \text{const.} \tag{2.214}$$

and, respectively:

$$\overrightarrow{E} = \frac{1}{2}\overrightarrow{E_0}e^{i(\omega t - kx)} + \text{const.} \tag{2.215}$$

where $\omega \approx \Omega$ was considered (note: Ω refers to a resonance frequency).

For a stationary regime, we have no time dependence, which means $\frac{\partial}{\partial t}(N_1 - N_2) = 0$, because $\omega T_1 \gg 1$, and this further means that:

$$\overrightarrow{P_0} = \frac{L}{\hbar}\frac{|\overrightarrow{d_{12}}|^2}{3}(N_1 - N_2)\frac{1}{(\Omega - \omega) + i\left(\dfrac{1}{T_2}\right)}\overrightarrow{E_0} \tag{2.216}$$

If we replace equation (2.216) in equation (2.212), we obtain the following expression for $\chi(\omega)$:

$$\chi(\omega) = \frac{\pi}{\hbar\varepsilon_0}L\frac{|\overrightarrow{d_{12}}|^2}{3}(N_1 - N_2)g_{\text{L}}^c(\omega, \Omega) \tag{2.217}$$

where $g_{\text{L}}^c(\omega, \Omega)$ is:

$$g_{\text{L}}^c(\omega, \Omega) = \frac{1}{\pi}\frac{1}{(\Omega - \omega) + i\left(\dfrac{1}{T_2}\right)} \tag{2.218}$$

and represents the Lorentz spectral line form function.

If, instead of equation (2.207) for isotropic media, we use equation (2.185) applicable for anisotropic media, then the electric susceptibility will be a tensor that correlates a certain polarization component with electric field components over different directions. In mathematical terms, this can be expressed as:

$$\vec{P}_{0i} = \frac{\pi}{\hbar}\left(\widetilde{d_i d_j^\dagger}\right)(N_1 - N_2)g_L^c(\omega, \Omega)\vec{E}_{0j}$$
$$= \varepsilon_0 \chi_{ij}(\omega)\vec{E}_{ij}$$

(2.219)

where χ_{ij} are the components of a second-order tensor.

Let us now analyze the absorption and dispersion properties of isotropic media, for frequencies close to the resonance frequency, Ω. If we look back at equations (2.213) and (2.217), while also considering the Lorentz spectral line form function, g_L^c (see equation (2.218)), we will obtain:

$$\chi_{dis}(\omega) = \frac{\pi}{\hbar\varepsilon_0}L\frac{|\vec{d}_{12}|^2}{3}(N_1 - N_2)\left[\frac{1}{\pi}\frac{(\Omega - \omega)}{(\Omega - \omega)^2 + \left(\frac{1}{T_2}\right)^2}\right]$$

(2.220)

and:

$$\chi_{abs}(\omega) = -\frac{\pi}{\hbar\varepsilon_0}L\frac{|\vec{d}_{12}|^2}{3}(N_1 - N_2)\left[\frac{1}{\pi}\frac{\frac{1}{T_2}}{(\Omega - \omega)^2 + \left(\frac{1}{T_2}\right)^2}\right]$$

(2.221)

There is a correlation between the linear susceptibility, $\chi(\omega)$ and the complex valued propagation constant, $k(\omega)$, which intervenes in the solutions for equations (2.214) and (2.215). If we insert these solutions into the field equation, equation (2.209), we obtain:

$$k^2(\omega) = \frac{n^2\omega^2}{c^2}\left[1 + \frac{\chi(\omega)}{n^2}\right]$$

(2.222)

If we consider that:

$$\chi(\omega) \ll \frac{1}{n^2}$$

(2.223)

which is a physically acceptable assumption, and we also consider the approximation:

$$\sqrt{1 + \frac{\chi}{n^2}} \approx 1 + \frac{\chi}{2n^2}$$

(2.224)

we will obtain the following expression for the propagation constant:

$$k = k_{dis} + ik_{abs} \approx \frac{n\omega}{c}\left[1 + \frac{\chi_{dis}(\omega)}{2n^2}\right] + i\frac{\omega\chi_{abs}(\omega)}{2nc}$$

(2.225)

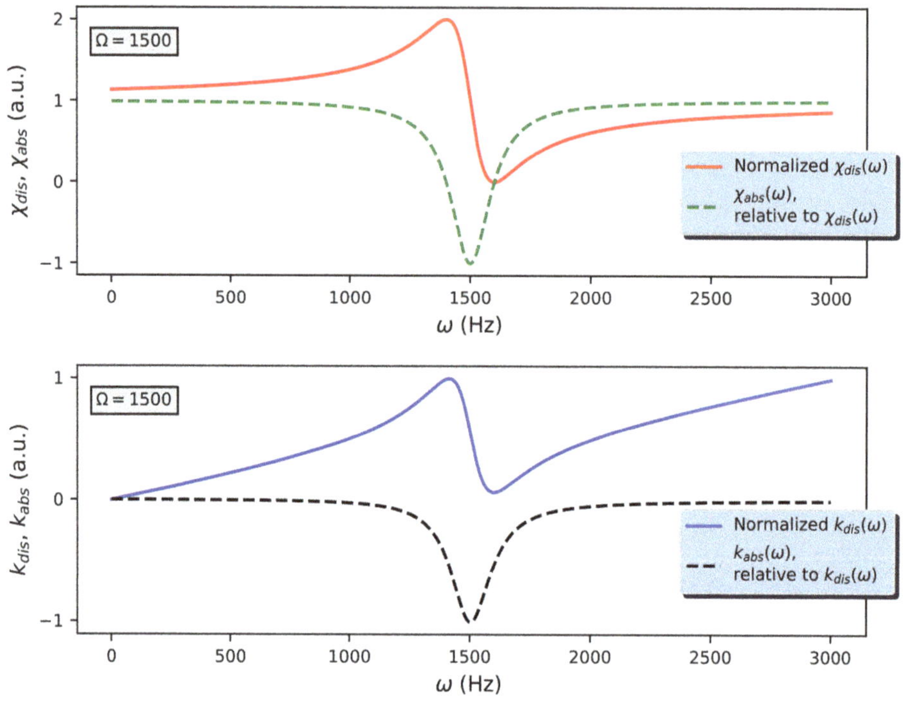

Figure 2.20. Susceptibility and propagation constant as functions of frequency, for both dispersion (real part) and absorption (imaginary part).

The frequency dependency of the susceptibility and the propagation constant, around a resonance (or transition) frequency, Ω, is shown in figure 2.20. For plotting the susceptibilities, $\chi_{\mathrm{dis}}(\omega)$ and $\chi_{\mathrm{abs}}(\omega)$, we used equations (2.220) and (2.221), and for plotting the real and imaginary parts of the propagation constant, $k_{\mathrm{dis}}(\omega)$ and $k_{\mathrm{abs}}(\omega)$, we used equation (2.225).

The propagation constant (real part) $k_{\mathrm{dis}}(\omega)$ describes the dispersive properties of the medium. These properties are determined by the susceptibility, $\chi_{\mathrm{dis}}(\omega)$. The frequency dependency of $\chi_{\mathrm{dis}}(\omega)$ emphasizes the anomalous dispersion phenomenon.

Absorption properties of the medium are described by the imaginary part of the propagation constant, $k_{\mathrm{abs}}(\omega)$. Light intensity is defined as:

$$I = \frac{nc\varepsilon_0}{2}\,|E_0|^2\mathrm{e}^{-ikz} \tag{2.226}$$

If we look at Lambert–Beer law (previously used in section 2.3 to define the gain threshold condition), intensity is written as:

$$I = I_0\mathrm{e}^{-\alpha z} \tag{2.227}$$

If we compare the two equations above, i.e., equations (2.226) and (2.227), we can obtain the expression of the absorption coefficient, α as a function of frequency:

$$\alpha = -2k_{\mathrm{abs}} = \frac{\Omega\pi}{\varepsilon_0 \hbar c n}L\frac{|\overrightarrow{d_{12}}|}{3}(N_1 - N_2)g_{\mathrm{L}}(\omega, \Omega) \tag{2.228}$$

In equation (2.228), the $g_{\mathrm{L}}(\omega, \Omega)$ function represents the imaginary part of the Lorentz spectral line form function, $g_{\mathrm{L}}^{c}(\omega, \Omega)$, and has the following form:

$$g_{\mathrm{L}}(\omega, \Omega) = \frac{1}{\pi}\frac{\dfrac{1}{T_2}}{(\Omega - \omega)^2 + \left(\dfrac{1}{T_2}\right)^2} \tag{2.229}$$

The frequency dependency of $g_{\mathrm{L}}(\omega, \Omega)$ is shown in figure 2.21.

The Lorentz spectral line form function usually appears in every analysis of optical absorption or resonance phenomena. It can be proven that $g_{\mathrm{L}}(\omega, \Omega)$ satisfies the normalization condition, i.e.,

$$\int g_{\mathrm{L}}(\omega, \Omega)\mathrm{d}\omega = 1 \tag{2.230}$$

Taking this into consideration (i.e., equation (2.230)), the width of a Lorentz spectral line is defined as:

$$B = 2\Delta\omega_{\mathrm{L}} \tag{2.231}$$

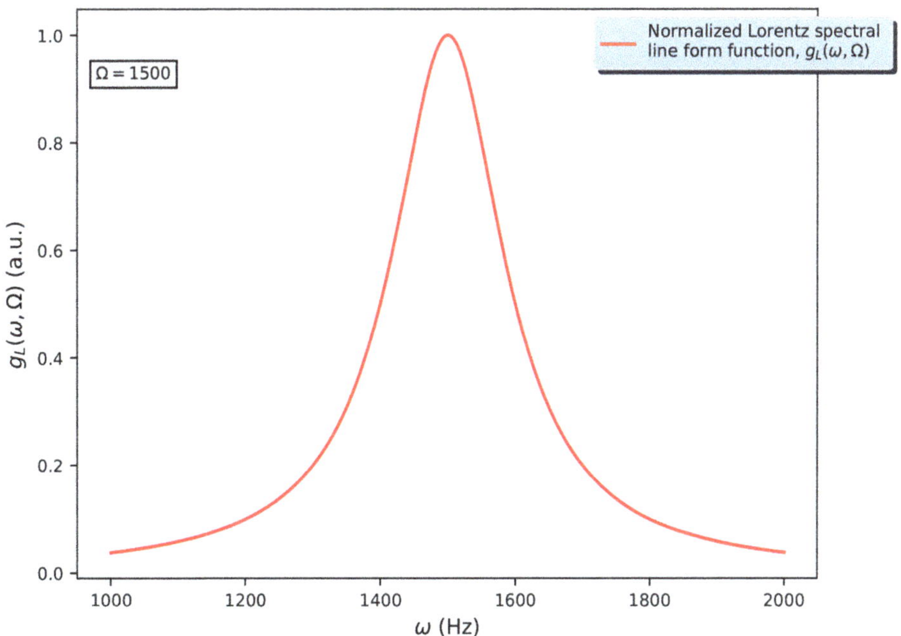

Figure 2.21. Real part of the Lorentz spectral line form function, as a function of frequency, around the transition frequency Ω.

where $\Delta\omega_L$ results from the condition that:

$$[g_L(\omega, \Omega)]_{\Delta\omega_L} = \frac{1}{2}[g_L(\omega, \Omega)]_{max} \tag{2.232}$$

or, simply put, B is the full width at half maximum (FWHM) for the function $g_L(\omega, \Omega)$. If we take a closer look at equation (2.229), we notice that:

$$\Delta\omega_L = \frac{1}{T_2} \tag{2.233}$$

where T_2 represents the spin–spin relaxation time of the medium.

From equation (2.228), we observe that the absorption coefficient, α, has a linear dependency on the population difference, $(N_1 - N_2)$. Moreover, we observe that N_2 has a negative contribution to the absorption coefficient, as it favors the energy transfer from the medium to the electromagnetic field, i.e., to emission processes. In the case where $N_2 > N_1$, the absorption becomes negative, and therefore the electromagnetic field passing through the medium is amplified (see section 2.3).

The absorption properties of a medium, or the transition strength between the involved atomic states, can also be described by the absorption cross-section per atom, or oscillator strength. The atomic absorption cross-section, σ_c is defined for an atom with two states, $N_1 = N$ and $N_2 = 0$, using the ratio between the absorbed power per atom and the incident power per unit area:

$$\begin{aligned}
\sigma_c &= \frac{P}{NI} \\
&= -\frac{1}{NI}\frac{dI}{dz} \\
&= \frac{\alpha}{N}
\end{aligned} \tag{2.234}$$

where I is the incident power per unit area (i.e., intensity) and P is the absorbed power per unit volume. We can observe, from equation (2.234), that the cross-section, σ_c can only be written as a function of the absorption coefficient, α, and the number of atoms per unit volume, N. This leads to an explicit form:

$$\sigma_c = \frac{\Omega\pi}{3\varepsilon_0\hbar cn}L\,|\vec{d_{12}}|^2 g_L(\omega, \Omega) \tag{2.235}$$

The atomic cross-section can be correlated with the oscillator strength, which is a physical quantity defined as the ratio between the quantum and classic cross-sections. As such, if we consider the states $\langle m|$ and $\langle n|$, involved in a transition, then the oscillator strength f_{mn} is given by:

$$\begin{aligned}
f_{mn} &= \frac{(\sigma_c)_{quantum}}{(\sigma_c)_{classic}} \\
&= \frac{2m\Omega_{mn}\,|\vec{d_{mn}}|^2}{3e^2\hbar}
\end{aligned} \tag{2.236}$$

which can have both positive and negative values (indicating absorption or emission), and where $\frac{e}{m}$ is the specific charge of the oscillating dipole.

Oscillator strengths are particularly useful for estimating the elements of the electric dipole matrix, whenever there is not enough experimental data, by using specific summing rules, depending on the nature of the involved physical processes. For example, in the case of transitions involving a single electron, the sum of all oscillator strengths must be unitary. However, if there are multiple electrons involved in a transition, the Kuhn–Thomas summing rule is used, which states that the sum of oscillator strengths that describe transitions on a given energy level is determined by the number of electrons involved in that transition.

2.8.2 Atomic saturation

We have deduced the dispersion and absorption properties of a medium, using the polarization equation, i.e., equation (2.207). Quantum equations that describe the behavior of substantial media (i.e., made of substance/matter, not fields or vacuum) must also account for saturation properties of mentioned media. Saturation, in this context, manifests through an unattenuated propagation of electromagnetic radiation through the medium, if the radiation has sufficient intensity. From a physical point of view, saturation properties can be explained as follows:

- **low intensity**—the energy absorbed by the atoms in the presence of electromagnetic fields that oscillate with frequencies that are equal to a transition frequency, is quickly released back (or yielded) to the environment through relaxation processes, and therefore absorption properties remain unchanged;
- **high intensity**—absorbed energy cannot be dissipated only through relaxation processes, and therefore a redistribution of populations takes place, between energy levels that participate in the transition. This population redistribution intervenes in the susceptibility, χ_{dis} and χ_{abs}, through the term $(N_1 - N_2)$, as to saturate the dispersion and absorption properties.

Saturation can be described using equation (2.194), in two cases:
- **in the absence of an electric field**, in stationary regime, the population difference $(N_1 - N_2)$ reaches an equilibrium state given by Boltzmann distribution, i.e.,

$$(N_1 - N_2) = (N_1 - N_2)^{\mathrm{eq}} \tag{2.237}$$

- **in the presence of an electric field**, where the stationary regime can also be reached, i.e., $\frac{\partial}{\partial t}(N_1 - N_2) = 0$, but where the equilibrium state is given by:

$$\frac{\hbar\Omega}{2} \frac{(N_1 - N_2) - (N_1 - N_2)^{\mathrm{eq}}}{T_1} = -\frac{i\omega}{4}\left(\overrightarrow{P_0}\overrightarrow{E_0}^* - \overrightarrow{P_0}^*\overrightarrow{E_0}\right) \tag{2.238}$$

The equilibrium value described by equation (2.238) corresponds to the case where the average power yielded by the atoms, through relaxation processes, is equal to the average power received from the electromagnetic field, through absorption processes. When this state is reached, the atoms stop absorbing the radiation, and therefore it passes unattenuated through the medium. If we consider that:

$$\overrightarrow{P} = \varepsilon_0 [\chi_{\text{dis}}(\omega) + i\chi_{\text{abs}}(\omega)]\overrightarrow{E} \tag{2.239}$$

then, from equation (2.238) we get:

$$N_1 - N_2 = \frac{(N_1 - N_2)^{\text{eq}}}{1 + \dfrac{I}{I_{\text{sat}}} \cdot \dfrac{\pi g_{\text{L}}(\omega, \Omega)}{T_2}} \tag{2.240}$$

where I_{sat} is the power per unit area of a wave which, at resonance, reduces the population difference to half of its unsaturated value. It is described by:

$$I_{\text{sat}} = \frac{n\varepsilon c}{\left(\dfrac{2T_1 T_2}{\hbar^2}\right)L\left(\dfrac{|\overrightarrow{d_{12}}|^2}{3}\right)} \tag{2.241}$$

The population difference drops as the intensity of the electromagnetic field rises. However, over a certain value, as a result of saturation phenomena, this is an unwanted effect in the case of lasers, where the highest population inversion is desired.

From equations (2.228) and (2.240) we can deduce that, when saturation phenomena become important, the absorption coefficient, α, becomes:

$$\alpha_{\text{sat}} = \frac{\Omega\pi}{\hbar\varepsilon_0 cn}L\frac{|\overrightarrow{d_{12}}|^2}{3}(N_1 - N_2)^{\text{eq}}\frac{1}{\pi}\frac{\dfrac{1}{T_2}}{(\Omega - \omega)^2 + \left(\dfrac{1}{T_2}\right)^2\left(1 + \dfrac{I}{I_{\text{sat}}}\right)} \tag{2.242}$$

Therefore:

$$\alpha_{\text{sat}}(\Omega) = \alpha(\Omega)\frac{1}{1 + \dfrac{I}{I_{\text{sat}}}} \tag{2.243}$$

meaning the saturation effect manifests by lowering the height of the absorption line. At the same time, the absorbed power per unit volume is:

$$P_{\text{sat}} = \frac{\hbar\Omega}{2T_1}(N_1 - N_2)^{\text{eq}} \tag{2.244}$$

2.8.3 Homogeneous and inhomogeneous broadening of a spectral line

The Lorentz spectral line, described by $g_L(\omega, \Omega)$ function (see equation (2.229)), is said to be homogeneously broadened, because the relaxation processes which define it act in the same way over all atoms of the medium, atoms that have the same transition frequency (noted as Ω throughout equations). Therefore, the frequency dependency of the macroscopic polarization of the material is the same as in the case of individual atoms. Typical processes that determine a homogeneous broadening are atom–atom collisions and coupling with lattice vibrations.

If, however, the atoms or molecules of the medium have different transition frequencies, Ω_i, then the spectral line emitted by the medium is different than the spectral line emitted by individual oscillators, therefore being inhomogeneously broadened. Common processes that determine or influence the distribution of frequencies, Ω_i, are the Doppler effect, crystal lattice defects, field inhomogeneities, etc.

For example, in the case of Doppler broadening (i.e., the medium we analyze is in gas phase), the movement of the gas molecules determines the modification of the transition frequency, $\omega_i \rightarrow \Omega_i$, as a function of their velocities, $\vec{v_i}$, for each oscillator. Therefore, in the interval $d\Omega_i$ we find transition frequencies for dN molecules out of N molecules per unit volume. Therefore:

$$dN = N g_G(\omega_0, \Omega_i) d\Omega_i \tag{2.245}$$

where:

$$g_G(\omega_0, \Omega_i) = \frac{\sqrt{\dfrac{4\ln 2}{\pi}}}{\Delta\omega_G} \exp\left[-4\ln 2\frac{(\Omega_i - \omega_0)^2}{\Delta\omega_G^2}\right] \tag{2.246}$$

is the Gaussian spectral line form function, which satisfies the normalization condition:

$$\int_{-\infty}^{\infty} g_G(\omega_0, \Omega_i) d\Omega_i = 1 \tag{2.247}$$

In order to evaluate the Doppler broadened inhomogeneous susceptibility, we have to take into consideration the fact that the polarization, $\vec{dP}$, as a function of frequency, ω, is obtained by summing all components of the form $d\chi(\omega, \Omega_i)$, therefore:

$$\chi(\omega) = \frac{\pi}{\hbar\varepsilon_0}L\frac{|\vec{d_{12}}|^2}{3}(N_1 - N_2)\int_{-\infty}^{\infty} g_L(\omega, \Omega_i)g_G(\omega, \Omega_i)d\Omega_i \tag{2.248}$$

From equation (2.248) we observe that $\chi_{dis}(\omega)$ and $\chi_{abs}(\omega)$ can be calculated.

References

[1] Beiser A 1973 *Concepts of Modern Physics* (Columbus, OH: McGraw-Hill)

[2] Boyd G D and Gordon J P 1961 Confocal multimode resonator for millimeter through optical wavelength masers *Bell Syst. Tech. J.* **40** 489–508

[3] Boyd G D and Kogelnik H 1962 Generalized confocal resonator theory *Bell Syst. Tech. J.* **41** 1347–69

[4] Ewart P 2019 *Atomic Physics* (San Rafael, CA: Morgan & Claypool)

[5] Fox A G and Li T 1961 Resonant modes in a maser interferometer *Bell Syst. Tech. J.* **40** 453–88

[6] Kostić S, Lazarević Z Z, Radojević V, Milutinović A, Romčević M, Romčević N Z and Valčić A 2015 Study of structural and optical properties of YAG and Nd:YAG single crystals *Mater. Res. Bull.* **63** 80–7

IOP Publishing

High Resolution Laser Microprocessing
Implementation and techniques
Bogdan Ştefăniţă Călin, Marian Zamfirescu and Niculae Puşcaş

Chapter 3

Laser systems for materials processing

3.1 Pulse generation: Q-switch and mode-locking

3.1.1 Q-switch

The amount of energy a simple laser system (active medium plus resonant cavity) can generate is governed by three characteristics: steady state gain, losses and threshold values. As previously stated, in order to generate laser radiation, the steady state gain of the system must go above the population inversion threshold, coupled with the overall energy loss as well. The energy loss of a laser cavity is generally described by a parameter called 'quality factor', Q. If we purposely introduce supplementary losses in the cavity, for example by removing or limiting reflections off one of the mirrors, we can significantly lower the quality factor, Q. As such, the laser medium can retain more energy than usual, from pumping, as it is not balanced out by the laser generation after reaching the threshold value. More specifically, the population inversion reaches much higher values in comparison to the threshold. Therefore, if we *suddenly* remove the purposely introduced losses, the difference in population inversion generates much higher radiation intensity inside the cavity, which results in a short and intense laser pulse, until the energy inside the system is depleted below the threshold. The time required for pumping to achieve population inversion is typically much longer than the lifetime of the cavity, θ_c, which more precisely means that the pulse duration is comparable to τ_c. Furthermore, population inversion is depleted after the generation of a pulse, which means the whole process needs to be restarted for the generation of another pulse. As the name implies, this method is known as 'Q-switching', as the key element is quickly removing the supplementary losses. Q-switching is schematically represented in figure 3.1.

As we can observe in figure 3.1, population inversion increases as long as cavity losses are high (i.e., the Q-factor is low). As losses are quickly removed, population inversion slows down and starts to drop shortly after. When population inversion starts to drop, a laser pulse begins to form (output intensity rises), until it reaches a peak intensity at the moment when population inversion goes below the threshold

doi:10.1088/978-0-7503-3239-2ch3　　　3-1　　　

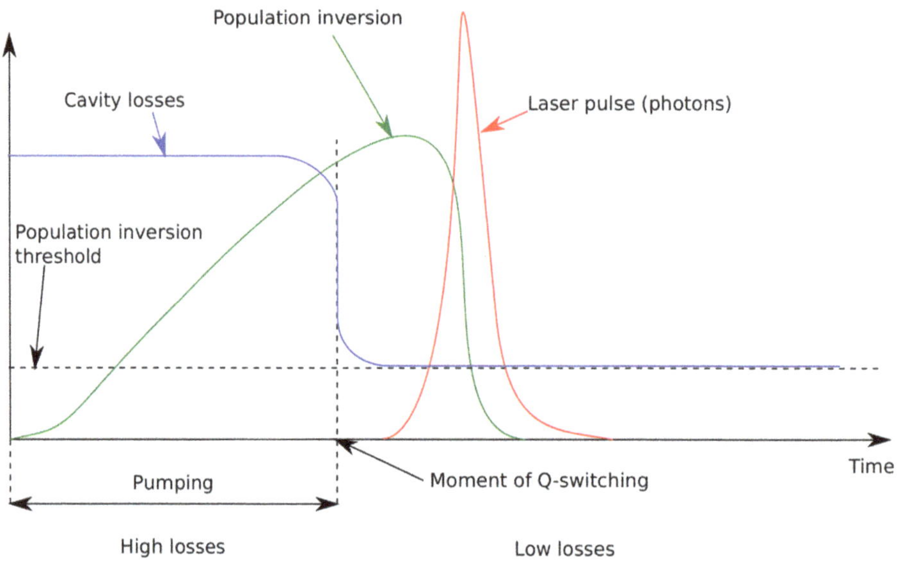

Figure 3.1. Schematic representation of Q-switching: cavity losses, population inversion and number of photons.

value. The resulting laser pulse power can reach values of up to 3–4 *orders of magnitude higher* than the equivalent power obtainable using the same system in continuous-wave mode.

In order to obtain Q-switching, three conditions must be met. Firstly, losses (both cavity and purposely inserted ones) must be higher than laser gain so that laser oscillation does not occur before the moment of switching the quality factor, Q. Secondly, removing the purposely inserted losses (i.e., switching the quality factor, Q) must happen quickly enough, typically below 10 ns, so that all energy is transferred to the single laser pulse that we intend to obtain. Thirdly, pumping and lifetime of upper energy levels must exceed the laser cavity lifetime, otherwise energy does not accumulate in the system.

Using the expression for the population inversion between the upper and lower levels at threshold, per unit volume:

$$\mathrm{d}N_t = \left(N_{\text{upper}} - \frac{g_{\text{upper}}}{g_{\text{lower}}} N_{\text{lower}} \right) \cdot \mathrm{d}V \tag{3.1}$$

as well as rate equations for upper/lower levels populations of the laser transition and photon density, we can obtain an expression for the maximum power, P_{max} achievable via Q-switching:

$$P_{\text{max}} = \frac{\hbar\omega}{2\tau_{\text{c}}} \left[N_t \ln \frac{N_t}{N} - (N_t - N_0) \right] \tag{3.2}$$

where N_0 is the initial population inversion (i.e., with low Q), N_{upper} is the population of the upper energy level (when transition happens, i.e., after Q-switching), N_{lower} is the population of the lower level, dV is the unit volume, h is Planck's constant, ν is the frequency of photons, τ_c is the lifetime of the cavity.

If we consider $N_0 \gg N_t$, then P_{max} becomes:

$$P_{max} \approx \frac{N_0 \hbar \omega}{2\tau_c} \tag{3.3}$$

Equation (3.3) indicates that approximately half of the population inversion is depleted in a duration equal to τ_c. Using this equation we can also calculate the maximum energy in the system, i.e.,

$$E_{sys} = P_{max}\tau_c = \frac{N_0 \hbar \omega}{2} \tag{3.4}$$

However, the maximum energy of the resulting laser pulse has to take into consideration the energy utilization factor, i.e., how much energy is transferred outside the system:

$$\begin{aligned} E_{max} &= \frac{N_0 - N_t}{N_0} \frac{N_0 \hbar \omega}{2} \\ &= \frac{\hbar \omega (N_0 - N_t)}{2} \end{aligned} \tag{3.5}$$

Since the basic definition of power is energy over time, the duration of a Q-switched laser pulse can be obtained using equations (3.3) and (3.5):

$$\begin{aligned} \tau_p &= \frac{E_{max}}{P_{max}} \\ &= \tau_c \frac{N_0 - N_t}{N_0} \end{aligned} \tag{3.6}$$

As difference between population inversions ratios (before and after Q-switching), resulting laser pulse duration drops and pulse extraction efficiency increases (see figure 3.2). Pulse duration cannot, however, get shorter than the cavity lifetime, τ_c.

Switching the quality factor, Q, fast enough to obtain a Q-switch regime can be obtained via various methods. One approach is splitting these methods into active and passive. Active methods are electronically controlled, common examples being the chopper/rotating mirror (mechanically removing light from the cavity via reflection or absorption), electro-optical modulator (changing optical characteristics inside the laser cavity), acousto-optical modulator (inducing a fast switching diffraction grating via acoustic waves). Passive methods involve the use of saturable absorbers (reusable) and thin films (consumable). While all of them function properly, most laser systems employ only electro-optical, acousto-optical or saturable absorber methods.

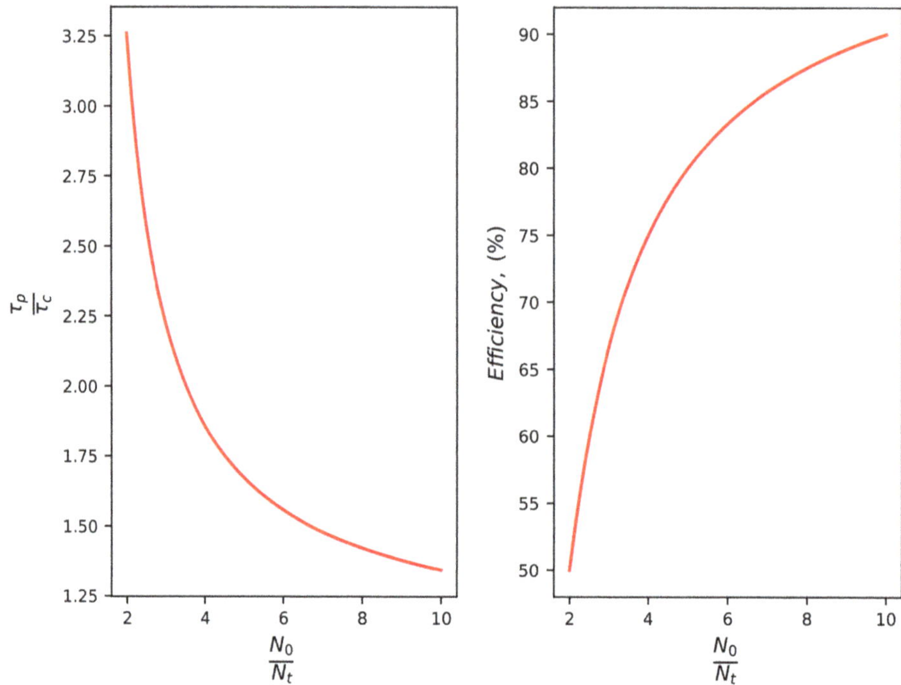

Figure 3.2. Pulse duration and photon extraction efficiency as functions of the initial inversion ratio.

A very common Q-switching method is to use an electro-optic modulator (EOM), which is a device built around a crystal that changes its optical properties when a high voltage is applied (as the name infers). This modulation can be either phase, intensity, polarization and frequency. The underlying physical effect is the Pockels effect, which describes a modification of the refractive index in nonlinear crystals.

Inside a laser cavity, an EOM is generally used to change the polarization of the beam. It is used together with a half-wave plate, polarizers or a window positioned at the Brewster angle, that works as an energy extraction device, i.e., lowers the quality factor, Q. Laser radiation inside a cavity is linearly polarized, due to the fact that photons emitted through stimulated emission retain all properties of the incident photon (phase, wavelength and polarization). A polarizer is usually used to ensure a specific polarization.

The main concept is that when the cell is in a certain state (either on or off), light propagating through the EOM changes its polarization, which in turn means that cavity losses are high, because light is not allowed to go through the active medium multiple times and get amplified, while also depleting the active medium of accumulated energy. The reason light does not go through the active medium is that it is either reflected by a window that is placed at the Brewster angle, or absorbed by a linear polarizer that does not allow that particular polarization direction to pass through. When the EOM changes state, it switches the polarization of light by 90°, allowing it to pass through the whole resonator.

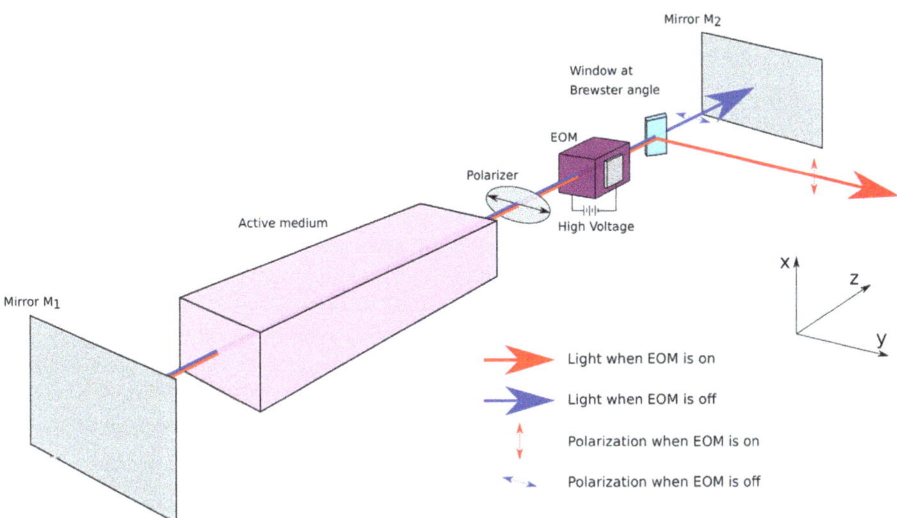

Figure 3.3. Active Q-switch resonator using an EOM, polarizer and window at Brewster angle.

For clarity, let us consider the setup presented in figure 3.3, where we have the cavity mirrors, M_1 and M_2, the active medium, A, a polarizer, P an *EOM* and a window, W_B, placed at the Brewster angle so as to reflect one of the polarization directions. Light that is generated by the active medium goes through the polarizer P_1 that forces the polarization direction inside the cavity. If the **EOM is on**, light changes polarization by 90° when it passes through the crystal of the EOM, and is reflected outside the cavity by the window, W_B, therefore generating high losses inside the cavity. With high losses, lasing conditions are different and determine the active medium, A, to store more energy before it can be released through stimulated emission processes. If, however, the **EOM is off**, light does not change polarization, and goes through the system, i.e., it passes through the window, W_B, it is reflected off the second mirror, M_2, goes back though the window, EOM, polarizer and active medium. Finally, it is reflected off the first mirror, M_1, and the process repeats. This means there is an oscillation and radiation is emitted out of the system, through mirror M_2. Therefore, we induce high losses so that the active medium stores more energy. When the active medium is saturated, we turn the EOM off and a giant short laser pulse is emitted, as the active medium is depleted of energy.

Pulse duration in the Q-switch regime is approximately 1 ns, but it depends on the pumping energy, resonator length and losses.

Another active method of achieving Q-switch regime is to use an acousto-optical modulator (AOM). As the name suggests, this device generates acoustic waves inside a crystal, with a very precise frequency. These waves represent variations in density, which in turn means variations in the refractive index. As such, these waves form a diffraction grating, which deflects (or, rather, diffracts) part of light away from its original path, which results in lower energy participating in the roundtrip inside the resonator, i.e., higher losses. Therefore, the active medium stores more energy.

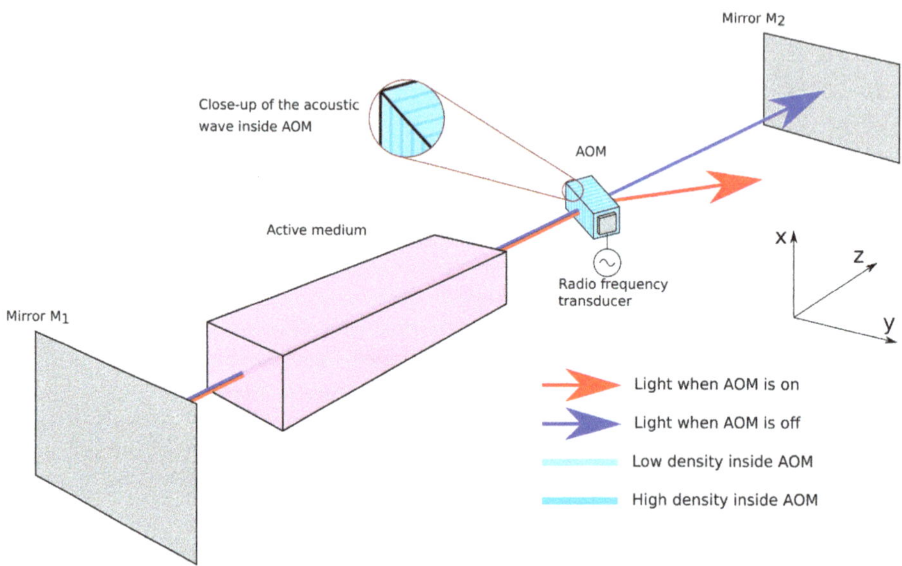

Figure 3.4. Active Q-switch resonator using an AOM.

When the wave is stopped, all energy is released in a giant pulse, in a similar fashion to the EOM Q-switch regime. A schematic of this method is presented in figure 3.4.

The AOM introduces low optical losses. The radio-frequency command circuit is rather simple and may be used for the Q-switch regime at kHz frequencies. Common AOMs are made of quartz, which is characterized by a switching time of around 100 ns (i.e., longer than EOM). AOMs are used especially for Nd^{3+}: YAG and small-gain continuous-wave lasers.

As mentioned above, passive Q-switching employs the use of materials that require no external control/intervention. One such case is represented by the use of a saturable absorber. More specifically, in this case a saturable absorber refers to an optical material that absorbs incident light until it reaches a certain threshold value, after which its optical transmission quickly increases. Most often a bleachable dye cell is used, placed inside the laser cavity, interrupting the optical path. It is important to note that at the beginning of the process, the dye cell is not opaque, but rather shows low transmission, i.e., it still permits roundtrips to happen inside the cavity and energy to accumulate, albeit at a lower efficiency. As the intensity of the light inside the cavity increases, it reaches the threshold value. At this moment, the optical properties of the dye cell change quickly, showing a significantly increased transmission. This method is schematically presented in figure 3.5.

In the case of bleachable dye, they are characterized by variable absorption coefficients. The absorbing dye molecules are excited by the incident laser radiation, going from the fundamental level, to an excited state in which they stop absorbing radiation. The absorption cross-section is high, $\approx 10^{-16}$ cm^{-2}. Therefore, oscillation threshold grows, due to increased losses. Initially, the dye absorbs low power laser light (below threshold) and allows only $\approx 1\%$ transmission. Oscillation occurs and

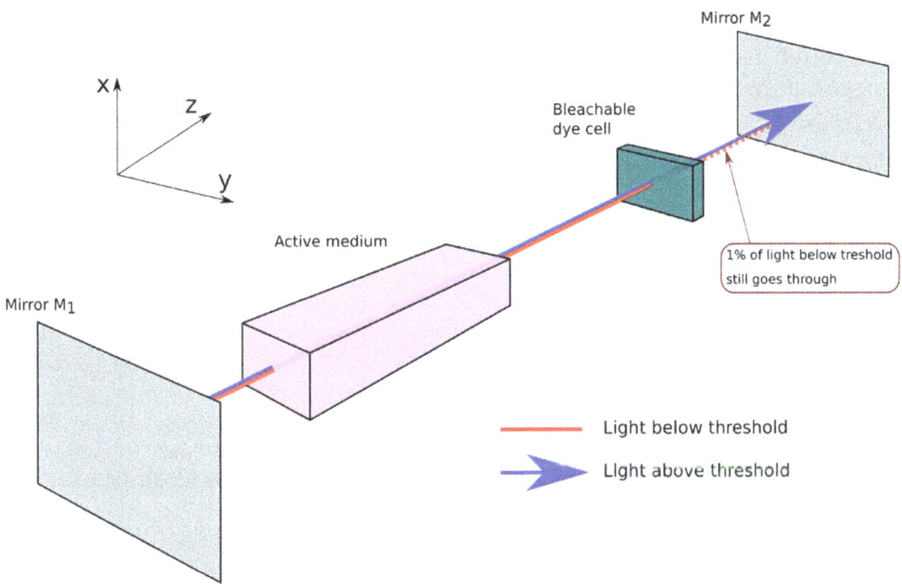

Figure 3.5. Passive Q-switch resonator using a bleachable dye cell.

the energy inside the resonator grows. When it exceeds the threshold, the dye becomes transparent and a high intensity pulse is emitted. The resonator is depleted of energy, and therefore the dye molecules relax back to the fundamental state, after which the process is repeated. This type of Q-switching is rather simple and has no control circuits. The dye can be prepared as to not have any chemical or photo-chemical degradation during the operation of the laser. Although this method has some compelling advantages, the 'opening' of the dye cell is uncontrollable, which determines a low spatial quality of the emitted beam.

3.1.2 Mode-locking

A laser system can use the energy in a frequency band where laser medium gains overcome system losses (see figure 2.9, more precisely oscillation modes **above** gain threshold and **below** laser line profile). As indicated in figure 2.9, a laser cavity usually has several longitudinal modes that are found under its usable frequency band. If the laser works under several transverse modes, then the resulting laser beam is comprised of several electromagnetic components with different frequencies, i.e., the sum of fields corresponding to each oscillation mode. Each oscillation mode is independent and therefore their amplitude and phase vary in time because the optical path is not identical for all the modes due to physical phenomena such as dispersion in the active medium. These variations determine nonlinear interactions between modes. In other words, the total electromagnetic field inside the laser cavity fluctuates randomly in time. The inverse of the frequency spectrum/band indicates the characteristic time of these fluctuations. The energy of all oscillation modes is useful for laser generation, but their phase relation determines a random energy fluctuation overall. As such, if we can induce a stable, fixed phase relation between all oscillation modes, then we can obtain a precise time-variation of the electromagnetic field inside the cavity, and the resulting laser field, implicitly. **Mode-locking** refers to the process of fixing the phase relation between all oscillation modes inside a laser cavity.

Consecutive oscillation modes are differentiated by:

$$\omega_{i+1} - \omega_i = \omega = \frac{\pi c}{L} \tag{3.7}$$

where L is the length of the laser cavity.

The electric field of a multimode cavity oscillation is defined by:

$$E(t) = \sum_n E_n e^{i[(\omega_0 + n\omega)t + \phi_n]} \tag{3.8}$$

The period of the electric field in the above equation, equation (3.8), is:

$$T = \frac{2\pi}{\omega} = \frac{2L}{c} \tag{3.9}$$

which represents the duration of a roundtrip in the laser cavity.

As mentioned previously, different modes oscillate randomly. Fixing the phase relation between all modes, for example $\Delta\phi_n = 0$, would yield necessary conditions in order to obtain a well-defined electromagnetic field. Let us consider N oscillating modes, of equal amplitude, for simplicity. The electric field is:

$$\begin{aligned} E(t) &= E_0 \sum_{\frac{N-1}{2}}^{\frac{N+1}{2}} e^{i(\omega_0 + n\omega)t} \\ &= E_0 e^{i\omega t} \frac{\sin\left(\dfrac{N\omega t}{2}\right)}{\sin\left(\dfrac{\omega t}{2}\right)} \end{aligned} \tag{3.10}$$

Therefore, the intensity of the field is:

$$I(t) \approx E(t)E^*(t) \approx \frac{\sin^2\left(\dfrac{N\omega t}{2}\right)}{\sin^2\left(\dfrac{\omega t}{2}\right)} \tag{3.11}$$

Equation (3.11) can be used to describe a train of pulses with a peak power that is equal to N (number of modes) times the average power that is distributed over all oscillation modes. In figure 3.6 we can observe a simplified example of six oscillation modes with equal amplitude, on the left side with random phase difference (that also varies, in real cases), and on the right side with a fixed phase difference, i.e., *mode-locked*.

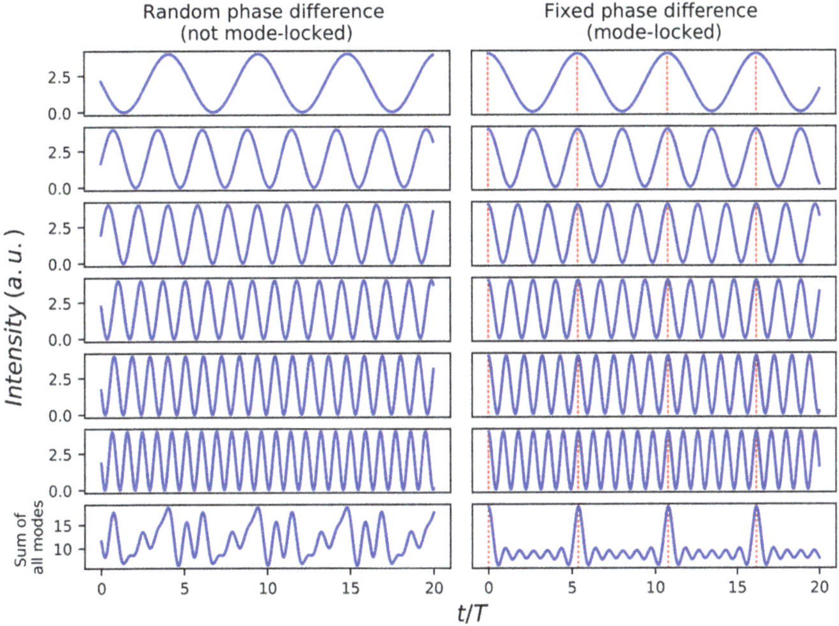

Figure 3.6. Schematic representation of optical field amplitude for a resonator with six oscillating modes, with and without mode-locking. The red dotted line indicated the temporal position where the phase difference is zero (i.e., mode-locked) for all oscillation modes.

Without mode-locking, the peak power of a laser system is the sum peak powers of each and every oscillation mode, i.e., proportional to $(2N + 1)E_0^2$. In the mode-locking regime, the peak power becomes proportional to $(2N + 1)^2 E_0^2$. This is not to be confused, however, with the average output power, which remains unchanged. Mode-locking can be thought of as setting a precise time-related order to oscillation modes generated inside the cavity, i.e., generating very short pulses with high peak power. It is not a method of producing more average power, or energy, out of a given system.

Let us set the spectral width of the laser line above threshold as ω_{th}. The number of oscillation modes that are phase-locked is, therefore, $N_l \approx \frac{\omega_{th}}{\omega}$. The temporal pulse width, t_p, is defined as:

$$
\begin{aligned}
t_p &= \frac{T}{N_l} \\[2mm]
&= \frac{\dfrac{2\pi}{\omega}}{\dfrac{\Delta\omega_{th}}{\omega}} \\[2mm]
&= \frac{1}{\Delta\nu_{th}}
\end{aligned}
\tag{3.12}
$$

In the case of mode-locked laser systems, the duration of the resulting pulse is inversely proportional to the bandwidth of the laser cavity. However, it is important for the bandwidth of the laser cavity to not exceed the gain bandwidth. For example, in the case of a Nd^{3+}: YAG laser system delivering the fundamental wavelength of 1064 nm, the frequency bandwidth is $\Delta\nu_{th} = 1.2 \times 10^{10}$ s^{-1}, which indicates a pulse duration of $t_p = 80$ ps.

Until now we have discussed what mode-locking represents in the context of laser generation. On the other hand, the technical solutions for obtaining mode-locked systems are similar to Q-switching, i.e., it uses AOM, EOM or saturable absorbers. In other words, we have to introduce supplementary cavity losses (similar to Q-switching), but only for specific oscillation modes, not the overall radiation inside the cavity.

Before discussing how we can obtain mode-locking on purpose using a device or component added inside the laser cavity, we need to mention the special case of self-locking. When certain conditions are met, nonlinear effects exhibited by the active medium can actually induce a fixed phase difference between modes. For example, Kerr-lens mode-locking uses the Kerr effect (change in refractive index as a function of light intensity) in combination with an appropriately placed aperture, in order to favor only light above a certain intensity threshold to remain stable inside the laser cavity. However, this method is rarely used due to being highly sensitive to operational conditions such as temperature. In the case of a modulator, be it AOM or EOM, it is electronically controlled at a frequency equal to $\delta\nu = \frac{c}{2L}$, which is the frequency separation between two consecutive longitudinal oscillation modes, as well as a frequency corresponding to a cavity roundtrip. As such, light inside the laser cavity will suffer induced losses (either absorbed, deflected or diffracted), for all but one oscillation mode.

For simplicity, let us consider mode-locking in the time domain. As mentioned above, the modulator introduces supplementary losses driven at specific frequencies. If the modulator-driven losses are synchronized to the laser cavity roundtrip, then a particular pulse will be favored for doing multiple roundtrips inside the laser cavity, while other pulses/oscillations will be attenuated. Frequency mode-locking is realized using an EOM, but in this case unfavorable oscillations are not attenuated, but rather frequency-shifted. If the modulation frequency is synchronized to the cavity roundtrip, then pulses that arrive too early will suffer a down shift, while those arriving late would be up-shifted. After repeated roundtrips inside the cavity, all modes will oscillate in phase with each other, as well as in phase with the EOM.

3.2 Ultrafast laser amplifiers

Laser microprocessing has been used in industry since the invention of pulsed Nd:YAG lasers. However, the applications were limited by thermal effects and low throughput of the nanosecond Q-switched lasers available at that time. In the 1980s a worldwide technological competition stimulated the invention of a new type of lasers with pulse duration in the picosecond and femtosecond range. Chirped pulses amplification (CPA) [28] was the disruptive technology that promoted the femtosecond laser systems in many

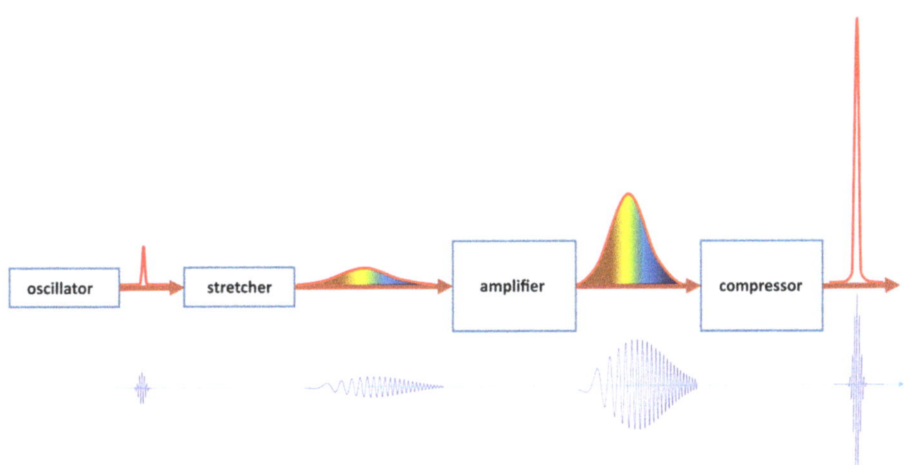

Figure 3.7. CPA scheme.

fields, from photochemistry to laser particle accelerators. The fundamental concept of CPA is presented in Figure 3.7. For the huge impact of ultrafast laser amplifiers on science and technologies, the Noble Prize in Physics was awarded in 2018 to the inventors of CPA, Donna Strickland and Gerard Mourou.

In the 1990s, ultrafast laser processing was rapidly adopted since the ultrafast laser pulses demonstrated many advantages over nanosecond pulses. When a laser beam is focused on the material surface, the laser energy absorbed by the laser irradiated material is quickly transferred to electrons from the solid lattice. In the case of long laser pulses of the order of ns, the free electrons generated by the laser pulse scatter on the material lattice, inducing the heating of the irradiate substrate. Locally, at the focus spot, temperatures much higher than the melting point are reached. The surface temperature profile depends on the laser characteristics: pulse energy, beam shape, lasers wavelength, and material properties: surface reflectivity, absorption coefficient and the material thermal conductivity. When the pulse duration is longer than the thermal diffusion time, which is at the order of tens of ps for most of the materials, the heat will propagate far from the laser interaction area. The heat-affected zone (HAZ) is much larger than the irradiated area. Then, the nanosecond laser processing has a poor precision and is not adapted for application where pre-existing surface features could be damaged by heat. In opposition to the long pulses regime of laser–material interaction, for pulse duration shorter than the thermal diffusion time, the laser will induce a multiple-ionizations process, the material is ablated by Coulomb explosion and most of the heat induced by laser absorption is removed from the surface. The substrate remains unaffected and the dimensions of the structures produced by femtosecond pulsed lasers could be of the order of tens to hundreds of nanometres, below the size of the focused laser spot (see Figure 3.8).

Ultrafast pulses are desired in order to minimize the HAZ at the laser interaction area on the material to be processed [2]. Laser oscillators emitting ultrafast pulses at

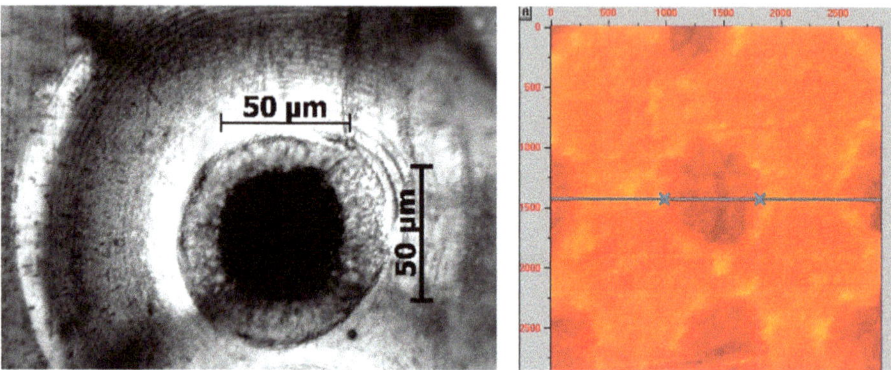

Figure 3.8. Comparison between two laser-processed surfaces with nanosecond and femtosecond pulses. (a) The optical microscopy image of the nanosecond laser irradiated surface reveals a large heat-affected zone far from laser-ablated area. (b) The atomic force microscopy image shows gold films processed by femtosecond laser pulses. The hole diameters are about 800 nm, below the laser focused spot.

MHz repetition rate do not offer enough peak power to induce laser ablation of materials. However, such lasers can be used in processing methods where the interaction and material modification are based on cumulative effects, such as photochemical polymerization. Two-photon photopolymerization is a commonly used photochemical process for 3D laser lithography. Such laser processing allows for 3D structuring with resolution down to 100 nm, opening an entire spectra of applications: 3D scaffold for tissue engineering, optical waveguides, micro optic devices, micro-optoelectro-mechanical systems (MOEMS), etc. A femtosecond laser oscillator delivers usually pulses at repetition rates of tens of MHz and energy per pulse of the order of nJ. Laser beam parameters at this order of magnitude are ideal for the above-mentioned applications. On the other hand, the laser material processing based on laser ablation requires focused femtosecond laser beams with pulse energy of the order of µJ. The pulse energy delivered by femtosecond laser oscillators is much below the ablation threshold of most of the solid materials, even if the laser beam is tightly focused. A laser amplification system is needed in order to increase the pulse energy. However, the femtosecond pulses cannot be directly amplified in the ultrashort pulse regime. During amplification, the nonlinear optical Kerr effect generated by high intensity pulses may induce beam distortions and even filamentation followed by damage of the amplification crystal. In order to prevent optical damage in the amplification chain, the pulses have to be amplified in the picosecond regime rather than femtosecond.

The main advance in laser physics was done by invention of the CPA technique based on a Ti:Sapphire crystal as the active medium. This material has the fluorescence emission wavelength from 650 nm to 1100 nm, with a maximum band around 800 nm. Amplification of the pulse energies is done from nJ to mJ and even tens of J in PW-class laser systems used in laser–plasma particle acceleration. Due to the intrinsic quantum nature of the light and as consequence of the Heisenberg uncertainty principle, the pulse duration and the spectral broadening are directly linked by the relation

$\Delta \tau \Delta E \geqslant \hbar$. Then, the femtosecond pulses have broad spectrum. As an example, a laser with 200 fs pulse duration centered at 800 nm, has about 5 nm spectral width at full-width-half-maximum (FWHM) and a 10 fs laser pulse is about 95 nm broad in spectrum. This spectral property of the laser beam allows a stretcher setup (an optical setup based on spectral dispersion by gratings) to stretch the pulses in time from femtosecond to picosecond pulses. Laser peak power is therefore decreased several orders of magnitude, below the threshold of nonlinear optical effects. The laser beam is safely amplified in the picosecond regime on several amplification stages. After each amplification stage the beam is expanded in order to keep the laser irradiance under the damage threshold of the optics in the propagation path.

There are two main amplification schemes of femtosecond pulses: *regenerative* and *multipass* amplifiers. The gain and efficiency are the main characteristics that distinguish between the two types of amplifiers. The input beam intensity relative to saturation intensity of amplification setup is the parameters that define which type of amplification is the most appropriate. Details are presented in the following subsections.

3.2.1 Gain and extraction efficiency. Frantz–Nodvick equation

The laser amplification process does not differ much at atomic scale from the laser emission process in a laser oscillator [17]. In a four-level laser system such as the Ti: sapphire active medium represented in figure 3.9, the optical pumping from the fundamental level E_0 to highest level E_3 has to be done at the maximum of the absorption band of the Ti:sapphire crystal in the green spectral range. Mainly two types of laser systems are available to pump at this spectral domain. One is the argon ion (Ar^+) gas laser with emission at 514.5 nm and typical output power of 10–20 W. The second and most used laser pump system for Ti:sapphire crystals is the frequency-doubled Nd:YAG (yttrium–aluminium–garnet crystal) laser with fundamental emission at 1064 nm and the second harmonic at 532 nm obtained from nonlinear crystals such as potassium titanyl phosphate (KTP), potassium dideuterium phosphate (KDP) or beta barium borate (BBO). Solid state lasers, including the diode-pumped solid state lasers (DPSSLs) are more preferred as pumping sources for Ti:sapphire lasers because of their compactness and more energetic optical output compared with the ion gas laser. However, for pumping beams at energies of tens to hundreds of J per pulse as is the case of very large scale PW-class lasers, Nd:glass instead of Nd:YAG active medium is used in pump lasers,

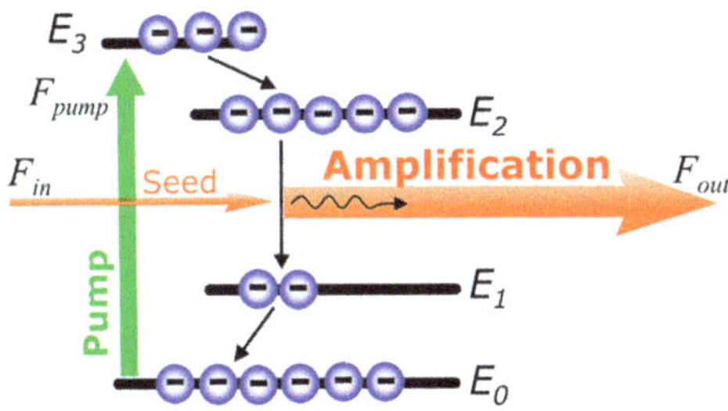

Figure 3.9. A four-level amplification process.

in order to manage better the thermal loading and homogeneity of the large size crystals. Emission is at 1.054 μm (for phosphate glasses) which is not much different to the Nd:YAG crystal.

The long fluorescence lifetime of Ti:sapphire active medium at a wavelength of around 800 nm corresponding to the metastable level E_2, as well as the fast transitions τ_{32} and τ_{10} allow for inversion of population between levels E_2 and E_1 ($N_2 > N_1$). The main difference between a laser oscillator and an amplifier is that the amplifiers do not have a resonant cavity. The amplification process is initiated by a seed beam originating from a laser oscillator or from an intermediate amplification stage. A laser pulse traveling trough the pumped active medium will induce stimulated emission, then the energy density (laser fluence) of the traveling pulse will increase up to a maximum possible value representing the saturation level F_{sat} of the active medium:

$$F_{sat} = \frac{h\nu_L}{\sigma_a} \tag{3.13}$$

where h is the Plank constant, ν_L is the laser frequency and σ_a is the amplification cross-section.

In a practical situation one may be interested to extract from the amplification device as much energy as possible. From this point of view there are two main parameters that quantify the amplification process: the gain G and the extraction efficiency η_{extr}. The gain is defined as the ratio between output laser fluence F_{out} and input laser fluence F_{in}:

$$G = \frac{F_{out}}{F_{in}} \tag{3.14}$$

while the extraction efficiency takes into account the energy density F_{ac} accumulated in the active medium:

$$\eta_{extr} = \frac{F_{out} - F_{in}}{F_{ac}} \tag{3.15}$$

The accumulated fluence F_{ac} for a four-level laser depends on the pump energy density F_{pump}, absorption efficiency of the pumping radiation η_P, fluorescence quantum efficiency η_F, and the quantum defect $\eta_Q = \lambda_{pump}/\lambda_L$:

$$F_{ac} = \eta_P \eta_F \eta_Q F_{pump} \tag{3.16}$$

Output energy density extracted after a single pass of the laser pulse through the amplification medium depends on energy density of the input laser pulse, pump energy density, and is limited by the saturation fluence F_{sat}, as described by the Frantz–Nodvick equation [10]:

$$F_{out} = F_{sat}\ln\left\{1 + \left[\exp\left(\frac{F_{in}}{F_{sat}}\right) - 1\right]G_0\right\} \tag{3.17}$$

Then, the gain for a single path is:

$$G = \frac{F_{\mathrm{sat}}}{F_{\mathrm{in}}}\ln\left\{1 + \left[\exp\left(\frac{F_{\mathrm{in}}}{F_{\mathrm{sat}}}\right) - 1\right]G_0\right\} \tag{3.18}$$

G_0 is the gain at low value of the input fluence:

$$G_0 = \exp(\sigma_a n_v l_a) \tag{3.19}$$

where l_a is the length of the active medium and n_v is the inversion population density.

From the above equations, the extraction efficiency is:

$$\eta_{\mathrm{extr}} = \frac{GF_{\mathrm{in}} - F_{\mathrm{in}}}{F_{\mathrm{ac}}} = \frac{1}{\ln G_0}\left[\ln\left(1 + \left[\exp\left(\frac{F_{\mathrm{in}}}{F_{\mathrm{sat}}}\right) - 1\right]G_0\right) - \frac{F_{\mathrm{in}}}{F_{\mathrm{sat}}}\right] \tag{3.20}$$

The gain and the extraction efficiency cannot be simultaneously high, and the maximization of one or the other will decide a certain amplification regime.

Laser amplification in low input energy regime. If we consider the input fluence $F_{\mathrm{in}} \ll F_{\mathrm{sat}}$, the gain will be:

$$G \approx G_0 = \exp(\sigma_a n_v l_a) \tag{3.21}$$

In this regime the input energy is much lower than the saturation level and the laser fluence increases exponentially while propagating through the amplification medium: $\frac{dF}{dz} = gF$. The depopulation is insignificant at the time scale of the temporal pulse width. Then, enough accumulated energy density remains after the first pass of the laser pulse. Hundreds of passes are possible before extracting the entire accumulated energy. An optical configuration is necessary to be set in order to provide multiple passes, for each pass the output fluence becomes input for the next amplification round.

Laser amplification in high input energy regime. When the input laser fluence is higher than the saturation level $F_{\mathrm{in}} \gg F_{\mathrm{sat}}$, the increase of laser pulse fluence is linear while the pulse propagates along the z direction of the amplification crystal: $\frac{dF}{dz} = gF_{\mathrm{sat}}$. For a total amplification length l_a the output fluence will be:

$$F_{\mathrm{out}} = F_{\mathrm{in}} + h\nu_L n_v l_a \tag{3.22}$$

The gain becomes:

$$G = 1 + \frac{F_{\mathrm{sat}}}{F_{\mathrm{in}}}\sigma_a n_v l_a \tag{3.23}$$

In this case the extraction efficiency is close to 1.

Figure 3.10 shows the graphic representation of equations (3.18) and (3.20). It is clear that at low input fluence F_{in} the highest gain can be achieved, but for a low extraction efficiency, while the extraction efficiency is high for an input fluency close to saturation value. As such, two distinct amplification regimes can be distinguished, i.e., the 'high gain' regime and the 'high efficiency' regime (see Figure 3.11).

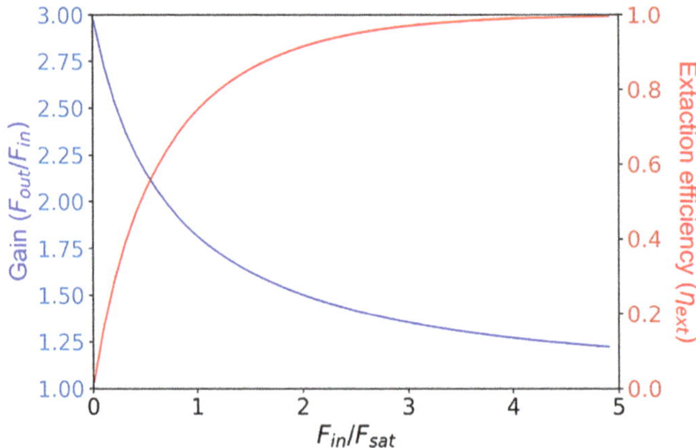

Figure 3.10. Gain and extraction efficiency.

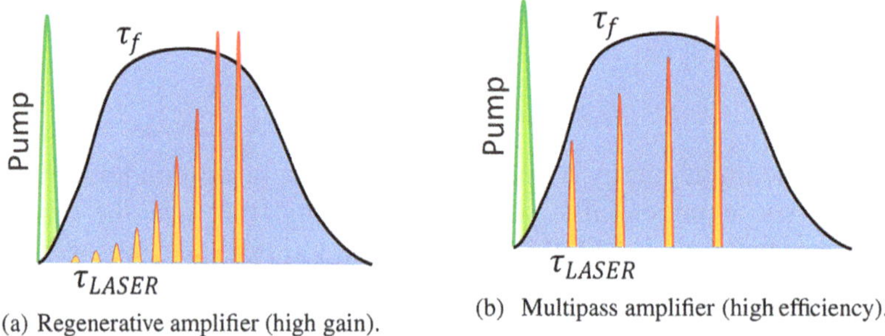

(a) Regenerative amplifier (high gain).

(b) Multipass amplifier (high efficiency).

Figure 3.11. High gain versus high extraction efficiency regime. In regenerative amplifiers (high gain) the pulse energy increases exponentially (a). For multipass amplifiers (high efficiency) the pulse energy increases linearly (b).

3.2.2 Regenerative laser amplifiers

Most of the femtosecond lasers systems in material processing are using the so-called regenerative amplifiers with output energies of hundreds of μJ up to a few mJ. The amplification configuration corresponds to the 'high gain' amplification regime. The energy of the input pulse is only few nJ as emitted by laser oscillators. As we will describe later in this chapter, the femtosecond pulses are not directly amplified in ultrashort pulse regime. Because of the huge intensity at the order of $\mathbf{MW\ cm^{-2}}$ that corespondents to an amplified femtosecond laser pulse, laser damage of the amplification crystal or beam distortion could occur because of the nonlinear optical effects induced by high peak power pulses. In order to prevent such effects, a temporal stretcher is introduced between laser oscillator and the regenerative amplifier. The pulse from the oscillator is stretched in time from femtoseconds to tens or hundreds of picoseconds time duration, therefore lowering several orders of magnitude the laser intensity, much below the laser damage threshold.

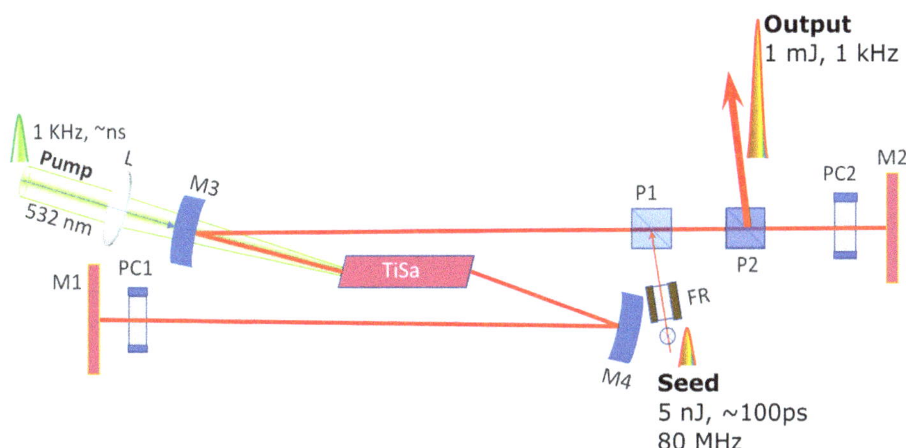

Figure 3.12. Regenerative amplifier.

Figure 3.12 shows the sketch of a typical regenerative amplifier formed by a folded optical cavity defined by two end mirrors (M1 and M2) with 100% reflectivity of the entire spectral band of the initial pulse. Injection and extraction of the laser pulse is done by optical polarizers (P1 and P2) and Pockels cell (PC1 and PC2). The Pockels cells are synchronized with the pumping laser pulses. Their control units define precisely the seeding time τ_1 and the extraction time τ_2.

The seed beam is injected in the cavity from the polarizer P1. The Faraday rotator (FR) has the role of optical isolator and prevents any damage of the optical components from the previous modules in the laser system (laser oscillator and stretcher). At time τ_1 a high voltage $V_{\lambda/4}$ is applied to the Pockels cell PC1 corresponding to a quarter haveplate ($\lambda/4$). At the same time, the second Pockels cell has no voltage applied. A vertically polarized beam injected in the cavity will be directed by P1 toward the M1 end mirror, it passes through Ti:sapphire crystal and PC1 Pockel cell, it hits the M1 mirror and return to the other M2 end mirror. After passing two times the Pockels cell PC1, the polarization of the injected pulse is rotated from vertical to horizontal. Then, at P1 the beam is transmitted and remains in the cavity. The length of the optical path in the cavity L_c is about 1–2 m, resulting a roundtrip of the pulse of the order of few nanoseconds ($\tau_p = 2L_c/c$). Before returning the laser pulse back to M1 mirrors, the voltage of the PC1 ramps up to $V_{\lambda/2}$. We have to take into account that usually the seed beam is at 80 MHz, as from a typical Ti:sapphire laser oscillator. This corresponds to about 12 ns between each pulse. By ramping up the voltage of PC1 to $V_{\lambda/2}$ in less than the roundtrip τ_p of the pulse inside the cavity, the pulse returning from the M2 mirror, after hitting the M1 mirrors it passes again two times the PC1 and rotates it polarization with 180°. Thus, the pulse preserves its horizontal polarization and remains in the cavity. By opposite, all the subsequent vertically polarized pulses at 80 MHz will keep their vertical polarization and will be rejected from the amplification cavity at P1. This mechanism provides simultaneously a pulse selection method and keeps the selected pulse inside the cavity. At τ_2 the voltage of PC2 is set at $V_{\lambda/4}$. After passing two times through the PC2 Pockels cell, the polarization of the amplified pulse is rotated from horizontal to vertical and is extracted from the cavity by the P2 polarizer. All the process is repeated synchronously with the next pumping pulse, at the repetition rate of the pump laser. The typical repetition rate of a regenerative amplifier is of the order of kHz up to hundreds of kHz.

Regenerative amplification is at high gain, but with low efficiency extraction for a single pass. Therefore, it is desired that the traveling pulse is kept inside the cavity long enough to extract as much energy as possible from the available inversion of population. The fluorescence lifetime for Ti:sapphire is 3.2 µs allowing many passes of the selected pulse before extraction. The number of passes N of the amplified pulse in the cavity is of the order of tens to hundreds, where $N = (\tau_2 - \tau_1)/\tau_p$, and it is controlled by the injection and extraction time as set from the electronic units of the Pockels cells.

Extraction time τ_2 has to be set at the moment of saturation of the gain, when the last two subsequent pulses in the cavity have the same amplitude, and practically the cavity losses can no longer be compensated by the gain (figure 3.13(b)). The time settings can be verified by a fast photodiode placed behind one of the end mirrors of the regenerative cavity. The photodiode signal represents a series of the intracavity laser pulses amplified after each roundtrip. For a correct measurement of traces the oscilloscope has to be set in low impedance input mode at 50 Ω.

At a later extraction time, after the saturation time, any subsequent pulses traveling in the cavity will lose energy as shown in figure 3.13(c). An earlier extraction time, before reaching the saturation regime, will have an impact on the energy stability of the extracted laser pulse (figures 3.13(a) and (d)), unsuitable for a sensitive experiment such as laser microprocessing at ablation threshold. A fine tuning of the extraction time is also important in order to prevent the output of double or multiple pulses as shown in figure 3.13(f). In a practical experiment, the output of the laser is measured by a powermeter, which is a slow measuring device that integrates the signal over a certain time (1–2 s). If a double or multiple pulse situation occurs, the measured power could be apparently high enough. However,

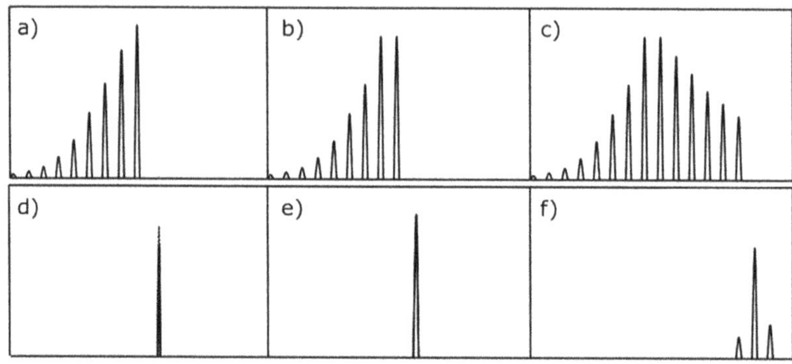

Figure 3.13. Laser pulses inside the regenerative cavity (a, b). Extracted pulses (d, e).

the misleading measurement will produce experimental artifacts since the laser pulse may have replicas, a very poor intensity contrast. Replicas of the output laser pulse have to be checked by a fast photodiode placed outside the cavity laser. Examples of the oscilloscope traces corresponding to the output beam at different delays configurations are presented in figure 3.13(d)–(f). It is to be mentioned that the width of the measured traces do not rely on the laser pulse width since the femtosecond optical signal is too fast compared with the rise time of the any fast photodiode.

Injection time τ_I is related to arrival time of the pump laser pulse when the inversion population occurs. A bad injection time is usually observed in the saturation traces as an important background signal and poor contrast of the peaks. The output laser pulse has in such a situation two temporal components, a fast optical signal of the order of femtoseconds on the top of a long signal with temporal width of the order of nanoseconds corresponding to the amplified spontaneous emission (ASE).

3.2.3 Multipass laser amplifiers

Laser pulse energy of the order of tens of mJ and more can be extracted from amplifiers only in a high efficiency amplification regime. For ultrashort lasers based on Ti:sapphire crystal, the active medium is pumped by a ns pulse laser, usually at 532 or 527 nm obtained from the second harmonic of a Nd:YAG or Nd:glass laser system. The green laser is directed to one of the end faces of the amplifiers crystal. In some pumping configuration, two or more laser beams can simultaneously pump the crystal from both sides. A seed beam with a long pulse of hundreds of ps is directed to the active medium synchronously with the pump beams. The laser beam passes several times through the active medium in a bow-tie configuration (figure 3.14).

Due to the energy accumulated in the crystal, the initial pulse is amplified several times per pass. However, only a few amplification steps are possible. After a certain limit the increase in number of passes will induce loses or can even induce damage inside the laser crystal or at the crystal faces if the laser-induced damage threshold is

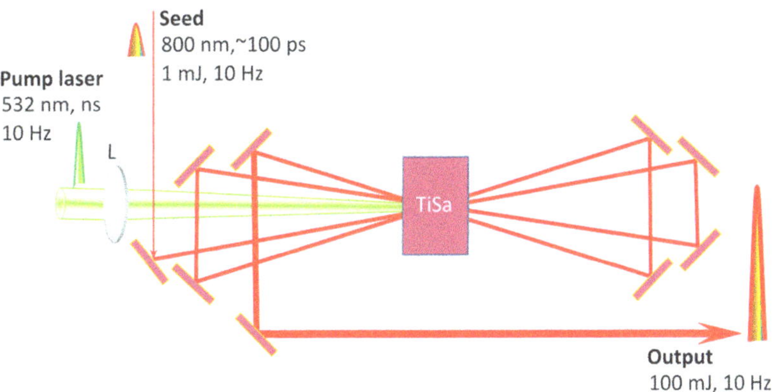

Figure 3.14. Multipass amplifier. High efficiency regime.

reached. In order to get more output energy, an amplification chain is designed with increased beam diameter from one amplification stage to another. The design of a multipass amplifier takes into account the seed pulse energy, the size of the seed beam, the laser damage threshold of the amplifier crystal, the pump energy and the absorption efficiency. These input parameters will decide the number of passes and the maximum possible output power.

In a numerical example we consider the pump beam with the following parameters: wavelength $\lambda_p = 532$ nm, beam diameter $d_p = 5$ mm, pulse energy $E_p = 0.5$ J and the seed beam with the parameters wavelength $\lambda_{seed} = 800$ nm, beam diameter $d_{seed} = 5$ mm, pulse energy $E_{seed} = 0.6$ mJ. For a Ti:sapphire crystal the typical damage threshold is 4 J cm^{-2}, the saturation fluence is 0.9 J cm^{-2}, the absorption efficiency of the pumping radiation is $\eta_p = 0.92$, fluorescence quantum efficiency is $\eta_F = 0.8$, and the quantum defect is $\eta_Q = \lambda_{pump}/\lambda_L = 0.665$.

The accumulated fluence F_{ac} in the amplification crystal is given by equation (3.16). In a multipass laser amplifier the output of a previous amplification pass i becomes the seed for the next pass $i + 1$, then $F_{in}^{i+1} = F_{out}^i$. The output energy E_{out} of each pass can be obtained from the Frantz–Nodvick equation (3.17), where the output fluence is given by the relation $F_{out} = \frac{4E_{out}}{\pi d_p^2}$.

Figure 3.15 shows the output energy and the gain after each pass in the case of the numerical example mentioned before. From the graphic representation it is clear that the maximum extracted power is reached around 210 mJ after six passes. If more passes are considered, the gain becomes lower than a critical level, and no longer compensates the losses and the output energy decreases. Then, for the given input parameters, the optimal number of passes is only six. Of course, different results are obtained depending on the size of the pump beam at the input face of the crystal, pump pulse energy, or seed pulse energy, but the number of passes is at the order of few, as a characteristic for high efficiency–low gain amplifiers.

In some applications, such as laser–plasma particle acceleration or generation of secondary sources, much higher energies are required. In this case a chain of amplifiers is designed. For each amplification stage the laser beam is expanded

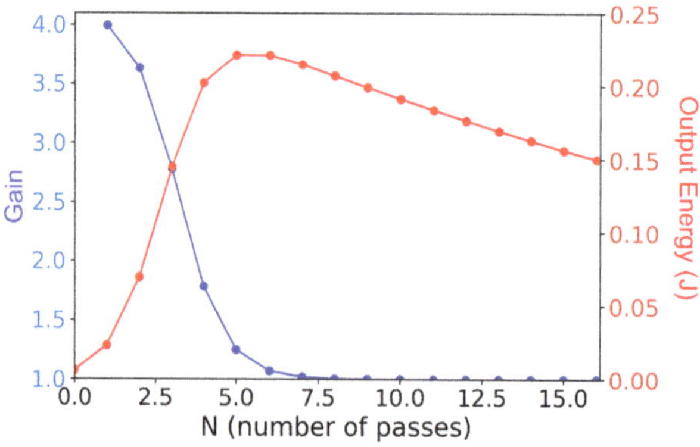

Figure 3.15. Output energy and gain after each pass in a multipass amplifier.

in order to keep the fluence below the damage threshold of the optical components in the amplifier. As a consequence, the size of the mirrors and the crystal will increase accordingly. High output power requires also pump lasers with high pulse energy. Because of technological constraints, such lasers operate at relatively low repetition rates of the order of few Hz or even below. However, it is to be mentioned that in laser microprocessing by ultrafast laser pulses the regenerative amplifiers are the most commonly used, one of the important parameters being also the laser repetition rate that strongly influences the throughput.

3.3 Characterization of laser pulses

Lasers have become versatile tools used in research as well in industry. For all applications, the beam characterization is very important for validation of the experimental results or for some technological process. The metrology of laser beams usually includes the measurement of laser power, pulse duration and laser pulse energy for pulsed lasers, spatial measurement of laser beam profile and quality factor M^2. In some cases where the wavelength of laser emission can vary, the spectral measurements also have to be taken into account. For ultrafast lasers additional parameters are very important to measure and control: the spatial and spectral phase. Most of the time we are referring to laser beam intensity as the main parameter that defines a certain interaction of laser with materials. From a radiometry point of view, the term of *intensity* has to be considered with some precaution in order to avoid confusion. Laser irradiance or power density measured in W cm^{-2} is in fact the proper term. In laser material processing the term *laser intensity* is readily used and refers to the optical power per unit area and the measurement unit is W m^{-2} or from practical reasons W cm^{-2}. This is a relevant parameter in laser–matter interaction, but cannot be directly measured. Behind this parameter we can find two others that in practice can be determined: the laser power (W) and the laser beam diameter or the beam area (cm^2). Correct measurement of beam area is not a simple task and we will refer to that in a dedicated subsection (see section 3.3.2). For the moment we will focus on measurement of laser power and laser pulse energy.

3.3.1 Energy and power measurement

A pulsed laser can be characterized by pulse energy E measured in joules (J) and by average power P_{av} measured in W. These parameters are obtained by devices like energy meters and power meters. Another parameter, the peak power P_{peak}, characterize the laser pulses as ratio between pulse energy E_{pulse} and pulse duration τ (s). At this point, the terms can be confusing and the most common question when energy meters or power meters are used is: *how do we avoid the laser damage of the optical sensor?* First of all, no matter what kind of measuring device is used, the first golden rule is to expand the laser beam in order to cover the sensitive area of the detecting device, the second rule is not to expand too much in order to avoid the clipping of the beam. Ideally, for a beam with Gaussian profile the beam diameters has to be 1/3 of the sensor diameter (see Figure 3.16).

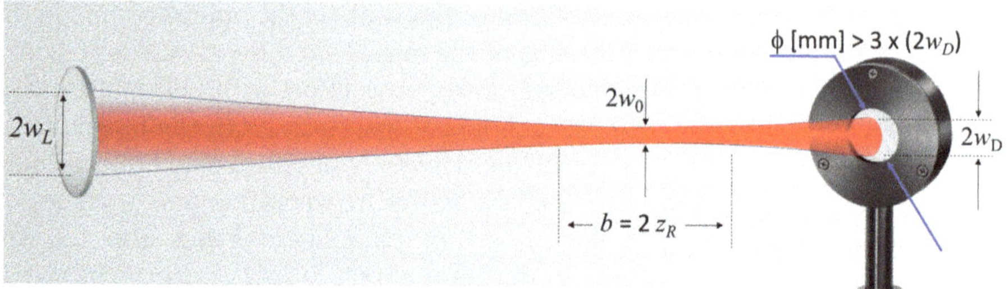

Figure 3.16. Accomodation of the laser beam on the active area of the detector. The size of the sensor has to be about three times larger than the beam size. If a lens with long focal length is used to reduce the beam size, it is critical to avoid positioning the detector close to the focus spot.

In radiometry the optical power refers to the radiant flux, and it is related to the energy $E(J)$ carried by photons per time unit. The units of optical power are joules per second (J s^{-1}) or watt (W). For a continuous-wave laser, the number of photons is constant in time, then the laser power measurement is more intuitive. However, what we can measure for a continuous-wave laser, apart from the laser power, is the power fluctuations. Two type of fluctuations can be defined, the fast fluctuation or the laser power noise, which is the power variation in a short period of time, in the range of ms up to a few seconds, and long-term fluctuations which refers to power variation over tens of minutes or hours. Usually, the long-term fluctuation is related to thermal drifts of the components inside the laser device, or of the device housing, because of the environment thermal drift. Most often, a laser system is thermally stabilized using a temperature-controlled breadboard.

For a pulsed laser, the power as measured by a powermeter refers to an *average power* and has the same significance as for a continuum wave laser. The difference is that the total energy per second is carried in burst of pulses. If the repetition rate of the laser is ν measured in hertz (Hz), the average power of a pulsed laser is given by the relation:

$$P_{av}\ (W) = E_{pulse}\ (J)\nu\ (s^{-1}) \tag{3.24}$$

where E_{pulse} represents the energy carried by a single bunch of photons, or a single laser pulse (see Figure 3.17). From here we can define the *peak power*, which is the energy carried by a single pulse over the pulse duration τ:

$$P_{peak}\ (W) = \frac{E_{pulse}\ (J)}{\tau\ (s)} \tag{3.25}$$

For ultraintense lasers the peak power is the relevant parameter as it includes also the information of pulse duration. Usually, in one experiment the laser beam is tightly focused on the sample in a spot with a few micrometers in diameter. Then, we refer to laser intensity (or laser irradiance), measured in W cm^{-2}. The huge number of photons arrives at the same time in a narrow space. In this irradiation regime, nonlinear optical effects occur from two-photon absorption to multiphoton ionization. As a numerical example, a regenerative amplifier with 1 kHz repetition rate that emits pulses with energy of 1 mJ and 100 fs pulse duration has a peak power of

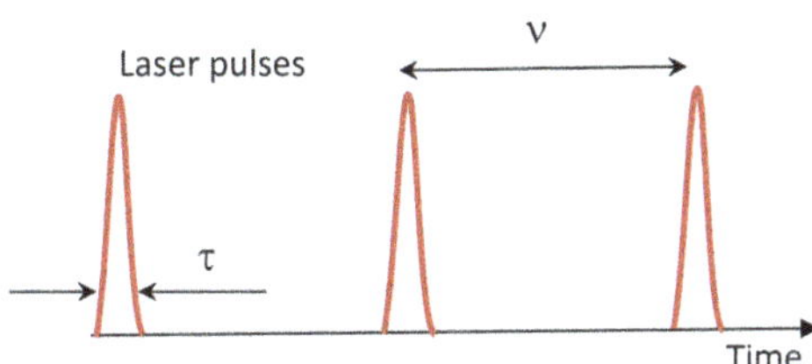

Figure 3.17. Burst of laser pulses.

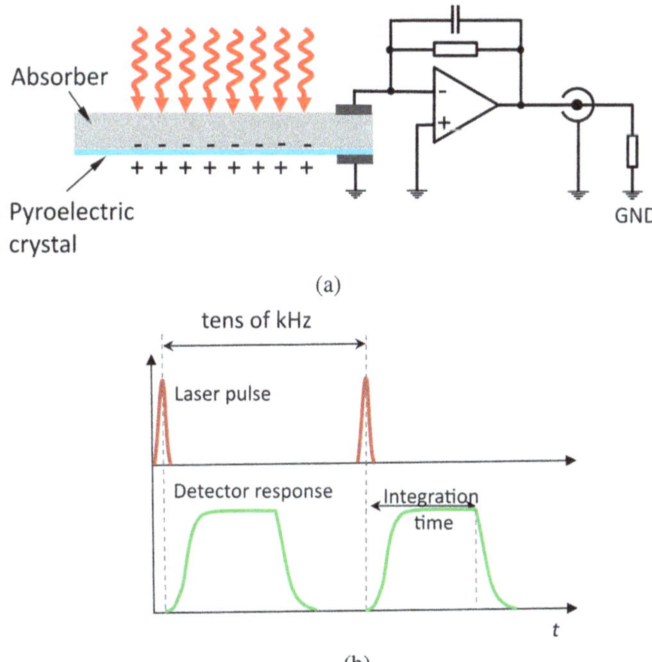

Figure 3.18. (a) Structure of a pyroelectric sensor. (b) Time response of the pyroelectric sensors is slow, at the order of microseconds compared with duration of laser pulses.

10 GW. If the beam is focused on a spot with beam diameter of 10 μm the laser intensity of the focused beam is 12.7×10^{-15} W cm^{-2}. In a practical situation we can directly measure the pulse energy of the pulsed laser or the average power, then the laser intensity is indirectly determined.

For a correct measurement of energy or power, some consideration has to be taken into account, related to the beam repetition rate and the technical limitations of the existing devices. Usually, the energy meters available on the market for the range of energies from nJ to tens of mJ are based on pyroelectric sensors. These sensors uses a pyroelectric crystal such as lithium tantalate (LiTaO$_3$). On the top of the pyroelectric materials an absorber assures the heat transfer from the laser beam to the active material (see Figure 3.18).

Spontaneous polarization $P_{SP}(T)$ of the active material strongly depends on temperature. Then, a laser pulse produces a local heating of the surface and by pyroelectric effect induces a local electric field. The electronic circuit of the sensors transforms the detected signal in electric pulses with amplitudes proportional to the laser pulse energy. However, due to the relatively slow response of the sensor, the pyroelectric energy meters are limited to laser beams with repetition rate of the order of kHz or below. These sensors are ideal for single-pulse measurements and characterization of pulse-to-pulse temporal stability.

If necessary, the pulse-to-pulse energy can be in principle evaluated by pyroelectric sensor, but for laser frequency of a few kHz and below, as usually supported by such devices. Figure 3.19 shows a typical pulse-to-pulse energy measurement over time and the energy distribution around the average energy. The energy stability can be evaluated by the standard deviation of the laser energy as follows:

$$\sigma = \sqrt{\frac{1}{N}\Sigma(E_i - \bar{E})^2} \tag{3.26}$$

where $\bar{E}$ is the average energy measured over N pulses. For a typical pulsed laser the standard deviation of the pulse-to-pulse laser energy is about 1–5%.

For lasers with high repetition rates, from tens of kHz to MHz, the pulse energy could be measured by fast photodiodes, but with some precautions. The photodiode signal is in principle proportional to the laser pulse energy, however, the photodiode response is no longer linear over a certain laser intensity and the photodiode is quickly saturated.

At high repetition rate, the beam can be characterized by measuring the average power using a powermeter based on a thermoelectric sensor or thermopile (see Figure 3.20). The Seebeck effect is used to convert the heat into electricity at the junction of different types of wires. The structure of the sensor consists of a series of thermopiles arranged in a circular geometry with the hot junction oriented towards the center of the active area and the cold junction oriented outward. The active area is an absorber selected for a specific range of laser wavelengths. The time response of the thermopile is very slow, of the order of a few seconds, and is sensitive to any infrared source or improper manipulation,

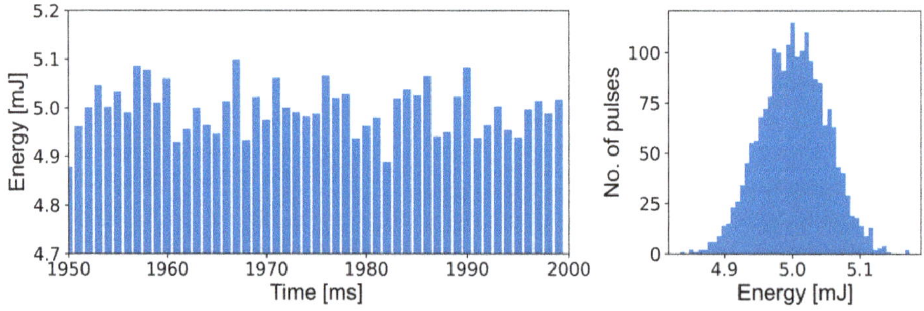

Figure 3.19. Pulse-to-pulse energy stability and the pulse energy distribution.

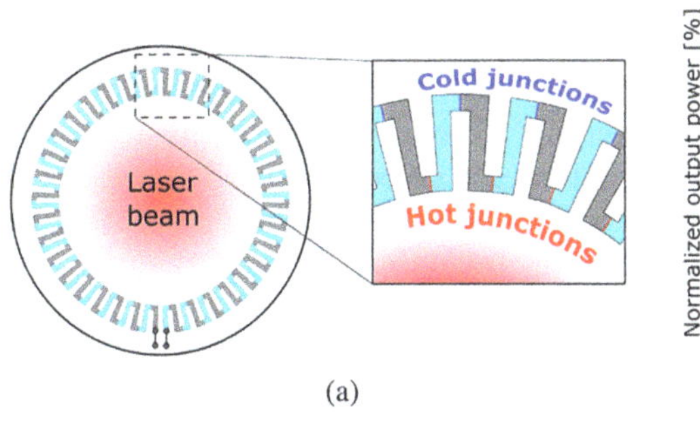

(a)

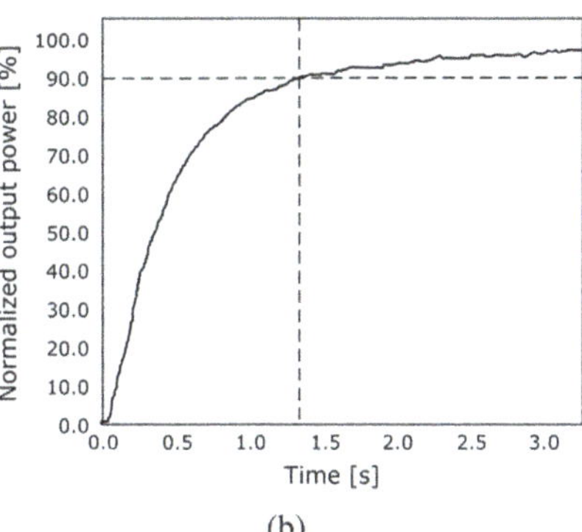

(b)

Figure 3.20. (a) Thermoelectric sensor (thermopile). (b) Time response of the thermopile is very slow, at the order of a few seconds.

including the hand direct contact of the operator with the device housing. This type of sensor is the most used for lasers with power of a few kW.

When the thermopile sensor is used to measure the average power of a high repetition rate laser, the pulse energy can be recovered using the equation (3.24). However, the pulse-to-pulse energy stability cannot be evaluated by such a device due to its very slow response; eventually only the long-term stability of the laser can be evaluated by this means.

A good practice in measuring the power or energy of the laser beam is to carefully estimate the laser irradiance or laser fluence at the sensor surface (see section 3.3.3). The laser intensity should not be too close to the damage threshold of the sensor active area, the beam diameter has to uniformly cover the sensor if the beam profile is flat-top, and has to be roughly 1/3 from the sensor size if the spot is Gaussian, as already mentioned before in this section. If the beam size is larger than the sensor active area, the beam can be accommodated by using a beam reducer with known demagnification, or by a simple convergent lens with long focal length. However, using a lens is very risky since the detector could be placed by mistake too close to the focus spot. If a lens is used, the proper calculation of beam diameter along the beam propagation path has to be carefully done using the equations described in section 3.3.4.

A common mistake in estimation of the maximum measurable power with respect to the laser damage threshold of the sensors is to not take into account some beam parameters such as pulse duration, repetition rate or laser wavelength. First of all, laser wavelength has to be set on the detector acquisition software interface. The sensor head gives different electronic output at the same optical power depending on the laser wavelength. This is related to the wavelength-dependent quantum efficiency $\eta(\lambda)$ and is taken into account by the calibration curve of the detector that gives its the spectral responsivity $R(\lambda)$ (A W^{-1}). If the detector is properly calibrated as usually delivered from the factory, once the right wavelength is set, the correct power or energy value is displayed by the detector interface.

Fluence damage threshold is commonly provided by the detector's producers for several of the most common lasers, at 1064 nm, 532 nm, 266 nm, etc. The measurement conditions are also provided, such as the pulse duration and laser repetition rate. If the pulse duration of the laser to be characterized differs much from the sensor characteristics provided, for example a subpicosecond laser is to be measured and the sensor datasheet present the damage threshold fluence in the ns regime, the damage threshold for the ultrashort regime can be estimated by the following relation [29]:

$$F_{th}(\tau) = F_{th}^{(ns)} \sqrt{\frac{\tau}{\tau^{(ns)}}}$$

(3.27)

Similarly, if the laser wavelength differs from the provided data, the laser fluence can be adjusted by the equation below:

$$F_{th}(\lambda) = F_{th}^{(\lambda_0)} \sqrt{\frac{\lambda}{\lambda_0}}$$

(3.28)

In the case of a very different repetition rate, as for example the laser to be characterized is running at hundreds of kHz, the pulse energy is usually estimated from the average power and repetition rate (see equation (3.24)). Evaluation of the laser damage threshold fluence from this energy is definitely a mistake since a high repetition laser could induce heating of the surface and decrease of the damage threshold compared to the low repetition rate case as usually listed in the datasheets of commercial detectors. The proper estimation of maximum measurable power has to take in this case the average power and irradiance (W cm^2), which also includes the laser repetition rate.

3.3.2 Spatial measurement

Laser beam diameter is one of the important parameters to measure when laser intensity or the laser fluence has to be evaluated. First of all, it is to be mentioned that sometimes the laser beam diameter is incorrectly considered as the spot generated by the laser beam on a sensitive paper or the ablated spot on some materials. This type of measurement can give an immediate but rough evaluation of the laser transversal profile; it can be useful for fine tuning alignments in laser systems during maintenance, but this is quite inaccurate and has nothing to do with the real beam diameter. The size of the spot generated at the interaction of the laser beam with some surfaces is strongly dependent on the intensity threshold of the effect and on the laser intensity.

The dedicated device for measurement of laser profiles and evaluation of laser beam diameter is the laser profilometer with CCD or CMOS sensors (see Figure 3.21). The sensors are attentively selected and characterized by producers in order to assure uniform response over the entire matrix of pixels, large dynamic range and linear responsivity, as well as low dark noise. Usually, a beam intensity is much over the damage threshold of the sensor's pixels. Then, some precautions have to be considered when a laser profilometer is used in order to avoid damage to the sensor's sensitive

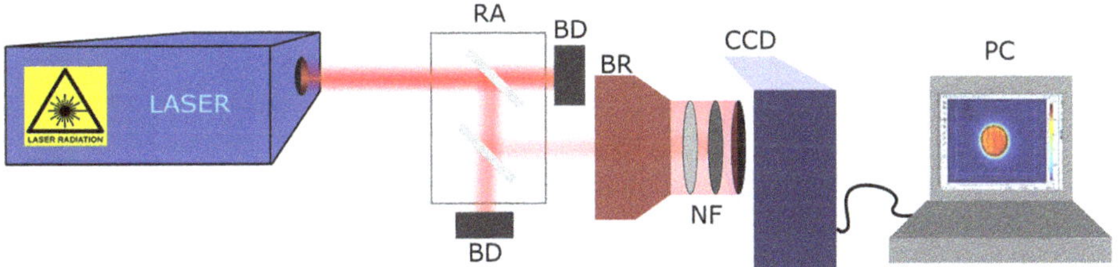

Figure 3.21. Laser beam attenuation for laser profilometry measurements. RA—reflective attenuator; BD—beam damp; BR—beam reducer; NF—neutral filters; CCD—laser beam profilometers with CCD; PC—computer with software for acquisition and data processing.

area. The laser beam can be easily attenuated with neutral filters with different optical density. However, such attenuation may induce beam profile distortions, especially in the case of high-power lasers and ultraintense lasers because of the filters heating or the nonlinear optical effects induced by the laser beam into the filters' glass. If neutral filters are used, an important rule to be applied is to set the neutral filters ordered by increasing the optical density in the direction of beam propagation, from low attenuation value to the highest attenuation, toward a CCD sensor. In this way, the beam distortion is reduced and the risk of damaging the filters is minimized. An alternative method to attenuate an intense laser beam is to use a cascade of reflective glasses as attenuators, using the 4% reflectance of each glass surface. Sometimes, the beam diameter has to be adapted to the camera aperture. Then, a laser beam reducer with known demagnification can be included in the optical path.

The beam profile is recorded and analyzed by dedicated software that gives us information such as beam diameter on two orthogonal directions and beam ellipticity. The measured parameter is the laser irradiance, or the laser intensity distribution in a transversal plane. For a symmetric beam with Gaussian or super-Gaussian profile the theoretical radial distribution of the irradiance I is given by:

$$I(r) = \frac{2P}{\pi w^2} \exp\left[-2\left(\frac{r}{w}\right)^n\right] \tag{3.29}$$

with $I_0 = \frac{2P}{\pi w^2}$, where r is the radial coordinate, I_0 is the maximum intensity, P is the laser power, w is the beam radius and n is the super-Gaussian factor. For $n = 2$, the above relation becomes the classical Gaussian distribution of the beam profile, for $n > 2$ the profile is super-Gaussian, and for $n \gg 2$ the profile is flat-top (or top-hat).

It should be mentioned that, despite other cases where the width of a Gaussian shape is considered at FWHM, as for example on spectral lines or for temporal pulse shapes, the diameter of laser beam profile is defined at the pixels' position where the beam irradiance is decreased to $1/e^2$ or 13.5% from maximum irradiance. This is consistent with the definition of beam transversal profile as a Gaussian function described by equation (3.29). Also, the normalization factor in equation (3.29) is chosen so, by integrating $I(r)$ over the entire beam surface, the result is exactly the optical power P.

If the beam transversal irradiance is asymmetric and far from a Gaussian profile, as a real laser beam it is, the profiles cannot be fitted correctly by a Gaussian function. The first- and second-order moment can be used to define the centroid and beam radius, respectively [23]. Since the beam profiles are measured by digital cameras with finite pixel size and pixel numbers, the irradiance for each pixel is $I(x_i, y_i)$. Beam centroid (x_c, y_c) is given by the first moment:

$$x_c = \frac{\sum\limits_{i,j=0}^{N} x_i I(x_i, y_i)}{\sum\limits_{i,j=0}^{N} I(x_i, y_i)}; \quad y_c = \frac{\sum\limits_{i,j=0}^{N} y_i I(x_i, y_i)}{\sum\limits_{i,j=0}^{N} I(x_i, y_i)} \tag{3.30}$$

The second moment is used for estimation of the beam radius on x- and y-axis:

$$w_x{}^2 = 2\frac{\sum\limits_{i,j=0}^{N} (x_i - x_c)^2 I(x_i, y_i)}{\sum\limits_{i,j=0}^{N} I(x_i, y_i)}; \quad w_y{}^2 = \frac{\sum\limits_{i,j=0}^{N} (y_i - y_c)^2 I(x_i, y_i)}{\sum\limits_{i,j=0}^{N} I(x_i, y_i)} \tag{3.31}$$

The image acquired by the digital camera is sent to a computer and the matrix corresponding to intensity map $I(x_i, y_i)$ needs several processing steps before computing the centroid. First of all, the acquired image has to be not saturated. For example, an 8-bit depth image, as for standard webcams sensors, has a maximum intensity at 255 counts, or a 16-bit depth image, as for many qualitative sensors, has 65 536 counts on its maximum intensity at saturation. To avoid image saturation, the laser power should be attenuated, or the sensor exposure time could be adjusted. The main issue on beam laser profile measurement is the noise that accumulates when the irradiance is numerically integrated over the entire number of pixels as the sum of $\sum_{i,j=0}^{N}$. In order to avoid this inconvenience, one step is to chose a mask (circle, ellipse or rectangular region of interest) and select only the intensities of pixels inside the mask area. As a rule of thumb, the size of mask has to be roughly three times larger than the diameter of the beam. The intensity of pixels outside the mask will be set at zero in order remove their contribution to the sum. This step is necessary, but not sufficient to remove the noise on image. Prior to beam profile measurements, the background has to be measured and subtracted from the data. Since the background data is also noisy, subtracting these values from low intensity pixels of the profile matrix could produce artifact and negative intensities towards the wedge of the beam profile altering the integration and computation of centroids. Then, a threshold can be imposed in the algorithm in order to remove the intensities below some negligible values. With these steps accurate values for the center of mass of the beam irradiance are obtained (see Figure 3.22). First- and second-order moments are largely used in laser beam profilometry and adopted as per the international standard ISO 11146-1 [15].

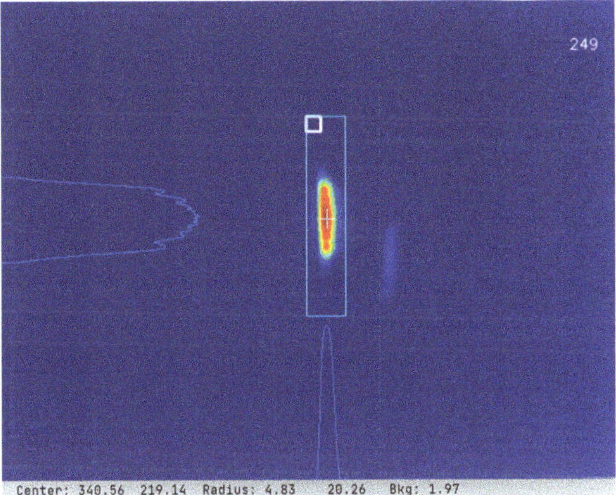

Figure 3.22. Computed centroid and beam size for transversal intensity of a laser beam emitted by a standard laser diode. Image was generated from a Python script with the algorithm as described in text.

3.3.3 Power density and energy density

With optical power, pulse energy and beam size measured as mentioned in the previous sections, parameters such as power density I (or irradiance) measured in W cm^{-2} and the energy density F measured in J cm^{-2} (or laser fluence) can be estimated. These are important values to know in various part of the experiments, from design of the laser multipass amplifier when the number of passes are computed based on pump power density (see section 3.2.3), to optical components in order to avoid their damage, and down to the level of sample, in order to estimate the laser–matter interaction regime. The laser energy density (or laser fluence) is usually expressed in mJ cm^{-2} or J cm^{-2}, and the formula depends on the shape of the beam profile. A flat-top beam has the irradiance spatial distribution evenly over the entire beam surface. Then, the expression of the laser fluence is given by:

$$F = \frac{E_{\text{pulse}}}{\pi w^2} \tag{3.32}$$

where w is the radius of the beam measured as described in the previous section. If the beam surface is elliptic, the beam area should be calculated as $A = \pi w_x w_y$, taking into account the beam ellipticity. In most of the cases, the beam profile has a Gaussian spatial distribution. If we consider two extreme cases, a flat-top beam and a Gaussian beam with the same optical power and the same beam diameter (as seen in figure 3.23(b)), we observe that the maximum intensity for the Gaussian case is two times higher compared to the flat-top beam. Because of this factor of 2, the fluence of a Gaussian beam is expressed as:

$$F = \frac{2 \cdot E_{\text{pulse}}}{\pi w^2} \tag{3.33}$$

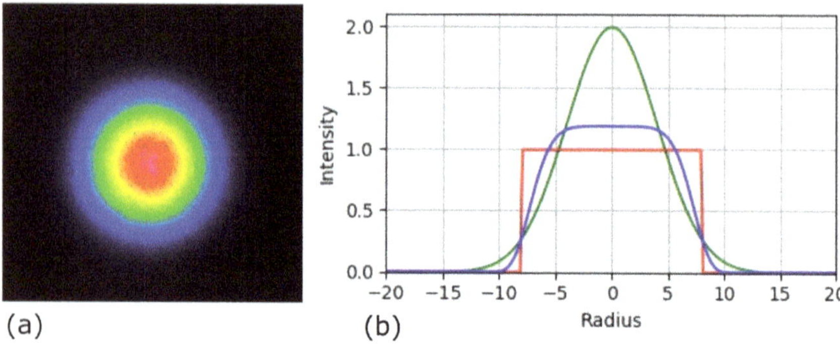

Figure 3.23. (a) Transversal intensity distribution of a laser beam as measured by a beam profiler. (b) The irradiance radial distribution for three type of profiles: Gaussian, super-Gaussian and top-hat.

3.3.4 Beam propagation. Beam quality factor M^2

Usually in laser processing the laser beam is tightly focused on the sample. Laser fluence on the focus spot cannot be directly measured, but involves the characterization of energy and beam profile far from focus and correct estimation of the beam diameter at focus. Characterization of the laser spot in focus is not trivial since the spot size can be comparable with the pixel resolution of digital cameras. However, the beam spot size can be estimated from Gaussian beam propagation theory (see section 2.6). A real Gaussian beam that propagates in the z direction after a lens, has the beam radius estimated by:

$$w(z) = w_0 \sqrt{1 + \frac{z\lambda M^2}{\pi w_0}} \qquad (3.34)$$

where λ is the laser wavelength in the propagation medium, w_0 is the minimum spot diameter at focus, known as beam waist, and M^2 is the beam quality factor.

M^2 is 1 for an ideal Gaussian beam and >1 for any real laser beam. As examples, a He–Ne gas laser provides the best quality beam profile close to TEM_{00} propagation mode with round and Gaussian transversal irradiance distribution and M^2 about 1.05–1.1. A solid state laser such as Nd:YAG or Ti:sapphire emits beams with a slight elliptical profile close to Gaussian, and sometimes slightly astigmatic, with M^2 about 1.5–2.5 depending on laser geometry, where M^2 could differ on the X- and Y-axes, depending on astigmatism. In ultraintense lasers, especially the pump laser systems, the beam profile is flat-top and M^2 is even higher as for the excimer lasers, exceeding values of 10–20. M^2 is given by the relation:

$$M^2 = \frac{\pi w_0^{\text{real}} \theta}{\lambda} \qquad (3.35)$$

where θ is the beam divergence and w_0^{real} is the beam waist for a real Gaussian beam, that differs from the beam waist w_0^{ideal} of an idea Gaussian beam (see figure 3.24).

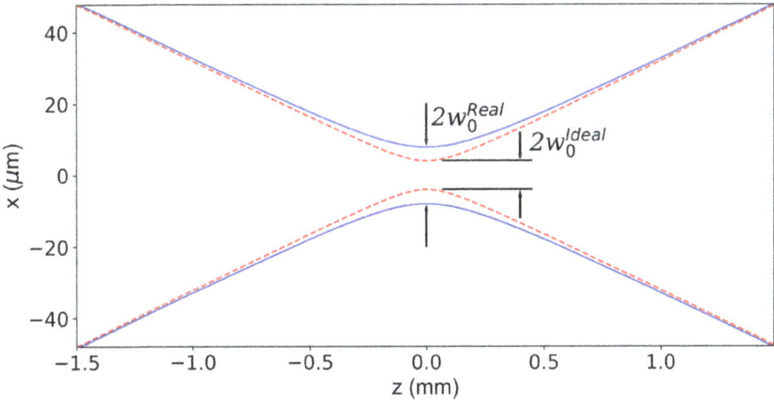

Figure 3.24. Beam propagation and focusing of a real beam.

For a collimated beam with beam diameter w_L that is focused by a focusing optics with focal length f_L, the beam divergence after focusing optics is given by $\theta = w_L/f_L$. Then, the beam waist of a real Gaussian beam, or the beam spot radius at focus, w_0 is defined by:

$$w_0^{real} = \frac{\lambda f_L M^2}{\pi w_L} \tag{3.36}$$

Measurement of M^2 is necessary in order to have a trusted estimation of laser beam spot on focus and a correct value for the laser fluence on the sample. As a numerical example, for a laser wavelength at $\lambda = 400$ nm, such as the second harmonic generation (SHG) from a Ti:sapphire laser, a collimated beam with diameter $w_L = 10$ mm, with beam quality factor $M^2 = 2$, focused by a lens with focal length $f_L = 100$ mm, will produce a beam spot at focus with radius $w_0^{real} = 8\mu$m, two times larger than for a beam spot estimated by neglecting the M^2 value in equation (3.36). This error will propagate on determination of laser fluence, with a four times overestimation of the real value for the given example.

The most straightforward method to experimentally evaluate M^2, is to measure the beam profile at several positions around the focus spot. The plot of beam diameter $w(z)$ versus z position can be easily fitted by equation (3.34) with fitting parameters w_0 and M^2.

3.3.5 Temporal measurement

Time characteristics of a nanosecond laser, or pulse duration, can be easily measured by a fast photodiode and an oscilloscope that has enough bandwidth to record the fast response obtained by the photodiode. As a general rule, for a correct measurement of nanosecond laser pulses, the rise time of the photodiode has to be faster than the pulse duration, and the oscilloscope has to be set on the measurement channel with termination of 50 Ω. In laser micro- and nanoprocessing, ultrafast lasers have a

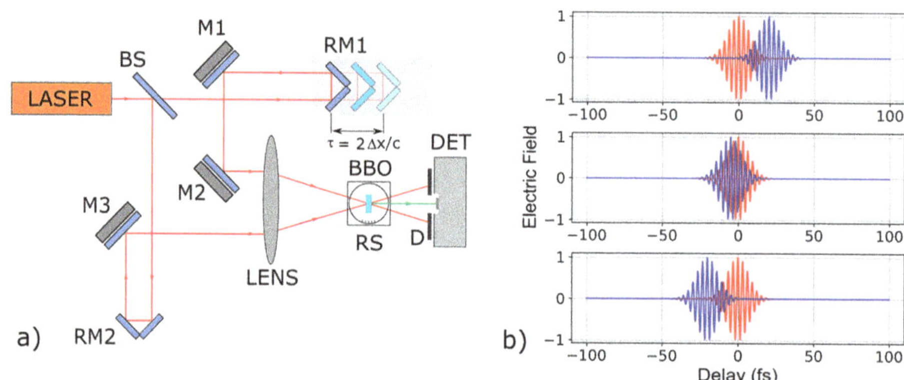

Figure 3.25. (a) Autocorrelator configuration. The components and their role are explained in the text. (b) Electric field overlap as function of delays, reconstruct the autocorrelation function.

pulse duration from tens of femtoseconds to a few picoseconds. Commercially available photodiodes have a much longer rise time, of the order of nanoseconds. Then, trying to use a photodiode to measure ultrashort pulses, the recorded trace on the oscilloscope gives in fact the photodiode response, and not the real temporal laser pulse shape. Femto- and picosecond laser pulses are measured by indirect methods, usually by employing the autocorrelation technique based on nonlinear optical effects such as SHG.

In figure 3.25(a), a typical arrangement of autocorrelator setup is presented. The laser beam to be characterized is split in two beams with equal intensities $I_1 = I_2$ using a beam-splitter (BS). The autocorrelator structure has two arms, a mobile arm with roof mirror RM1 returning the beam at 180° and a fixed arm, the reference, with roof mirror RM2. Roof mirror RM1 is positioned on a translation stage with positioning accuracy at the order of the resolution expected for the pulse measurement. The pulse is delayed with respect to the second arm by a time delay $\tau = 2\Delta x/c$, where Δx is the stage displacement and c is the speed of light. The delayed and reference beams are spatially superposed on an optical nonlinear crystal, such as for example barium borate (BBO) with thickness and crystal orientation optimized for the pulse duration and laser wavelength. A lens can be used in order to obtained the so-called non-colinear arrangement, as presented in figure 3.25(a). Also, the focusing lens is needed in the case of long pulses measurement in order to increase the laser fluence and consequently the conversion efficiency of the crystal. However, the laser fluence has to be kept low enough, below the damage threshold of the crystal. The crystal is mounted on a rotation stage in order to finely tune the angle between principal axis of the nonlinear crystal and beam directions. In non-colinear configuration, the phase-matching condition $\vec{k_1} + \vec{k_2} = \vec{k}_{\mathrm{SHG}}$ and maximum of intensity I_{SHG} are obtained along the angle bisector determined by the two beams incident on the BBO crystal. The incident beams emitted from a Ti:sapphire laser at wavelength of 800 nm are converted by the BBO crystal in a radiation at 400 nm. SHG represents only a few percents from the incident radiation, as the conversion efficiency depends on the beam intensity, pulse duration, thickness of the crystal, phase-matching angle, and laser polarization orientation with respect to the principal axis of the crystal. If all the preliminary conditions are optimized, the maximum SHG intensity is obtained when the two beams I_1 and I_2 are perfectly synchronized on the crystal. The intensity I_{SHG} is measured by a slow detector (DET) as function of time delay between the two pulses. The residuals of the fundamental beams are blocked by a diaphragm (D) that allows the SHG beam to reach the detector.

Figure 3.25(b) represents the electric fields of the two superposed beams as function of delays. At delay $\tau = 0$ corresponding to a perfect synchronization on the crystal, I_{SHG} is maximum as the time overlap of the two electric fields is maximum, and decreases as the positive or negative delay increases. The intensity variation with time delay $I_{SHG}(\tau)$ represent the intensity autocorrelation function and it is recorded experimentally by scanning the position of the roof mirror RM1.

The mathematical description of the intensity autocorrelation function is given by the following relation:

$$A(\tau) = \int_{-\infty}^{\infty} I_1(t)I_2(t - \tau)\mathrm{d}t \tag{3.37}$$

where τ represents in the experiment the delay between the reference pulse and its replica. The intensity I_{SHG} is proportional to the autocorrelation $A(\tau)$ mentioned above and it is a function of delay τ. Scanning the position of roof mirror RM1 over a range corresponding to delays larger than the pulse duration will reconstruct the profile of $I_{SHG}(\tau)$ with FWHM proportional to the laser pulse duration up to a factor that depends on the temporal pulse shape.

Figure 3.26 represents an autocorrelator setup built as based on the schematic from figure 3.25. The laser characterized in the following example was a Nd:YAG picosecond system (HyperRapid 50 from Lumera GmbH) with emission at 1064 nm and 500 kHz

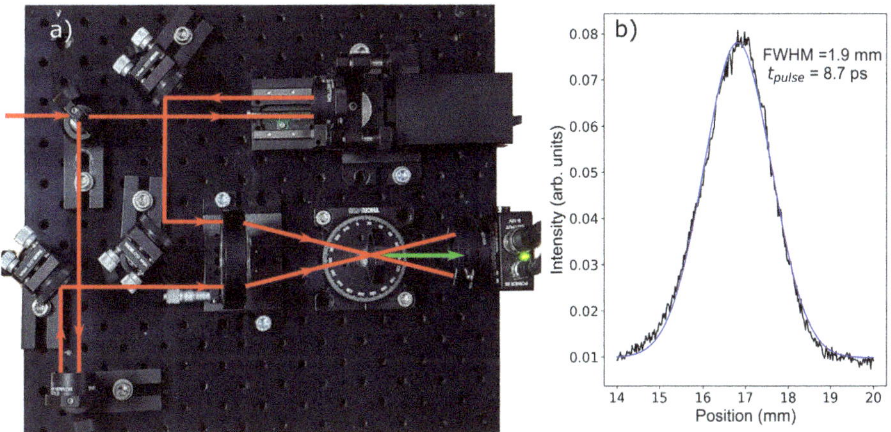

Figure 3.26. (a) Autocorrelator setup. (b) Reconstruction of the autocorrelation function and the data fit for laser pulse with 8.7 ps pulse duration.

repetition rate. The BBO crystal is 2.00 mm thick, and cut at an angle of $\theta = 23.3°$ for optimal conversion from fundamental wavelength to SHG at 532 nm. A photodiode connected to an oscilloscope measures the SHG signal for each position of the motorized stage. Translation and oscilloscope are controlled from computer by a Python script. Data collected are fitted with a Gaussian function or by a hyperbolic secant squared (sech^2) shape. A Gaussian shape gives the best fit in the example from figure 3.26(b). The width of the curve is $\Delta x^{\text{FWHM}} = 1.9$ mm. The conversion from position to time is given by the relation $\tau = 2\Delta x/c$. A factor 2 is taken into account in the expression above in order to consider the roundtrip of the beam when the stage is moving with Δx. Then, the conversion from distance to delay can be obtained by multiplying the displacement by a factor of 6.6 ps mm^{-1}. A final adjustment takes into account the shape of the temporal pulse. For a Gaussian temporal shape the autocorrelation function is larger than the pulse overlap by a factor of $\sqrt{2}$. For other pulse shapes this factor is different. For example, a pulse with sech^2 shape has the correction factor of 1.54:

$$\tau_P^{(\text{Gauss})} = \frac{1}{\sqrt{2}}\tau^{\text{FWHM}}, \quad \text{and} \quad \tau_P^{(\text{sech}^2)} = \frac{1}{1.54}\tau^{\text{FWHM}} \tag{3.38}$$

After all corrections, the measured pulse duration is 8.7 ps. The main source of error in this measurements is caused by positioning precision specific to the translation stage. For the stage used in the experiment (model MTS50/M-Z8 from Thorlabs) the minimum repeatable incremental movement is 0.8 μm, then the absolute error on pulse duration measurement is 3.7 fs, or 0.04% relative error.

Another configuration of setup for measurement of ultrafast pulses is the interferometric autocorrelator. In this version the two beams are colinear as in a Michelson interferometer and can provide information about phase modulation in the case of non-bandwidth limited pulses. The measured signal shows interfringes that depend on the phase modulation of spectral chirp of the pulse.

It is to be mentioned that the method of autocorrelation has several limitations. The retrieved shape is always symmetric, even if the real pulse in asymmetric. This is a possible issue especially in the case where we expect a pulse with a certain temporal shape, with double or multiple pulses. Also, in the case of intensity autocorrelators, the result is only a pulse envelope, and it shows no information about the spectral chirp that may occur in ultrashort pulses (see figure 3.27). More details about spectral chirp and group delay dispersion (GDD) are presented in section 3.4. The method described above involves scanning of the delays in the range of the pulse duration. Scanning can be done by displacement of a roof mirror or retro-reflector, rotating mirrors or prisms, fast scanning by coil driven mirrors etc, as different producers propose in their products. In all cases, the laser has to be at a certain repetition rate and does not work for single pulses such as in experiments with ultraintense lasers. However, in a single-pulse autocorrelator scheme that involves thicker nonlinear crystals, the information on pulse duration is given from the image of the SHG beam obtained in non-colinear incidence.

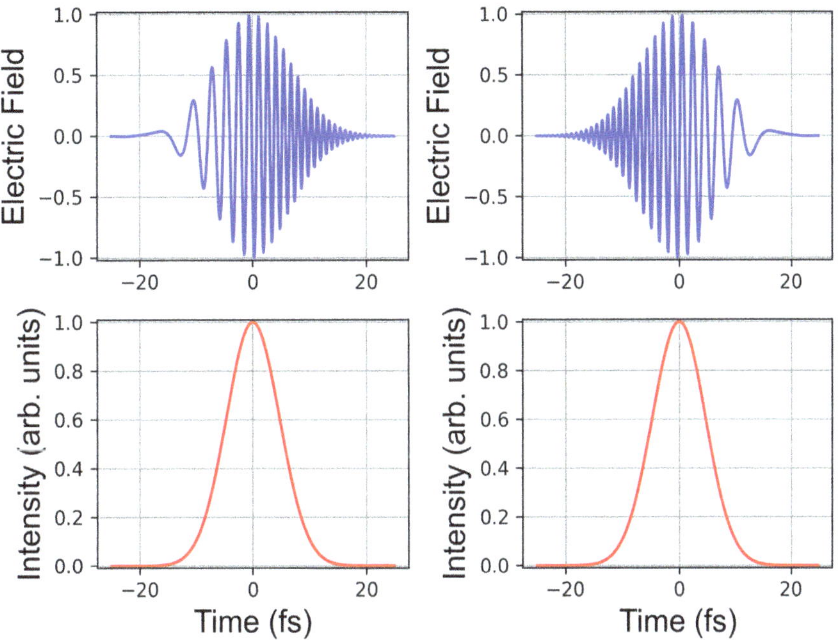

Figure 3.27. Negative and positive spectral chirp give the same temporal pulse envelope in second-order autocorrelation.

If the temporal characterization requires more information about the ultrashort pulses, such as pulse shape, spectral phase, or contrast, other techniques will be used, like: FROG, SPIDER, WIZZLERS, TUNDRA, etc.

3.3.6 Spectral measurement

Lasers systems have well-known emission bands with central wavelength depending on the type of active medium. Several examples of most common lasers and their emission at fundamental wavelengths are listed below: Nd:YAG—1064 nm; Ti: sapphire—around 800 nm; Nd:YLF—emission at 1047, 1053, 1313, 1324 and 1370 nm; Yb:YAG—1030 and 1050 nm; Er:YAG—1.55 and 2.94 μm, thulium-doped fiber: 1.9, 1.95 and 2.00 μm. However, several cases involve knowing the spectral distribution of the emission band as for the optical parametric oscillators (OPOs) and optical parametrical amplifiers (OPAs). Since the shortest pulses depend fundamentally on the spectral bandwidth, spectral monitoring is necessary in the process of amplification in ultraintense laser systems. In some cases, the spectral manipulation of large bandwidth is a flexible technique used in temporal pulse shaping. In these cases, spectral measurements together with spectral phase measurements provide useful information that can be correlated with the temporal shape of the pulse via Fourier transform.

The laser beams are intense enough to not require as complex spectroscopic equipment as the ones used in advance spectroscopy, such as photoluminescence or

Raman spectroscopy with very low intensity beams, where sensors with good control of dark noise are needed. Also, the spectral bandwidth of ultraintense lasers is large enough to allow spectroscopic measurement with poor spectral resolution larger than 1 nm. Then, the minispectrometers largely available on the market are good enough to get reasonable spectral behavior of the laser beam. When ordering a minispectrometer from a producer, several characteristics have to be analyzed and decided. Many options can be chosen in order to configure the device for a final application, but one has to keep in mind that the miniaturization and portability of such equipment come with the disadvantage that once the configuration is chosen, it remains fixed and cannot be modified after. The structure of a minispectrometer consists in the following main components:

- *Entrance aperture.* At the spectrometer aperture a pinhole with diameters of few μm up to tens of μm is placed in order to define the spectrometer resolution. Apertures with lower diameters may provide better spectral resolution, below 1 nm, as a trade off regarding the signal intensity that arrives on the sensor area. The light is coupled to a spectrometer by an optical fiber with SMA connector. Even if the optical fiber in not a component of the minispectrometers, the fiber characteristics can impact on the measurements quality.
- *Diffraction gratings.* Large dispersion angles and high intensity efficiency are obtained by holographic gratings with grooves density from 600 to 3600 lines mm^{-1}. A blazed grating can be chosen for a specific spectral range where the grating efficiency is optimized. The spectrometers can be configured in a such way that the spectra are recorded for both fundamental and second harmonic wavelength, for example, at 800 nm and 400 nm, and can be simultaneously measured. Larger spectral range can be obtained, but it requires grating with low grooves density around 300 lines mm^{-1} and it comes with sacrificing the spectral resolution that cannot be better than 1 nm in this case.
- *Sensor area.* The dispersed photons are detected by an array of pixels, usually uni-dimensional structure of 1024, 2048 or more pixels. The spectral sensitivity of the sensors are typical in the range of 200–1100 nm. However, the final spectral domain achievable depends on the grating configuration chosen.

Other parameters to take into account are related to electronic and software performances. The bit-depth value of pixel represents the analog-to digital-conversion resolution. An 8 bits resolutions is the lowest possible since this gives an analog signal on maximum 255 levels of intensities. A higher bit-depth of 10, 12 or even 16 bits is available on the market providing good performances in terms of intensity resolution. Cooling the sensor with Peltier element can significantly increase the signal-to-noise ratio, since the dark noise is lower as the sensor's temperature decreases. Sometimes, a triggering option is necessary for integrating the minispectrometer with a laser experiment. This is particularly useful when the measured signal has to be synchronized with the laser pulses in a time window of the order of micro- or miliseconds relative to the laser pulse. From

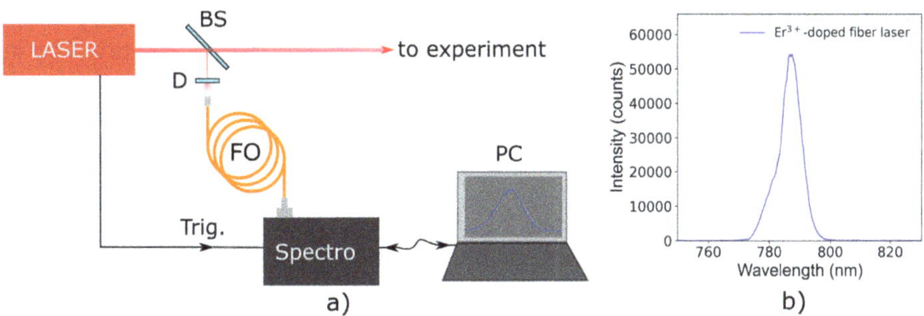

Figure 3.28. (a) Typical setup for laser spectral characterization of laser beam. BS—beam-splitter; D—diffuser; FO—fiber optic; PC—data acquisition system. (b) Emission spectra from an erbium-doped fiber laser.

the software point of view, a friendly and intuitive user interface is always desired. However, advanced functionalities provided by a Software Developing Kit (SDK) provides access to computer-to-device communication features, in order to integrate the minispectrometer in a more complex experiment with automatic algorithms for exposure, triggering and data recording. The libraries and functions provided are usually compatible with many programming languages such as C++, C#, Matlab, Labview, Python etc as preferred by the user. Figure 3.28(a) presents a simple setup for laser beam spectral characterization. Despite the simplicity of measuring the spectral shape of the beam, several precautions have to be kept in mind when such measurements are done. The first issue to avoid is the overexposure of the sensor, that will not only saturate the signal, but can even harm the detector. Fiber optics (FO) that collect the beam should be never exposed directly to the intense laser. A pick-up from the main beam path can be selected from a **BS** or even from the rear of a back polished mirror, with the precaution that the reflection or transmission is not clipping the spectral band. If necessary, the peak-up beam has to be more attenuated by using a diffuser (D) with neutral spectral transmission in the laser spectral range (ex. a piece of white paper, if the laser beam is red or near infrared). The optical fiber has usually a core diameter of few hundreds of μm. No coupling optics has to be used. The very small solid angle corresponding to the input at the core fiber will couple only a part of the diffused light, and the saturation of spectrometers is then avoided.

An important parameter to be correctly set is the exposure time. As the exposure time increases, normaly better signal-to-noise ratio is obtained. However, software saturation of the data signal could appear because of limited bit-depth available. For example, in the case of a 16-bit detector, the integrated signal is limited to 65 536 counts. Triggering can be useful in some experimental conditions, as for example when the laser runs at a repetition rate of 10 Hz and the exposure time is shorter than the time interval between two pulses (100 ms). In such conditions, in order to prevent an acquisition window with missing laser pulse, the synchronization of the spectrometer with laser electronics is necessary. Sometimes, a delay generator is included between laser and spectrometer, as well between laser and other equipment in experiment, in order to finely tune the start acquisition time. In general, the femtosecond lasers cannot be triggered externally. Then, the electronics of the laser is a master that provides the t_0 time in the experiment. All other parts of the equipment are slaves synchronized at different delays as imposed by the delay generator. The most common issue and probably the most neglected is the calibration of the spectrometer. Devices come calibrated from fabric. Bad manipulation and storage of the spectrometers could cause in time a drift of the measured peaks with respect to the calibration table. A misposition of the diffraction grating inside the spectrometer will shift the spectral bands with respect to the sensors pixels leading to an incorrect interpretation of the spectral data. Recalibration can be done at the factory, or in a metrology laboratory using some well-known emission sources and spectral etalons. Recalibration is necesary not only in wavelength, but also in intensity, in order to take into account the drift of the spectral response of pixels, that can be caused by accelerated aging of the sensor in overexposed conditions. Wavelength verification and recalibration could be in principle realized in any laboratory with optical sources that emits narrow peaks at known wavelength, the pixel positions of each peak are written in a calibration table, usually provided by the software interface of the spectrometer. Data are extrapolated by linear or polynomial function that transform the pixel position in wavelengths.

3.4 Dispersion of ultrafast laser pulses

A femtosecond laser is not monochromatic. A continuous-wave laser also has a finite spectral bandwidth, at the order of tens of picometers. The spectrum of a femtosecond pulse is much larger and it fundamentally depends on the pulse duration, since the spectral and temporal descriptions of the electromagnetic wave are correlated by the Fourier transform relation:

$$E(\omega) = \int_{-\infty}^{\infty} E(t)\mathrm{e}^{i\omega t}\mathrm{d}t \tag{3.39}$$

In other words, the time–frequency relations from quantum-mechanics $\Delta\tau\Delta E \geqslant \hbar$ are telling us that shorter pulses need larger emission bandwidth.

The Fourier limited pulses have the spectral bandwidth dependent on the spectral shape (*Gaussian* or sech^2, see Figure 3.29) and is given by:

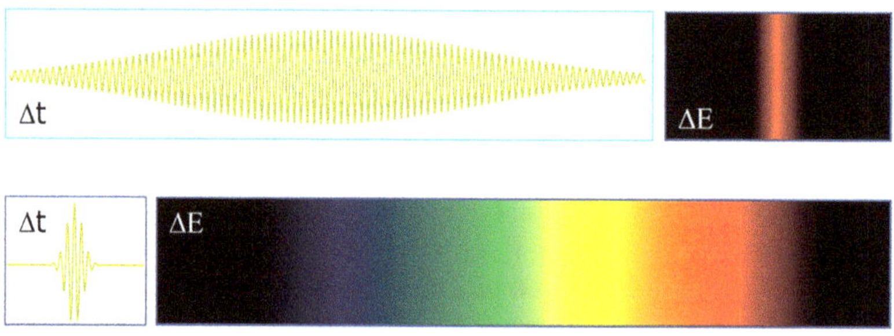

Figure 3.29. Time–frequency relations and Fourier limit or ultrashort pulses.

$$\Delta\nu^{\text{Gauss}} = \frac{0.441}{\tau_{\text{P}}^{\text{Gauss}}}, \quad \text{and} \quad \Delta\nu^{\text{sech}^2} = \frac{0.315}{\tau_{\text{P}}^{\text{sech}^2}} \tag{3.40}$$

In wavelengths, the relations from above are:

$$\Delta\lambda^{\text{Gauss}} = \frac{\lambda_0^2}{c} \cdot \frac{0.441}{\tau_{\text{P}}^{\text{Gauss}}}, \quad \text{and} \quad \Delta\lambda^{\text{sech}^2} = \frac{\lambda_0^2}{c} \cdot \frac{0.315}{\tau_{\text{P}}^{\text{sech}^2}} \tag{3.41}$$

3.4.1 Group delay dispersion

Ti:sapphire crystal has a very large spectral band of fluorescence, with more than 200 nm, from 700 to 900 nm. This is at the moment the only material capable of sustaining very short pulses below 10 fs. In a femtosecond laser pulse a very high number of longitudinal modes evolves as described in the chapter dedicated to mode-locking. The longitudinal mode is separated in frequency by $\delta\nu = c/2L$ which is at the order of 75 MHz for a typical cavity length $L = 2$ m. A bandwidth of $\Delta\lambda = 100$ nm at the central wave of $\lambda_0 = 800$ nm represents in frequency about $\Delta\nu = 46.8$ THz. The total number of modes evolving in the cavity is $N = \Delta\nu_0/\delta\nu$, about 6×10^5 modes. Spatial and temporal description of the total electric field of such a beam can be written as a linear combination of each spectral mode:

$$E(z, t) = \sum_{j}^{N} A_j e^{i(\omega_j t - k_j z)} \tag{3.42}$$

where A_j is the field amplitude, k_j is the wavevector, and ω_j is the angular frequency of the mode j. Propagation of the modes in any material depends on the dispersion relation:

$$|\vec{k_j}| = \frac{\omega_j}{c} n(\omega_j) = \frac{2\pi}{\lambda_j} n(\lambda_j) \tag{3.43}$$

where n is the refractive index of the material defined for the mode ω_j and wavelength λ_j.

Each mode propagates in the material with the *group velocity*:

$$v_{\text{G}} = \frac{d\omega}{dk} \tag{3.44}$$

From equations (3.43) and (3.44) the relation above can be written as:

$$\frac{1}{v_{\text{G}}} = \frac{d}{d\omega} k(\omega) = \frac{1}{c}\left(n + \omega\frac{dn}{d\omega}\right) \tag{3.45}$$

v_{G} corresponds to the velocity of the energy transport, and is generally different from the *phase velocity* ($v_{\text{G}} \neq v_{\text{P}}$), with:

$$v_{\text{P}} = \frac{\omega_0}{k_0} = \frac{c}{n(\omega)} \tag{3.46}$$

For pulses that propagate in a dispersive medium, a temporal phase shift $\varphi(\omega)$ is introduced between propagation modes, that potentially induce a pulse temporal deformation depending on dispersion. The phase change of a spectral mode ω_j with respect to the central frequency ω_0 is described by Taylor series as follows:

$$\varphi(\omega_j) = \varphi(\omega_0) + \left(\frac{\mathrm{d}\varphi}{\mathrm{d}\omega}\right)_0 (\omega_j - \omega_0) + \frac{1}{2!}\left(\frac{\mathrm{d}^2\varphi}{\mathrm{d}\omega^2}\right)_0 (\omega_j - \omega_0)^2$$
$$+ \frac{1}{3!}\left(\frac{\mathrm{d}^3\varphi}{\mathrm{d}\omega^3}\right)_0 (\omega_j - \omega_0)^3 + \cdots$$

$$(3.47)$$

In equation (3.47), the superior terms have the following significance:

- $(\frac{\mathrm{d}\varphi}{\mathrm{d}\omega})$– is the group delay (GD) and represents the time propagation of the pulse through the optical medium. If the pulse propagates in a non-dispersive media with constant refractive index n, the phase change after propagation distance L is $\varphi = \frac{\omega L n}{c}$. Then, the group delay is $\frac{\mathrm{d}\varphi}{\mathrm{d}\omega} = \frac{Ln}{c}$, and is at the order of picoseconds for optical paths of mm distance. If the pulse is spectrally narrow enough, or the optical medium is non-dispersive, the superior terms in the Taylor series are negligible and the pulse propagates with preserving the shape of the pulse envelope.

- $(\frac{\mathrm{d}^2\varphi}{\mathrm{d}\omega^2})$– is the second-order dispersion and corresponds to the group velocity dispersion (GVD). If GVD $= 0$, then $V_F = V_G$. But in almost any transparent material GVD $\neq 0$. At constant GVD, not depending on frequency, the dispersive propagation medium introduces a linear dispersion, or linear chirp. With propagation in longer dispersive media, the chirp accumulates and the pulse becomes longer and longer. From GVD, the GDD for the propagation medium is obtained as GDD $=$ GVD $\times L$. Two distinct cases can be discussed here. If GDD is positive, it induces a positive chirp to the spectral phase, then the short wavelengths are delayed with respect to the longer wavelengths. At negative chirp, the shorter wavelengths propagate in front of the beam (see figure 3.27). In optical systems a positive chirp can be in principle compensated by introducing in the optical path components with negative chirp, as for example pairs of prisms, to compress back the stretched pulse.

- $(\frac{\mathrm{d}^3\varphi}{\mathrm{d}\omega^3})$– is the third-order dispersion (TOD) and introduces a quadratic chirp. Quadratic chirp is more difficult to be controlled and compensated, and requires more complex systems such as a programmable AOM for fine tuning of the phase and amplitude of the spectral components.

In an optical system where the pulses are broad in spectrum and short in time, GDD and TOD have to be controlled and compensated. Femtosecond lasers systems or telecommunications on an optical fiber at very high bit-rates are good

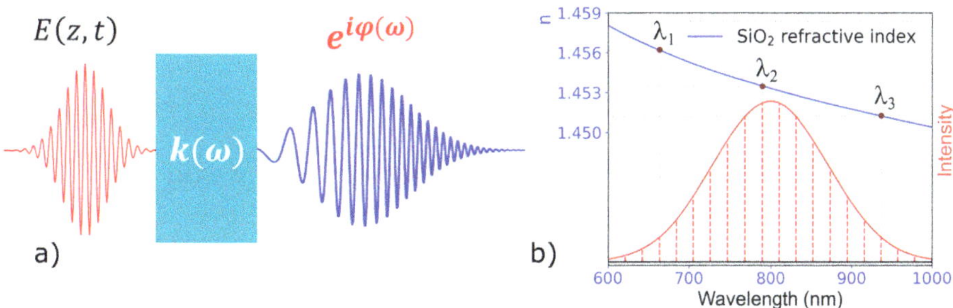

Figure 3.30. (a) GDD induces temporal stretching of ultrashort pulses at propagation in dispersive media. (b) Each spectral component propagates with different phase velocity due to the refractive index dependence on wavelength (chromatic dispersion).

examples. Practically, any optical component with chromatic dispersion (propagation through media with refractive index $n(\lambda)$) or angular dispersion (prisms and gratings) will introduce a certain phase shift $\varphi(\omega)$ for each spectral mode. If the phase has a quadratic dependence on frequency, the second and third derivative of $\varphi(\omega)$ are not null and introduce GDD and TOD that have impact on pulse propagation, changing the pulse duration and pulse temporal shape.

Each type of optical system (crystals, optical fibers, prism, gratings, thin films, mirrors etc) has a specific mathematical description of the frequency dependent phase shift introduced by the beam propagation geometry. For example, the phase change $\varphi(\omega)$ introduced by a slab of transparent material (see figure 3.30) is given by the frequency dependent refractive index $n(\omega)$ and the thickness L of the material:

$$\varphi(\omega) = \frac{\omega}{c} n(\omega) \tag{3.48}$$

or expressed in wavelength the phase $\varphi(\lambda)$ is:

$$\varphi(\lambda) = \frac{2\pi L}{\lambda} n(\lambda) \tag{3.49}$$

For practical reasons, since the refractive indices of materials are usually found in the literature as a function of wavelength [19, 22, 25], it is convenient to transform the differentials from $d\omega$ to $d\lambda$:

$$\frac{d}{d\omega} = \frac{d}{d\lambda}\left(\frac{d\lambda}{d\omega}\right) = -\frac{\lambda^2}{2\pi c} \cdot \frac{d}{d\lambda} \tag{3.50}$$

Then, the group velocity and phase velocity as a function of wavelength are:

$$v_{\mathrm{G}} = c\left[n(\lambda) - \lambda\frac{dn}{d\lambda}\right]^{-1}, \quad v_{\mathrm{P}} = \frac{c}{n(\lambda)} \tag{3.51}$$

In the expression of group velocity, the denominator is the *group index*:

$$n_G = n(\lambda) - \lambda \frac{dn}{d\lambda} \tag{3.52}$$

Similarly, the GVD and TOD as functions of wavelength are written as:

$$\text{GVD} = \frac{d}{d\omega}\left(\frac{1}{v_G}\right) = \frac{\lambda^3}{2\pi c^2} \cdot \frac{d^2 n}{d\lambda^2} \tag{3.53}$$

$$\text{TOD} = -\frac{\lambda^4}{4\pi^2 c^3}\left(3\frac{d^2 n}{d\lambda^2} + \lambda\frac{d^3 n}{d\lambda^3}\right) \tag{3.54}$$

We give below numerical examples and the steps on how to compute v_G, v_P, GVD, GDD and TOD, at the propagation of an ultrashort laser that propagates through optical components made of typical materials used in optics: fused silica and BK7 as found in lenses and mirrors; sapphire as the main material of the Ti:sapphire active medium in ultrafast lasers; and β-BaB$_2$O$_4$ (β-BBO) crystal used in optical nonlinear applications such as generation of harmonics. We consider an ultrashort laser beam with $\tau_P = 10$ fs pulse duration and $\lambda_0 = 800$ nm central wavelength. The minimal spectral bandwidth of a Gaussian shaped pulse is $\Delta\lambda^{\text{FWHM}} = 94.1$ nm, as given by the Fourier limited relation (3.40). This is a typical case of ultrashort laser beam used in multiphoton microscopy, or multiphoton laser processing systems, where pulses propagate through different components such as windows, cover glasses, filters, BSs, cube polarizer, etc, and are finally focused on the sample by microscope objectives with an important volume of glass.

The refractive index as function of λ included in the equations (3.53) and (3.54), can be obtained from database of optical constants [22] or computed using the Sellmeier equation [11]:

$$n^2(\lambda) = 1 + \frac{B_1 \lambda^2}{\lambda^2 - C_1^2} + \frac{B_2 \lambda^2}{\lambda^2 - C_2^2} + \frac{B_3 \lambda^2}{\lambda^2 - C_3^2} \tag{3.55}$$

Most of the programming languages offer mathematical libraries that help with computing of differentials, gradients, integrals, fast Fourier transforms (FFTs), polynomials etc. Python library numpy provides the function np.gradient() that helps to compute the values for:

$$\frac{d}{d\lambda}n(\lambda), \quad \frac{d^2}{d\lambda^2}n(\lambda), \quad \text{and} \quad \frac{d^3}{d\lambda^3}n(\lambda) \tag{3.56}$$

where $n(\lambda)$ is an array n_index of data obtained from equation (3.55), with Sellmeier parameters listed below. In order to avoid any confusion regarding the correct use of Sellmeier parameters, the values for C_j are provided in μm, not in μm^2 as found in databases for some materials.

Fused silica [19]:

B1 = 0.6961663,	B2 = 0.4079426,	B3 = 0.8974794,
C1 = 0.0684043μm,	C2 = 0.1162414 µm,	C3 = 9.896161 µm.

BK7 glass [35]:

B1 = 1.03961212,	B2 = 0.231792344,	B3 = 1.010469451,
C1 = 0.077464177μm,	C2 = 0.141484679 µm,	C3 = 10.17647547 µm.

Sapphire (o-ray) [35]:

B1 = 1.43134930,	B2 = 0.65054713,	B3 = 5.3414021,
C1 = 0.0726631μm,	C2 = 0.1193242μm,	C3 = 18.028251 µm.

Sapphire (e-ray):

B1 = 1.5039759,	B2 = 0.55069141,	B3 = 6.59273791,
C1 = 0.0740288μm,	C2 = 0.1216529μm,	C3 = 20.072248μm.

β-BBO (o-ray):[a][7]:

A = 2.7359

B1 = 0.01878,	B2 = 0.01822,	B3 = 0.01354,

β-BBO (e-ray):

A = 2.3753

B1 = 0.01224,	B2 = 0.01667,	B3 = 0.01516.

---

[a] For BBO crystal the Sellmeier equation has the form: $n^2(\lambda) = A + B_1/(\lambda^2 - B_2) - B_3\lambda^2$ [20].

From equations (3.51), (3.53) and (3.54) with refractive indices from (3.55) and gradients (3.56) computed with np.gradient() function from numpy in *Python* library, we obtain the v_G, v_P, GVD and TOD for fused silica, BK7 glass, sapphire and β-BBO, as plotted in figures 3.31 and 3.33. In the upper part of the plots the refractive indices are represented, including the indices for ordinary and extraordinary waves in the case of sapphire and β-BBO.

The plots in the middle represents the group velocity and phase velocity. Also, the spectral intensity distribution is represented (not at scale) for a Gaussian beam bandwidth with $\Delta\lambda^{\mathrm{FWHM}}$ = 94.1 nm (corresponding to a Fourier limited pulse with τ_P = 10 fs pulse duration) centered at λ_0 = 800 nm. A non negligible variation of the propagation parameters v_G, v_P is even visually observable in the range of the beam spectral bandwidth, for the spectral components of an ultrashort pulse that propagate in the optical media. Phase and group velocity slightly increase with wavelength, while the refractive index decreases. In the blue part of the spectrum the gap between v_P and v_G values becomes larger since the dispersion in this region increases (the gradient $dn/d\lambda$ of the refractive index is larger). It has to be mentioned that the group velocity corresponds to the group index n_G (see equations (3.51) and (3.52)) which is different from the refractive index shown in figure 3.31. When the relative difference between phase and group velocity is larger than 1% the pulses are significantly distorted at propagation.

In the bottom part, the plots represent the GVD and TOD of the materials mentioned above. In the spectral range from 1250 nm to 1350 nm the GVD changes

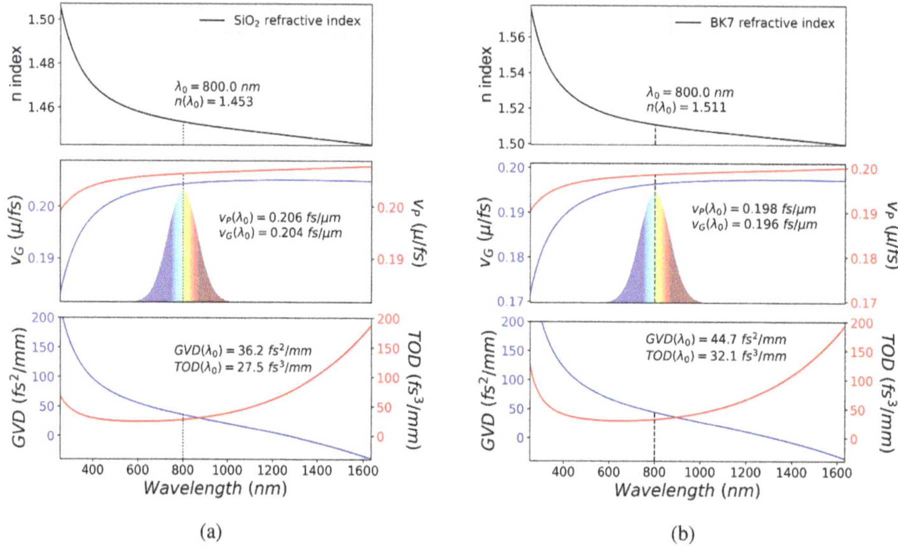

Figure 3.31. Group and phase velocity, GVD and TOD computed for: (a) fused silica and (b) BK7. Refractive index are obtained from Sellmeier equations as described in the text.

the sign for the exemplified materials. The positive GVD below this region is known as *normal dispersion*, while the negative dispersion is known as *anomalous dispersion*.

GVD has an impact on the pulse duration of a pulse traveling through the dispersive media and defines the GDD as a function of the length of the propagation path:

$$\text{GDD} = \text{GVD} \times L \quad (\text{fs}^2) \tag{3.57}$$

GDD is the parameter needed to compute the pulse broadening after traveling a certain optical path, no matter which kind of dispersion is used: a slab of dispersive material, optical fiber, chirped mirrors, pairs of prisms, gratings in a stretcher or compressor configuration etc. The output pulse duration at the exit of a dispersive system with known GDD is:

$$\tau_\text{P}^\text{OUT} = \tau_0 \sqrt{1 + \left(4\ln 2 \frac{\text{GDD}}{\tau_0^2}\right)^2} \tag{3.58}$$

Variation of the pulse duration τ_P^OUT for the elongated pulses as a function of initial pulse duration τ_0 for a series of different GDD values is presented in figure 3.32(a). At very short initial pulse, the output pulse become larger and large as the GDD increases. Then, in a laser experiment with ultrashort pulses one has to keep in mind to avoid as much as possible using optical components such as cube polarizer, filters, windows etc that introduce a large amount of glass that will consequently accumulate large GDD and induce an important pulse broadening. The same behavior is presented in figure 3.32(b). For a series of different initial pulse durations, the output pulse is

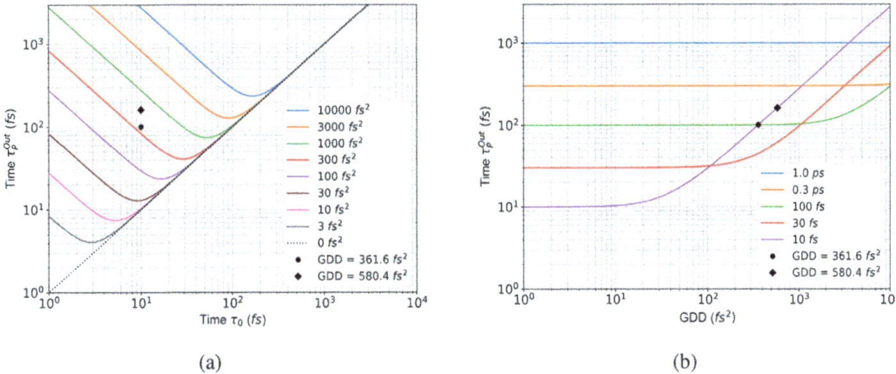

(a) (b)

Figure 3.32. Pulse broadening for: (a) series of GDDs as a function of initial pulse duration τ_0; (b) series of initial pulse duration as a function of GDD. The black dot and the diamond marker represent the pulse broaden to 101 fs, and, respectively, 161 fs for an initial pulse of 10 fs with GDD mentioned in the legend, introduced by a slab of 10 mm thickness of fused silica and sapphire.

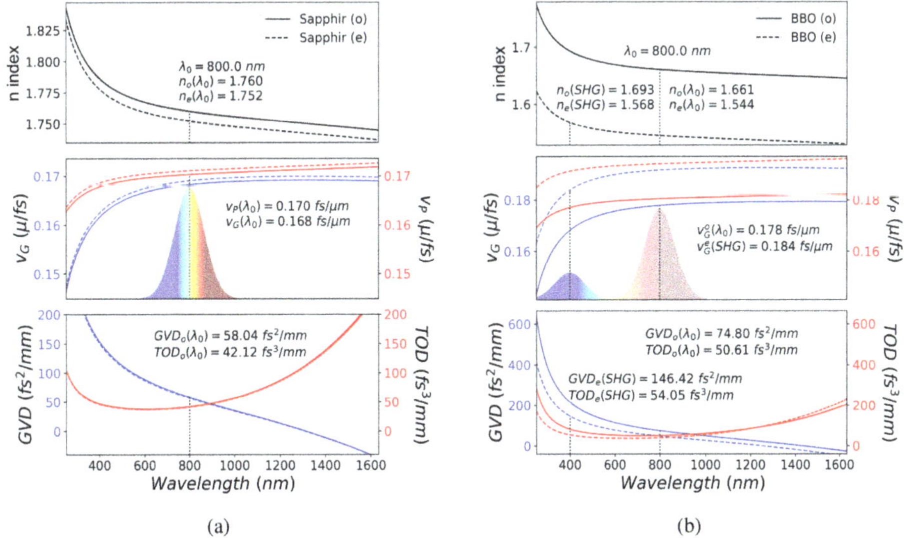

(a) (b)

Figure 3.33. Group and phase velocity, GVD and TOD computed for: (a) Sapphire and (b) β-BaB$_2$O$_4$ (BBO) crystal. The dispersion corresponding to ordinary and extraordinary rays are shown.

represented as a function of GDD. In both representations, it is observed that at GDD values (measured in fs^2) lower than the initial pulse duration τ_0 (fs) the propagation of pulses become insensitive to dispersion. As a consequence, initial pulse duration larger than 1 ps propagates practical non-stretched for a typical value of GDD below 10^4 fs^2 that corresponds to a total of hundreds of mm propagation path in fused silica.

Fused silica has at 800 nm the GVD $= 36.2$ fs^2 mm^{-1} (from figure 3.31(a)). After distance $L = 10$ mm of propagation, the accumulated GDD $= 361.6$ fs^2. Then, in the case of an incident pulse with initial duration $\tau_0 = 10$ fs, the elongated pulse duration is $\tau_p^{OUT} = 101$ fs for that GDD.

Much stronger GVD at 800 nm is induced by a sapphire crystal in a femtosecond laser oscillator (see figure 3.33(a)). If we consider a Ti:sapphire rod with 10 mm length, the GDD introduced by the active medium itself is about 580 fs^2. A laser pulse of initial $\tau_0 = 10$ fs is broadened to 161 fs. For this reason, the cavity of a Ti: sapphire laser oscillator includes always dispersion compensation components, such as a pair of prisms or chirp mirrors, in order to obtain the shortest pulses possible.

Next, it is worth commenting on the dispersion of the laser pulses propagated during the generation of the SHG in nonlinear optical crystals. We clearly notice a difference between the group velocities of the fundamental wavelength at 800 nm and its second harmonic at 400 nm generated in BBO (β-BaB$_2$O$_4$) crystals (figure 3.33(b)). If we consider $v_{G_{800}}$ and $v_{G_{400}}$ the group velocities of the two pulses traveling inside the crystal, we define the *group velocity mismatch* (GVM) or the temporal *walk-off* as:

$$\text{GVM} = \frac{1}{v_{G_{\text{blue}}}} - \frac{1}{v_{G_{\text{red}}}} \quad (\text{fs mm}^{-1}) \tag{3.59}$$

For a crystal with thickness L the GD is obtained as:

$$\text{GD} = \text{GVM} \cdot L \quad (\text{fs}) \tag{3.60}$$

which represents the traveling time of the pulse through the optical medium.

In the numerical example from figure 3.33(b), the BBO crystal has at 800 nm GVD $= 74.8$ fs^2, $v_G = 0.178$ fs mm^{-1}—for the ordinary ray, and at 400 nm (SHG) GVD $= 146.42$ fs^2 and $v_G = 0.184$ fs mm^{-1} for the extraordinary ray. This corresponds to a configuration of type I crystal, where the output SHG beam has the polarization orthogonal to the incident fundamental beam [4]. For long pulses in the range of ns, the SHG crystals are typically several mm in length. If we consider a BBO crystal with thickness of $L = 10$ mm, then the GDD at is 748 fs^2 at fundamental wavelength of 800 nm, and, respectively, 1464 fs^2 at SHG wavelength of 400 nm. If the initial pulse duration is 10 fs and 800 nm wavelength, after propagation in 10 mm of material it will be elongated to 208 fs. The equivalent of the SHG at 400 nm could be elongated to 406 fs, but the SHG efficiency decreases dramatically. First of all, the Intensity (W cm^{-2}) drops down 20 times just because of the pulse stretching by a factor of $\times 20$. Also, the two waves travel in the crystals at different group velocities that induce a temporal walk-off that become important as the thickness of the crystal is larger, preventing the formation of the SHG at 400 nm. Equation (3.60) leads to a GD of 1.8 fs, that apparently is not as much as compared to the pulse duration, but represents a phase difference of 0.7π that means a complete out of phase between the two waves. For short pulses, SHG is possible only for crystals with thickness of the order of fractions of mm, in order to minimize the group delay mismatch.

In summary, the steps to compute the propagation of a short laser pulse (pulse temporal broadening, pulse spectral chirp etc) in dispersive environments are:

- **Phase**. Find the equation of phase change as a function of wavelength. For usual dispersive systems (slabs of dispersive materials, prisms, chirped mirrors, gratings etc) the phase has a known formula that is specific from system to system. Equation (3.49) is valid only for a slab of dispersive material;
- **Phase derivatives**. Obtain the first-, the second- and the third-order derivative of the phase change in order to have the analytical expressions of v_G, GVD and TOD ((3.51)–(3.54)). Keep in mind that since these corespondent to superior orders of phase derivatives, they will have different expressions for each type of dispersive system. Transform their relations form frequency to wavelength (equation (3.50)) if the wavelength-dependent dispersion of refractive index is used;
- **Refractive index**. Compute the Sellmeier equation (3.55) or retrive the experimental refractive index from databases. If the Sellmeier equation is used, kccp in mind that different references use slightly different notations and parameters;
- **Refractive index derivatives**. Compute numerically the refractive index gradients (3.56). Software libraries can be used such as np.gradient().
- v_G, **GVD and TOD**. Compute numerically the group velocity, GVD, and TOD using the analytical expressions obtained from the phase derivatives and the refractive index gradient for a whole spectrum, or at a desired wavelength.
- **GDD**. The group velocity delay represents the accumulated GVD in the dispersive system over the entire optical path. For well-known dispersive systems the analytical expressions can be found in literature.
- **Pulse broadening or compressing**. Once the GDD is known for a given dispersive system, the pulse duration can be computed depending on the initial pulse width. Equation (3.58) can be used for a Fourier limited initial pulse width. A dispersive system with negative GDD can temporally compress the pulse that was already stretched in time.

Several most used examples of dispersive systems are shortly described below, presenting their configurations, GVD and TOD equations and their applications in laser experiments.

3.4.2 Pairs of gratings

Pairs of gratings are commonly used in stretchers and compressor lasers systems in order to manipulated the pulse spectral chirp. These dispersive configurations are important in CPA lasers in order to stretch the pulse before amplification and to compress back to the shortest possible pulse duration. CPA lasers have complex optical paths that need fine tuning of dispersion for best compression of pulses. Also, the pulse duration variation, from tens of femtoseconds at best compression regime, up to tens of picoseconds in stretched pulses regime, is needed in experiments. The gratings dispersive configurations are preferred due to their very large values of GDD (positive for stretchers and negative for compressors) that can be precisely

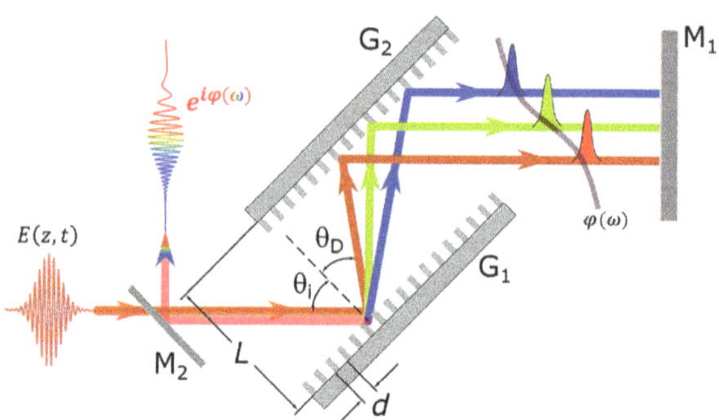

Figure 3.34. Dispersion controlled by a two-gratings pair.

tuned by adjusting both GVD and TOD, just from geometrical arrangement of the assembly, using motorized stages and automation protocols to control positions and incident angles of the optical components.

Figure 3.34 shows a compressor arrangement with two parallel gratings positioned at distance L one from the other. The incident pulse hits the first grating G_1 at incidence angle θ_i. The spectral modes are spatially dispersed according to the well-known gratings equation:

$$m\lambda = d(\sin\theta_i + \sin\theta_D) \tag{3.61}$$

where d is the grating's grooves pitch, m is the diffraction order, and θ_D is the angle of diffraction. From this relation, the angle of diffraction is deduced:

$$\theta_D = \arcsin\left(\sin\theta_i + m\frac{\lambda}{d}\right) \tag{3.62}$$

After the first diffraction grating, each spectral mode travels on different optical paths and hits the second grating G_2. With the gratings set perfectly parallel, the emerging beam should remain collimated before and after reflection on mirror M_2. This is usually a roof mirror or retro-reflector that changes the beam height with respect to the optical table in order to be easily reflected at 90° by the mirror M_2. Because of a different optical path, a phase change $\varphi(\omega)$ introduced for each spectral component ω of the initial beam, proportional to the diffraction angle θ_D and distance between the parallel gratings L [31]:

$$\varphi(\omega) = -\frac{\omega}{c}2L\cos\theta_D(\omega) \tag{3.63}$$

When one compares the phase change induced by gratings with the case of dispersive materials (equation (3.48)) it is obvious that in the gratings case, there is no dispersion of refractive index $n(\omega)$ involved. The phase $\varphi(\omega)$ is controlled by the diffraction angle and distance between the gratings. Relations for v_G, GVD, GDD, and TOD, are obtained from the derivatives terms of Taylor expression 3.47, transformed from $d\omega$ to $d\theta$ as:

$$\frac{\mathrm{d}}{\mathrm{d}\omega} = \frac{\mathrm{d}}{\mathrm{d}\theta}\left(\frac{\mathrm{d}\theta}{\mathrm{d}\lambda} \cdot \frac{\mathrm{d}\lambda}{\mathrm{d}\omega}\right) = -\frac{\lambda^2}{2\pi c} \cdot \frac{\mathrm{d}}{\mathrm{d}\theta}\left(\frac{\mathrm{d}\theta}{\mathrm{d}\lambda}\right) \tag{3.64}$$

Then, the angular dispersion is need, and is defined as the derivative of equation (3.62):

$$\frac{\mathrm{d}\theta_\mathrm{D}}{\mathrm{d}\lambda} = \frac{m}{d} \cdot \frac{1}{\sqrt{1 - \left(m\dfrac{\lambda}{d} - \sin\theta_\mathrm{i}\right)}} \tag{3.65}$$

The GDD and TOD of a gratings dispersion setup are:

$$\mathrm{GDD} = -\frac{\lambda^3}{2\pi c^2} \cdot \frac{Nm^2 L}{d^2}\left[1 - \left(-m\frac{\lambda}{d} - \sin\theta_\mathrm{i}\right)^2\right]^{-3/2} \tag{3.66}$$

$$\mathrm{TOD} = -\frac{3\lambda}{2\pi c} \cdot \frac{1 + \dfrac{\lambda}{d}\sin\theta_\mathrm{i} - \sin_\mathrm{i}^2\theta}{1 - \left(\dfrac{\lambda}{d} - \sin\theta_\mathrm{i}\right)^2} \cdot \mathrm{GDD} \tag{3.67}$$

where N is the number of passes through the dispersion device. For a pair of gratings $N = 2$, and the diffraction order is $m = -1$.

Figure 3.35 represents the variations of GDD and TOD with modification of the double grating compressor's geometry. Fine tuning of GDD and TOD is achieved in the range of hundreds of kfs^2 for GDD and, respectively, kfs^3 for TOD by changing

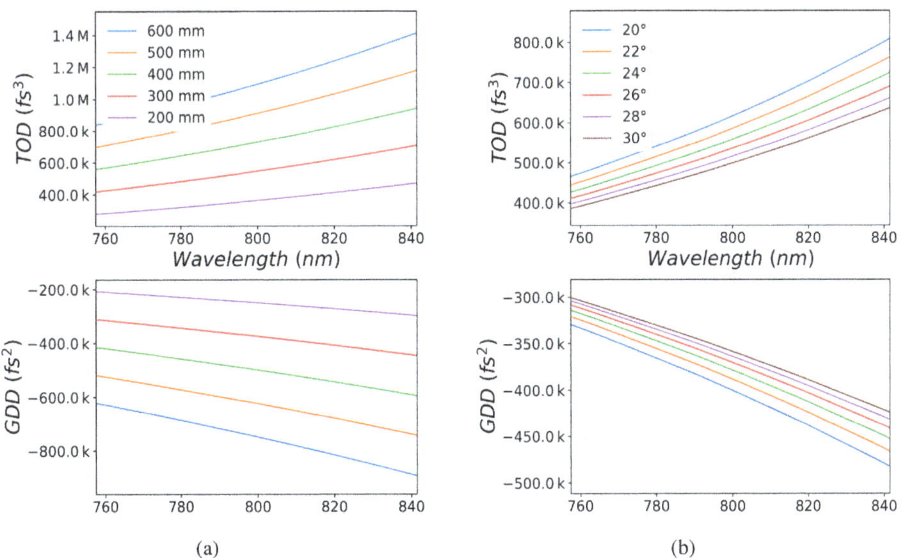

Figure 3.35. Tuning of GDD and TOD in a double grating compressor by modification of (a) distance L between gratings, and (b) input angle θ_i.

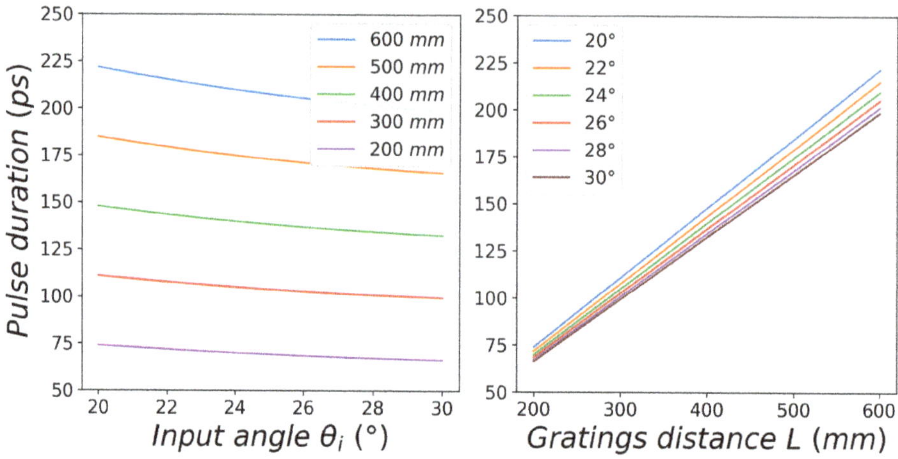

Figure 3.36. Pulse duration variation at the output of double grating compressor as function of input angle or gratings distance.

the distance L between gratings or by rotating the compressor angle relative to the input beam. In the left side graphics, the input angle θ_i is fixed at 25° and GDD and TOD are computed for a variation of L distance from 200 to 600 mm. On the right side, the distance is kept fixed at 300 mm and the input angle is varied from 20° to 30°. GDD values are negative and 3 orders of magnitudes larger than GDD induces by a slab of glass of a few mm thickness. The negative GDD is higher at large distance between gratings, and at lower incidence angles. The equivalent pulse elongation at the output of compressor for a Fourier limited laser pulse can be estimated from equation (3.58). In the example shown in figure 3.36, the initial pulse duration is $\tau_0 = 10$ fs, the grating groove is 800 lines mm^{-1} and the central wavelength is 800 nm. The grating distance L is modified as described above. The output pulse duration is typically from picoseconds up to the nanosecond range. A fine tuning of pulse elongation is obtained by changing the incident angle. The left side graphic shows the angle dependent pulse duration for different distance L. Large variation of pulse duration from tens to hundreds of picoseconds is achieved by changing the gratings distance. The right side graphic shows the gratings distance dependence of the pulse duration for different input angles.

If the input pulse is already chirped and has a positive chirp, the total GDD is eventually canceled and the pulse becomes shorter, with lowest possible pulse duration limited by the spectral bandwidth (see equation (3.41)). This is the case of optical compressor in CPA femtosecond laser systems. In some experiments, tuning the compressor can be a flexible method to obtain variable pulse duration, useful when optimization pulse duration or pulse shaping is necessary.

3.4.3 Chirped mirrors

Chirped mirrors are optical thin-films structures [30] designed to have simultaneously high reflectivity on a broad spectral band centered at the laser

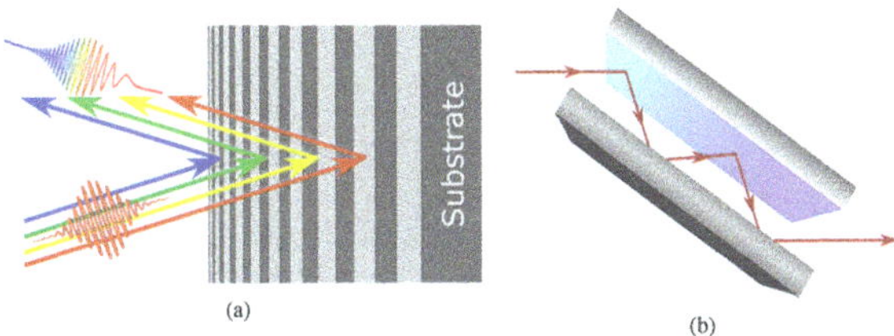

Figure 3.37. Chirped mirrors. (a) A gradient of layers thickness induces a phase shift on spectral frequencies. Higher frequencies are reflected from the top layers, while lower frequencies are reflected from the bottom layers. (b) Compressor based on multiple reflections on double chirped mirrors.

wavelength (see Figure 3.37), and a controlled GDD for the same spectral band. Dispersion is in general designed to have negative value in order to compensate the positive chirp introduced by other optical components. GDD is lower compared with other dispersive optical systems, such as double grating compressors, and has a fixed value on the order of tens to hundreds of fs^2. However, the compactness of such components makes it suitable for optical devices where multiple beam reflections will help to accumulate negative GDD large enough the compensate the positive dispersion in the system. Then, the total dispersion can be controlled both from the designed GDD at a single-reflection step and by the total number of passes in a multiple reflections process. This is typically the case of the modern femtosecond laser oscillators. Subpicosecond emission was possible only with control of dispersion introduced by the optical components, including the Ti:sapphire active medium (see the GVD from figure 3.33(a)). Pulses down to 10 fs were reached with chirped mirrors, replacing the traditional prism-pairs compressors [9], that cannot efficiently compensate for high order dispersion such as TOD.

Chirped reflectors are based on standard dielectric mirrors fabrication technology, where high reflectance is obtained on multiple thin-films with alternating transparent dielectrics coatings, such as TiO_2/SiO_2, Al_2O_3/HfO_2, AlAs/GaAs or any other combinations of similar materials. In Bragg reflectors, the alternating materials have the same optical propagation path of $\lambda/4$, but large contrast of the refractive index $\Delta n = n_1 - n_2$, resulting in a number of pairs with alternating physical thickness of the order of tens of nm. Chirped mirrors are designed with sequences of layer thicknesses decreasing from substrate to the top layer. As for Bragg reflectors, the design is computed from the transfer matrix method. For each layer i the transfer matrix M_i is written as [3]:

$$M_i = \begin{bmatrix} \cos \beta_i & -\dfrac{i}{q_i} \sin \beta_i \\ -iq_i \sin \beta_i & \cos \beta_i \end{bmatrix} \tag{3.68}$$

where $\beta_i = \dfrac{2\pi n_i(\lambda) d_i \cos(\theta_i)}{\lambda}$, λ is the wavelength, $n_i(\lambda) = n + ik$ is the wavelength-dependent complex refractive index of the layer i, including the extinction coefficient $k_i(\lambda)$ that accounts for absorption in the films, d_i is the thickness of the layer, θ_i is the angle of the ray relative to the normal incidence, and q_i is a polarization-dependent parameter as described later. After N layers the total matrix is:

$$M_{\text{tot}} = \prod_i^N M_i = \begin{bmatrix} m_{11} & m_{12} \\ m_{21} & m_{22} \end{bmatrix} \tag{3.69}$$

From the total matrix one can extract the reflectivity coefficient:

$$r(\lambda) = \frac{(m_{11} + m_{12} \quad q_S) \quad q_0 - (m_{21} + m_{22} \quad q_S)}{(m_{11} + m_{12} \quad q_S) \quad q_0 + (m_{21} + m_{22} \quad q_S)} \tag{3.70}$$

The parameters q_i, q_S and q_0 from relations (3.68) and (3.70) are defined for layer i, for substrate, and respectively for air, and are dependent on polarization. For oblique angle of incidence ($\theta_0 > 0$), these are substituted as follows:

$$\text{T E: } q_{(i,\,S,\,0)} \rightarrow p_i = n_i \cos \theta_i, \; p_S = n_S \cos \theta_S, \; p_0 = n_0 \cos \theta_0$$

$$\text{T M: } q_{(i,\,S,\,0)} \rightarrow s_i = \frac{n_i}{\cos \theta_i}, \quad s_S = \frac{n_S}{\cos \theta_S}, \quad s_0 = \frac{n_0}{\cos \theta_0} \tag{3.71}$$

p_i, p_S and p_0 are written for TE polarization, s_i, s_S and s_0 for TM polarization. The angles θ_i for each layer are obtained from the Snell–Descartes law of refraction:

$$\theta_i = \arcsin\left(\frac{n_0}{n_i} \sin \theta_0\right) \tag{3.72}$$

The spectral reflectance is obtained from reflectivity coefficient as follows:

$$R(\lambda) = |r(\lambda)|^2 \tag{3.73}$$

The reflectivity coefficient is a complex number and satisfies the relation:

$$r(\omega) = |r(\omega)|e^{i\varphi(\omega)} \tag{3.74}$$

then, the phase change for the Bragg mirror from relation (3.70) can be extracted as:

$$\varphi(\omega) = \arg[r(\omega)] \tag{3.75}$$

According to equation (3.50), the derivatives needed to obtain the group velocity and group delay dispersion are:

$$\frac{d}{d\omega}\varphi(\omega) = \frac{\lambda^2}{2\pi c}\frac{d}{d\lambda}\arg[r(\lambda)]$$

$$and \tag{3.76}$$

$$\frac{d^2}{d\omega^2}\varphi(\omega) = \frac{\lambda^3}{2\pi c^2}\frac{d^2}{d\lambda^2}\arg[r(\lambda)] = \text{GDD}$$

Reflectance $R(\lambda)$ and GDD are numerically computed from the transfer matrix formalism, and are exemplified in figure 3.38. The numerical design of the reflective structures considers the wavelength-dependent refractive index and extinction coefficient for pairs of TiO_2/SiO_2 layers [22], angle of incidence $\theta_i = 45°$, polarization

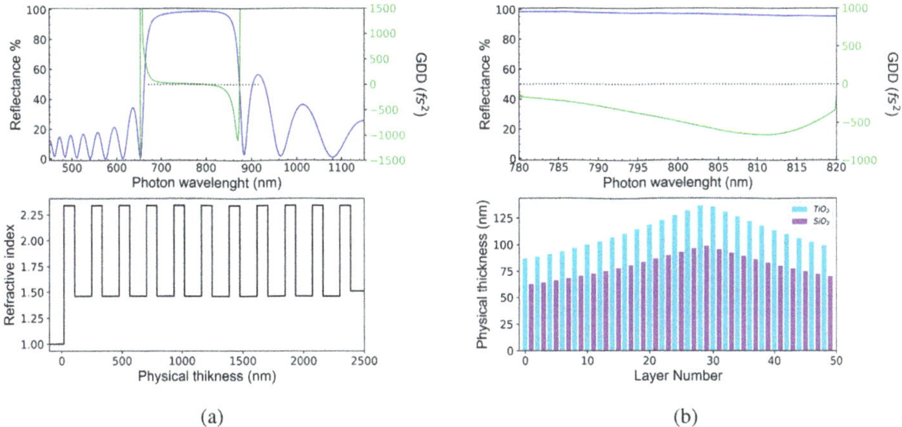

Figure 3.38. Reflectance and GDD calculated by transfer matrix method explained in the text, for: (a) simple Bragg mirror; (b) Bragg mirror with chirp.

p (vector of the electric field $\overrightarrow{E}$ parallel to the incidence plane). Figure 3.38(a) represents a typical reflectance of Bragg mirrors (10 pairs of TiO_2/SiO_2 layers) with a photonic stop-band in the spectral range 750–850 nm. The phase and GDD are almost constant in this region.

For chirped mirrors, the two materials with alternating low and high refractive indices, are designed to have a thickness gradient as shown in the bottom part of figure 3.38(b). Depending on the gradient profile and on the refractive index contrast, the phase and GDD can be engineered. Recent development of the chirped mirrors demonstrated that nonlinear gradient of thicknesses and quasi-symmetric structure help for maximization of the negative dispersion and reflectance at the same time. In the example presented here, the dependence of thicknesses gradient was defined by the following relation:

$$d_i = d_{\min} + \left[(d_{\max} - d_{\min}) \left(1 - \frac{N - i/q}{N} \right)^p \right] \tag{3.77}$$

where N is the total numbers of pairs, i is the layer index ($0 \leqslant i \leqslant N$), p and q are two parameters that can be used to tune the gradient shape from linear to quadratic ($p \in [1, 2]$), and from symmetric ($q = 1$) to simple Bragg structure without layers gradient ($q \rightarrow \infty$). For a simple Bragg mirror, GDD is almost zero in the spectral region of the photonic stop-band. In contrast, with proper tuning of the gradient layers thicknesses, the GDD for the chirped mirror presented above is down to -500 fs², in the spectral range around the central wavelength of 800 nm, typical for a Ti:sapphire laser.

The effect of chirped mirrors on frequency dependent phase can be visualized from the penetration depth of the light wavelengths in the mirrors layers, as represented from numerical calculation of the light field map shown in figure 3.39. Field maps for Bragg mirror and chirped mirror are both presented for better

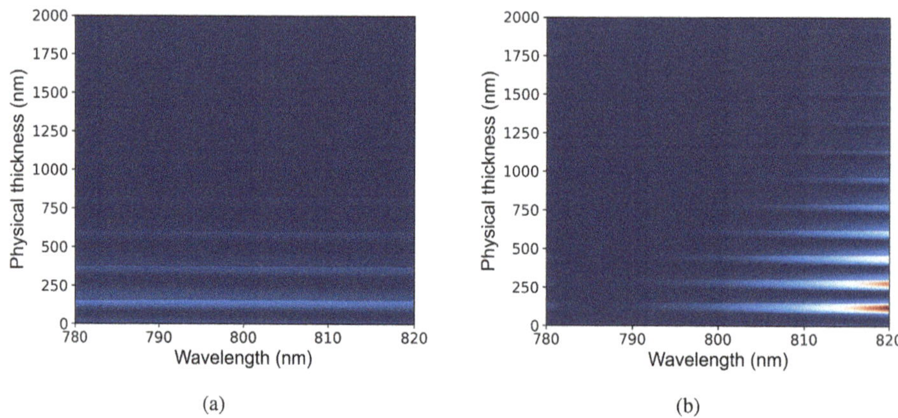

(a)   (b)

Figure 3.39. Field map shows penetration of light's wavelengths inside a multilayer structure: (a) Bragg mirror; (b) chirped mirror.

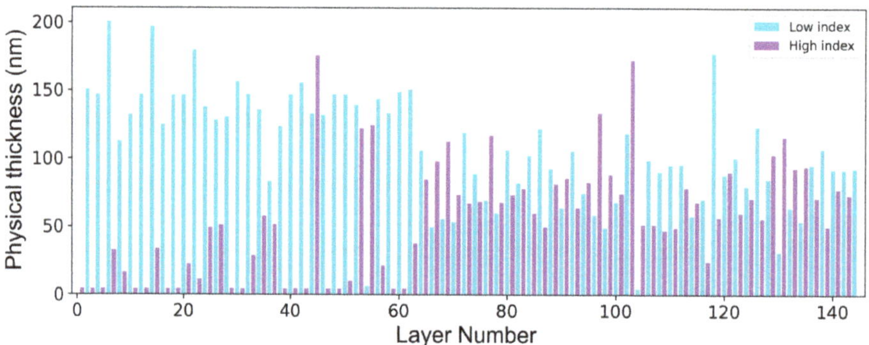

Figure 3.40. Index profile for high GDD values of chirped mirror optimized by genetic algorithm. (Data retrieved from [6].)

understanding this effect. In the spectral region of the stop-band, the reflectance of Bragg mirrors is high, $R > 98\%$. Then, the intensity of light's electric field drops abruptly after the first Bragg layers for all wavelengths. On the other side, for chirped structure, longer wavelengths penetrate deeper the mirror's layers compared with the lower wavelength, inducing a certain group delay between different laser spectral modes reflected by such structure.

Advanced optimizations use *genetic algorithms, needle optimization,* or *inverse design,* to obtain mirrors with broadband negative dispersion and high reflectivity [6, 34]. Figure 3.40 presents the optimization result for index profile obtained by the authors of reference [6]. A GaAs/AlAs structure of chirped mirror with GDD larger than -3200 fs^2 was designed by genetic algorithms method and demonstrated experimentally for a bandwidth of 7 nm centered at 1030 nm. In this case, the variation parameter in the optimization algorithm was the filling factor f as a ratio between GaAs and AlAs thickness.

3.4.4 Prisms compressor

Prisms pulse compressors have been introduced for the first time as dispersion compensation systems in femtosecond laser oscillators by J P Gordon and R L Fork [9, 14]. With the invention of chirped mirrors, the prism compressors are no longer used in modern commercial laser systems because of intrinsic positive dispersion introduced by prisms materials and the corresponding GVD and TOD that impact on the shortest pulse duration achievable in such systems. Nowadays, chirped mirrors are preferred because of compactness of the dispersive device, as already described in the subsection above. However, in some applications, the prisms compressor can still be used as an external dispersion compensating device for ultrashort lasers. Good examples are the multiphoton microscopy or two-photon laser processing of photopolymers with ultrashort laser pulses. The optical path of the microscopes incorporates a large amount of transparent glass, especially inside the objective microscope, that introduces a large amount of GDD and induces temporal pulse broadening (see section 3.4.1). In multiphoton microscopy it is critical to maintain the pulse duration as low as possible, typically below 100 fs in order to increase the quantum yield of fluorescence induced by nonlinear absorption of ultrashort pulse. A similar requirement is imposed in the two-photon photo-polymerization method used in 3D laser lithography, as discussed later in this book. In such applications, the device has to provide tunability of pulse dispersion in order to compensate for various optical path configurations (e.g. different microscope objectives) or various samples or sample thicknesses. Although chirped mirrors are preferred as an intracavity dispersion compensating device, they show only fixed GDD values at a certain photon wavelength. Dispersion of grating compressors can be easily tuned from incidence angles and gratings distance. However, the achievable dispersion is in the range of tens to hundreds of ps^2, two to three orders of magnitude over the typical GDD of glass optical components. In this case, the prisms compressor remains the most adapted solution satisfying the requirement of tunability in the range of hundreds of fs^2 to few ps^2 comparable with dispersion introduced by the total glass thickness present in a usual laser processing or microscopy setup. Figure 3.41 shows the configuration of a typical four-prism compressor design as originally introduced by Fork *et al* [9]. In the four-prisms version, the configuration is symmetrical. It uses identical prisms oriented such that their faces are parallel two by two. Due to the

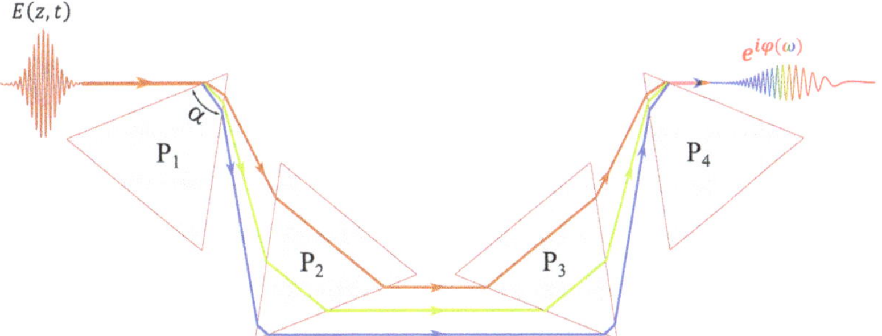

Figure 3.41. Four-prisms compressor.

symmetry of the structure, a more compact two-prisms design folded by a roof mirror (RM) can be used as it is more convenient from the point of view of alignment and compactness (figure 3.42).

The laser beam is incident on the first prism usually at Brewster's angle. Brewster configuration and p polarized incident beam are preferred in order to minimize the optical power losses caused by specular reflections on prisms faces. This is in particular one of the advantages of a prisms compressor over the grating compressors, that work at much lower diffraction efficiency for the designed diffraction order.

Under the condition of minimum deviation specific to prisms, the relationship between prism apex angle α and incident Brewster's angle θ_B is:

$$\alpha = 2\arcsin\left(\frac{\sin\theta_B}{n(\lambda_0)}\right) \tag{3.78}$$

where n is the refractive index of the prism's material and λ_0 is the reference wavelength or the central laser wavelength for which the refractive index is considered in apex angle calculation. The relation (3.78) gives us the geometric conditions regarding the prisms and setup design. Table 3.1 lists the Brewster's angle

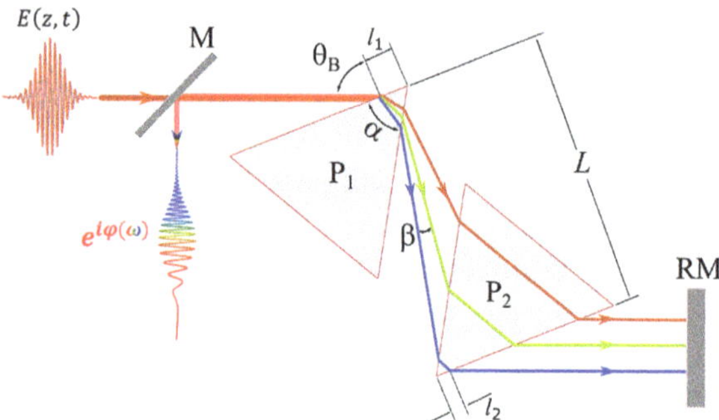

Figure 3.42. Two-prisms compressor.

Table 3.1. Brewster's angle, prism apex angle and intrinsic dispersions GVD and TOD for common glasses [24].

Material	Refractive index at $\lambda_0 = 800$ nm	Brewster's angle $\theta_B(°)$	Apex angle $\alpha(°)$	GVD (fs^2 mm^{-1})	TOD (fs^3 mm^{-1})
Fused silica	1.453	55.5	69.1	36.2	27.5
N-BK7	1.511	56.5	67	44.7	32.1
SF6	1.784	60.7	58.5	199.0	125.2
SF10	1.711	59.7	60.6	159.5	101.3
SF11	1.765	60.5	59.1	189.6	121.8
N-SF14	1.743	60.2	59.6	176.7	115.8

and prism apex angle for some materials currently used in prisms compressors. Depending on the entrance distance relative to the apex of the prism, the beam will travel a certain optical path in the prism material. This will inherently induce positive dispersion caused by the refractive index of prism material. The GVD and TOD for each material is provided in the same table.

The beam enters the compressor as close as possible to the apex of the first prism P_1 in such a way that the prism insertion distance l_1 is minimal. The prism introduces angular dispersion on the incident beam. On the exit face of P_1, the distribution of the angles of dispersion corresponds to the spectral bandwidth $\Delta\lambda$ of the laser beam. We consider the angle β_λ for a certain wavelength λ relative to dispersion angle of the reference wavelength. On the second prism, the short wavelengths travel closer to the apex of P_2, while longer wavelengths have longer optical path inside the prism glass. Then, the group delay of the spectral components has a negative dispersion and can compensate or even overpass the positive dispersion induced by the material itself. The input laser beam has a certain beam diameter D that involves a different propagation path inside the prism at different distances from apex. Then, the ray propagating deeper inside the prism with respect to the apex, will have larger group delay compared to the ray propagating closer to apex. After the first prism, the beam gets also a pulse front tilt (PFT). The role of the second prism P_2 is to compensate the front tilt introduced by P_1, to generate a parallel beams for each spectral component dispersed by the first prism and to introduce geometrically the negative dispersion larger than the intrinsic positive one. The spatial separation of the spectral components encountered at the exit of P_2 is removed by a second pass through a symmetric structure of prisms, or by reflecting back the beam on exactly the same way by a retro-reflector or an RM. The second travel will double the effect of the negative dispersion.

Phase shift $\varphi(\omega)$ has two components, one given by the optical propagation path (P^+) inside the glass inducing a normal positive dispersion, and a second component given by the optical path introduced by the spatial separation of the prisms (P^-), inducing an anomalous negative dispersion. Recalling the relations (3.49), (3.53) and (3.54), the phase is:

$$\varphi(\omega) = \frac{2\pi}{\lambda}(P^+(\omega) + P^-(\omega)) = \frac{2\pi}{\lambda}(P) \tag{3.79}$$

The GDD and TOD are expressed as:

$$\mathrm{GDD} = \frac{\lambda^3}{2\pi c^2} \cdot \frac{\mathrm{d}^2 P}{\mathrm{d}\lambda^2} \tag{3.80}$$

$$\mathrm{TOD} = -\frac{\lambda^4}{4\pi^2 c^3}\left(3\frac{\mathrm{d}^2 P}{\mathrm{d}\lambda^2} + \lambda\frac{\mathrm{d}^3 P}{\mathrm{d}\lambda^3}\right) \tag{3.81}$$

Propagation of the beams in a prisms system can be solved geometrically (figure 3.43) For solving this problem we have to define several geometrical parameters. In one experiment we can chose the setup geometry such as:

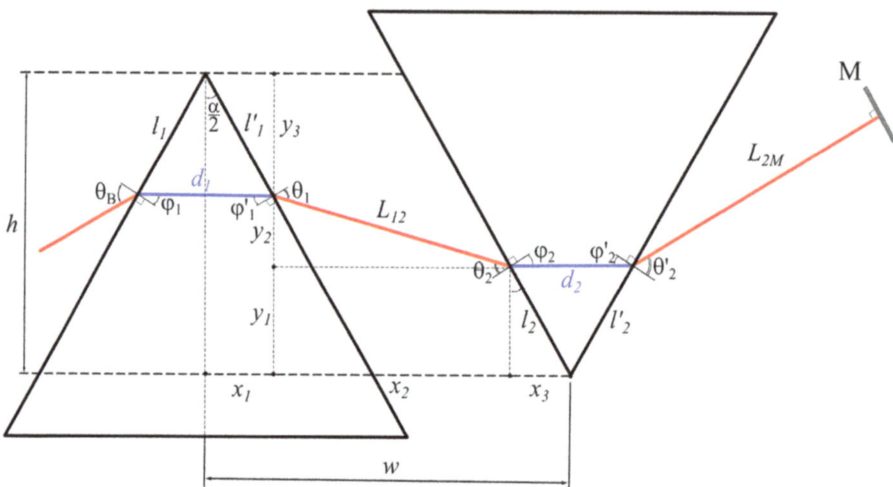

Figure 3.43. Geometry of rays propagation through prisms compressor.

- $\alpha-$ apex angle of the prisms.
- θ_B- Brewster's angle of incidence, where $\theta_B = \arctan n(\lambda_0)$, defined at the reference wavelength λ_0.
- $w-$ distance from the median to the median of each prism.
- $h-$ distance from one prism apex to the other prism apex measured in the median direction.
- l_1- the position insertion to the input surface of the first prism, relative to its apex.

Parameters w, h and l_1 are wavelength independent and define a certain configuration setup. Insertion l_1 has to be minimized and cannot be smaller that the beam diameter. In particular, parameters w and h can be tuned experimentally in order to have a fine control on the total GDD of the device. Other parameters that derive from the previous ones are:
- d_1 and d_2- physical propagation distance inside prisms.
- $L_{12}-$ prisms separation measured as physical distance from the exit surface of the first prism to the input surface of the second prism.
- $L_{2M}-$ physical distance from the exit surface of the second prism to the folding mirror.
- l'_1- the distance of the ray relative to the first prism apex at the exit surface.
- l_2- insertion to the second prism as the distance of the ray relative to the second prism apex at the input surface.
- φ_1, φ'_1, θ_1, θ_2, φ_2, φ'_2 – beam angles at input/output of the surfaces according to the refraction law.

It is obvious that these parameters depend on the wavelength of the beam and, respectively, on the refractive index of the material.

Snell's equations for each input end exit surface are:

$$\text{Prism} \quad P_1: \sin \theta_B = n \sin \varphi_1, \; n \sin \varphi'_1 = \sin \theta_1$$
$$\text{Prism} \quad P_2: \sin \theta_2 = n \sin \varphi_2, \; n \sin \varphi'_2 = \sin \theta'_2 \tag{3.82}$$

From geometric arrangement of the parallel prisms, the following equality are used:

$$\theta_1 = \theta_2$$
$$\alpha = \varphi_1 + \varphi'_1, \qquad \alpha = \varphi_2 + \varphi'_2 \tag{3.83}$$

Law of sines for each prism is expressed as:

$$\text{Prism} \quad P_1: \quad \frac{d_1}{\sin \alpha} = \frac{l'_1}{\sin(90 - \varphi_1)} = \frac{l_1}{\sin(90 - \alpha - \varphi_1)}$$

$$\text{Prism} \quad P_2: \quad \frac{d_2}{\sin \alpha} = \frac{l'_2}{\sin(90 - \varphi_2)} = \frac{l_2}{\sin(90 - \alpha - \varphi_2)} \tag{3.84}$$

We are looking to express the segments l'_1, L_{12} and l_2 as a function of experimental parameters w, h, l_1 and α. Then, we write the following relations:

$$w = x_1 + x_2 + x_3$$
$$h = y_1 + y_2 + y_3 \tag{3.85}$$

From trigonometrical formulas, the relations (3.85) become:

$$w = l'_1 \sin \frac{\alpha}{2} + L_{12} \cos \left(\theta_1 - \frac{\alpha}{2} \right) + l_2 \sin \frac{\alpha}{2}$$
$$h = l'_1 \cos \frac{\alpha}{2} + L_{12} \sin \left(\theta_1 - \frac{\alpha}{2} \right) + l_2 \cos \frac{\alpha}{2} \tag{3.86}$$

From the equations above we obtain the distance L_{12} between the two prism:

$$L_{12} = \frac{w - (l'_1 + l_2)\sin \dfrac{\alpha}{2}}{\cos \left(\theta_1 - \dfrac{\alpha}{2} \right)} = \frac{h - (l'_1 + l_2)\cos \dfrac{\alpha}{2}}{\sin \left(\theta_1 - \dfrac{\alpha}{2} \right)} \tag{3.87}$$

Then, from the second and third factors in the previous equations, the distance l_2 can be extracted as:

$$l_2 = \frac{h \cos \left(\theta_1 - \dfrac{\alpha}{2} \right) - w \sin \left(\theta_1 - \dfrac{\alpha}{2} \right)}{\cos \theta_1} - l'_1 \tag{3.88}$$

From equations (3.82) and (3.83):

$$\theta_1 = \arcsin \left[n \sin \left(\alpha - \arcsin \frac{\sin \theta_B}{n(\lambda)} \right) \right] \tag{3.89}$$

From the law of sines (3.84) and Snell's equations (3.82), the distances l'_1, d_1 and d_2 are expressed as:

$$l'_1 = l_1 \frac{\cos \varphi_1}{\cos(\alpha - \varphi_1)} = l_1 \frac{\cos\left(\arcsin\dfrac{\sin \theta_B}{n(\lambda)}\right)}{\cos\left(\alpha - \arcsin\dfrac{\sin \theta_B}{n(\lambda)}\right)}$$

$$d_1 = l_1 \frac{\sin \alpha}{\cos(\alpha - \varphi_1)} = l_1 \frac{\sin \alpha}{\cos\left(\alpha - \arcsin\dfrac{\sin \theta_B}{n(\lambda)}\right)} \tag{3.90}$$

$$d_2 = l_2 \frac{\sin \alpha}{\cos(\alpha - \varphi_2)} = l_2 \frac{\sin \alpha}{\cos\left(\alpha - \arcsin\dfrac{\sin \theta_2}{n(\lambda)}\right)}$$

The prisms have parallel opposite faces, then angles $\theta_2 = \theta_1$. With refraction law taken into account, $\sin \theta_2$ from 3.90 becomes:

$$\begin{aligned}
\sin \theta_2 &= n \sin \varphi'_1 \\
&= n \sin(\alpha - \varphi_1) \\
&= n \sin\left(\alpha - \arcsin\frac{\sin \theta_B}{n(\lambda)}\right)
\end{aligned} \tag{3.91}$$

Then, d_2 is reduced to:

$$d_2 = l_2 \frac{\sin \alpha}{\cos\left(\arcsin\dfrac{\sin \theta_B}{n(\lambda)}\right)} \tag{3.92}$$

Distance L_{2M} is deduced from the relation:

$$\frac{L_{2M} - c}{a \cos(90° - \theta'_2)} = \frac{a - l'_2}{a} \tag{3.93}$$

where a is a value related to the physical size of the prism and c is an arbitrary constant related to real distance from second prism to the mirror. Since a and c are constants, their values have no relevance on the derivatives of total optical path as resulting from equations (3.80) and (3.81), and have no impact on GDD and TOD. After replacement of parameters θ'_2 and l'_2 using the relations (3.82)–(3.84), L_{2M} becomes:

$$L_{2M} = \left[a - l_2 \frac{\cos \varphi_2}{\cos(\alpha + \varphi_2)}\right] n \sin(\alpha - \varphi_2) + c \tag{3.94}$$

At this point we have all the geometrical parameters describing the propagation of rays inside the prism compressor (d_1, d_2, L_{12} and L_{2M}) written as a function of

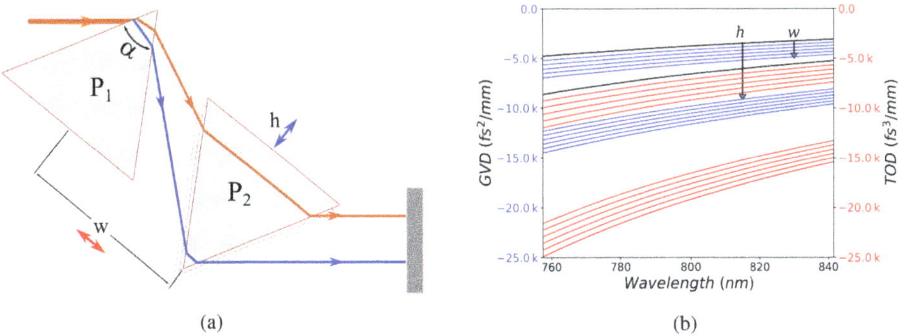

(a)

(b)

Figure 3.44. (a) Tuning of compressor dispersion by moving the prisms. (b) GDD and TOD of two-prisms compressor at different prisms distances.

experimental input parameters (α, w, h, l_1 and θ_B). Total propagation path $P = P^+ + P^-$ is given by:

$$P = 2n(\lambda)(d_1 + d_2) + 2L_{12} + 2L_{2M} \tag{3.95}$$

A factor of 2 is introduced in order to considered the double path propagation for a complete structure (symmetric four-prisms or two-prisms folded compressor).

It is worth noting that the pairs of prisms arrangement offers an easy way to adjust both positive and negative dispersion, by changing the distance w between the prisms and by varying the distance h by introducing the second prism further into the optical path (see figure 3.44(a)). Figure 3.44(b) shows the total GDD and TOD of an SF11 glass prism compressor for different w and h configurations. For two values of $h = 50$ mm and $h = 100$ mm the dispersions are calculated for values of w from 70 to 75 mm, and, respectively, from 150 mm to 155 mm in steps of 1 mm. It is evident that for a prism compressor the negative TOD is much larger than negative GDD. So, in a dispersion compensation system, even if GDD of the prism compressor will compensate for positive dispersion introduced by different optical components, the remaining TOD will be hard to be compensated and will induce a temporal pulse deformation with asymmetric shape. Some experiments are done by combining different dispersion systems such as prism with gratings (*grism*), or using active systems such as *acousto-optic modulators, spatial light modulators* (SLMs) etc.

3.4.5 Acousto-optic programmable dispersive filter

A CPA laser system uses dispersion of pulses, firstly to stretch the ultrashort pulses, emitted from a laser oscillator, from tens of fs to hundreds of ps (*stretcher*), and after amplification in a long pulse regime, the pulses are optically compressed back to the fs regime by using a negative dispersion compensation device (*optical compressor*). Stretching and compressing are currently realized with pairs of gratings systems as described early in this chapter. The laser beams travels during amplification stage a long chain of optical components that introduces GDD, TOD and even superior order of dispersion, especially in the nonlinear optical regime of the high intensity

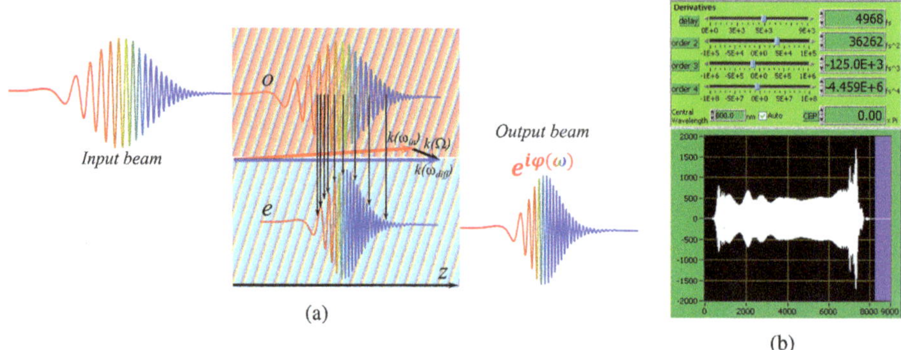

(a)

(b)

Figure 3.45. (a) AOPDF process. The extraordinary beam is generated by diffraction of the ordinary beam on an acoustic wave under the phase-matching condition for each spectral mode. (b) Acoustic wave shown on software interface of DAZZLER from Fastlite.

amplified beam. The role of optical compressor at the end of ultrashort laser amplifier is to compensate the positive dispersion introduced by stretcher (positive chirp) and eventually to fine tune the GDD and TOD in order to compensate the contribution of other optical components in the laser system. Because of high order dispersions that cannot be canceled by the grating compressor, the final pulse duration will not be compressed back to the Fourier transform limited pulse duration. The temporal pulse profile will be larger and asymmetric compared to initial pulse. In modern ultrafast laser systems with pulse duration of tens of fs, the superior order dispersion is controlled by an acousto-optic programmable dispersive filter (AOPDF) that manipulates the spectral phase in a flexible computer controlled way [16] (see Figure 3.45). AOPDE, also known from its commercial version as DAZZLER, is usually placed inside a stretcher–compressor chain just after the stretcher stage, at low pulse energy (nJ to μJ) and low beam size in order to fit the aperture of the commercially available equipment. The active dispersion control device is used firstly for very precise tuning of pulse duration, and in some condition it can be also used for temporal pulse manipulation, to generate a particular pulse shape or train of pulses.

In AOPDF, the chirped pulse travels inside a nonlinear crystal (typically TeO_2) colinear with an acoustic wave generated by an RF signal (around tens of MHz) with variable amplitude and frequency. As in any AOM, the RF signal produces a grating of instantaneous modulation of refractive index in the nonlinear crystal. The interaction of optic and acoustic waves is described by a three-wave mixing process with phase matching and frequency conservation conditions described below:

$$\vec{k}_{\text{diff}}(\omega_{\text{diff}}) = \vec{k}_{\text{in}}(\omega_{\text{in}}) + \vec{k}_{\text{ac}}(\Omega_{\text{ac}})$$
$$\omega_{\text{diff}} = \omega_{\text{in}} + \Omega_{\text{ac}}$$

(3.96)

where ω_{in} and ω_{diff} are the frequencies of the input and diffracted optical beams, and Ω_{ac} is the frequency of the modulated acoustic RF signal. During the interaction process, the input ordinary polarized optical wave is diffracted to an extraordinary polarized wave. This process take place only for a certain spectral component of the laser beam that satisfies the phase matching condition in interaction with one specific acoustic wave of the 'polychromatic' RF signal. The diffraction from ordinary to extraordinary wave takes place in the crystal at a precise position $z(\omega)$ that depends on the pulse chirp and spectral distribution of the acoustic wave. After propagation through the crystal with thickness L (tens of mm), a group delay occurs between ordinary and extraordinary waves due to the different group velocity v_G^o and v_G^e for each wave [21]:

$$\tau(\omega) = \frac{z(\omega)}{v_G^o} + \frac{L - z(\omega)}{v_G^o} \tag{3.97}$$

Due to temporal chirp of the stretched pulse, each spectral component will have a different group delay at the exit of the crystal depending on interaction with frequencies of the 'polychromatic' acoustic wave. This is traduced to a spectral phase modulated by the acoustic RF signal.

Because of limited length of crystals available, amplitude, frequency and resolution of the RF signal that can be generated, the device has inherent limitations in terms of magnitude of positive and negative dispersion that can be introduced ($\mathrm{GDD} = \pm 10^5 \ \mathrm{fs}^2$), pulse shaping capability (few ps) spectral resolution of the phase manipulation (down to 0.1 nm) and working spectral range (from UV to near IR). However, in tandem with stretcher–compressor systems, the dispersion can be well controlled, providing a convenient and flexible tool for manipulation of pulse dispersion, spectral phase and temporal pulse shape.

3.5 Pulse shaping

In many recent femtosecond laser processing experiments, the effect of temporal pulse shape on the quality of the micro- and nanostructuring have been studied and explored. It is well known that when focusing of ultrashort pulses on material surface at intensity exceeding the ablation threshold, different characteristics of the ablated spot can be achieved not only as a function of the pulse duration, but also as a function of asymmetric temporal shape of pulses, trains of double or multiple pulses and pulse dispersion with negative or positive chirp. As an example, asymmetric temporal pulse envelopes produce smaller hole diameters, below the diffraction limit compared to the non-chirped pulse [8]. This is explained by the nonlinear interaction of a photon with the lattice's electrons that needs a certain transient absorption profile in order to confine the energy of photons and the ionization process in a narrow volume.

3.5.1 Temporal pulse shaping

Dispersion of ultrashort pulse impacts directly on pulse temporal shape, as already discussed in section 3.4. Spectral phase with superior dispersion order $\mathrm{d}^2\varphi/\mathrm{d}\omega^2$ (GDD) and $\mathrm{d}^3\varphi/\mathrm{d}\omega^3$ (TOD) broadens the temporal shape and produces asymmetric laser pulses, respectively. These effects suggest that manipulation of spectral phase provides a flexible tool for temporal shape engineering by *Fourier synthesis technique*. Spectral distribution of the electric field of light is linked to its temporal distribution by the Fourier transform:

$$E(\omega) = \mathcal{F}[E_{\text{in}}(t)] = \int_{-\infty}^{\infty} E(t)e^{-i\omega t}dt \tag{3.98}$$

The reversed relation between $E(t)$ and $E(\omega)$ is given by the inverse Fourier transform:

$$E(t) = \mathcal{F}^{-}[E_{\text{in}}(\omega)] = \frac{1}{2\pi}\int_{-\infty}^{\infty} E(\omega)e^{i\omega t}d\omega \tag{3.99}$$

The spectral distribution $E(\omega)$ includes also the therms of phase and intensity that can be experimentally manipulated by a technique known as filtering. Then, an output spectral pulse $E(\omega)$ is written as follows:

$$E_{\text{out}}(\omega) = E_{\text{in}}(\omega)H(\omega) \tag{3.100}$$

where $H(\omega)$ is the filtering function that can act on spectral phase $\varphi(\omega)$ and on spectral intensity $S(\omega)$. E_{in} is the initial spectral distribution. It is important to mention that for correct computation of the Fourier transform, one has to know both **phase** and **intensity** spectral distribution. If the initial pulse has a Gaussian temporal shape, then E_{in} is:

$$E_{\text{in}}(\omega) = \frac{1}{2\pi}\int_{-\infty}^{\infty} E_0 e^{-\frac{t^2}{2\tau_P^2}}e^{-i\omega t}dt \tag{3.101}$$

where τ_P is the Fourier limited pulse duration given by relations (3.41). For a Gaussian spectral shape the shortest pulse duration is:

$$\tau_P^{\text{Gauss}} = 0.411\frac{\lambda_0^2}{c\Delta\lambda^{\text{Gauss}}} \tag{3.102}$$

The spectral filtering therm $H(\omega)$ can be simply the effect on spectral phase and spectral intensity as the pulse propagates through a dispersive medium:

$$H(\omega) = \sqrt{(S(\omega))}\,e^{-i\varphi(\omega)} \tag{3.103}$$

The temporal filtering therm $h(t)$ can be also written as inverse Fourier transform:

$$h(t) = \frac{1}{\sqrt{2\pi}}\mathcal{F}^{-}[H(\omega)] = \frac{1}{2\pi}\int_{-\infty}^{\infty} \sqrt{(I(\omega))}\,e^{-i\varphi(\omega)}e^{i\omega t}d\omega \tag{3.104}$$

A controlling dispersion device can manipulate the spectral phase or spectral intensity, or both simultaneously in order to obtain the desired pulse shape by Fourier synthesis technique. Let's consider a general form of spectral phase as reconstruction of the Taylor series (3.47) using the second and third dispersion terms:

$$\varphi(\omega, \omega_0) = \frac{1}{2}\text{GDD}\cdot(\omega - \omega_0)^2 + \frac{1}{6}\text{TOD}\cdot(\omega - \omega_0)^3 \tag{3.105}$$

The equation above is written in a simplified form since the linear phase has no effect of the temporal shape of the pulse, because its first derivative is a constant number.

The spectral shape of a filtered input beam is expressed as:

$$E_{\text{out}}(\omega) = \mathcal{F}[E_{\text{in}}(t)]I(\omega)e^{-\mathrm{i}\varphi(\omega)} \tag{3.106}$$

By inverse Fourier transform the desired temporal shape is obtained:

$$E_{\text{out}}(t) = E_{\text{in}}(t)\mathcal{F}^{-}[I(\omega)e^{-\mathrm{i}\varphi(\omega)}] \tag{3.107}$$

In a practical situation, the above relations are not solved analytically, but computed numerically by FFT, which is a good approximation of the analytical solutions. All programming languages have implemented dedicated libraries (such as numpy or scipy from Python) for computing FFT (scipy.fft) and inverse FFT (scipy.ifft). Figure 3.46 presents numerical calculations of temporal pulse shape for several dispersion cases. It has to be noticed that in all examples the spectral intensities (red curves) have identical spectral bandwidth at central wavelength of 800 nm.

The first example on the left of figure 3.46 shows the non-dispersed pulse, with zero GDD and TOD, at constant spectral phase (green curve). The pulse (blue curve) is compressed to its Fourier limited pulse width of 10 fs. The next example shows the pulse temporal broadening for GDD $= -500$ fs^2 at zero TOD, as negative linear chirp. Spectral modes at higher frequencies travel faster than lower frequencies (anomalous dispersion). With a positive GDD, not shown in these examples, the temporal pulse envelope has identical broadened shape, but the spectral modes at higher frequency will be slower than lower frequencies, positive linear chirp, as shown in figure 3.27 from section 3.3.5. In both cases, with positive and negative linear chirp, the elongated pulse duration is given by the relation (3.58) and is $\tau_{\text{P}}^{\text{OUT}} = 139$ fs^2.

In the last two cases on the right side of figure 3.46, negative and positive quadratic chirp, with TOD of $\pm 10\,000$ fs^3, is presented. The temporal pulse envelopes show asymmetric shapes with oscillations toward positive and negative delays, depending on TOD sign. Pulse duration of pulses with quadratic chirp is

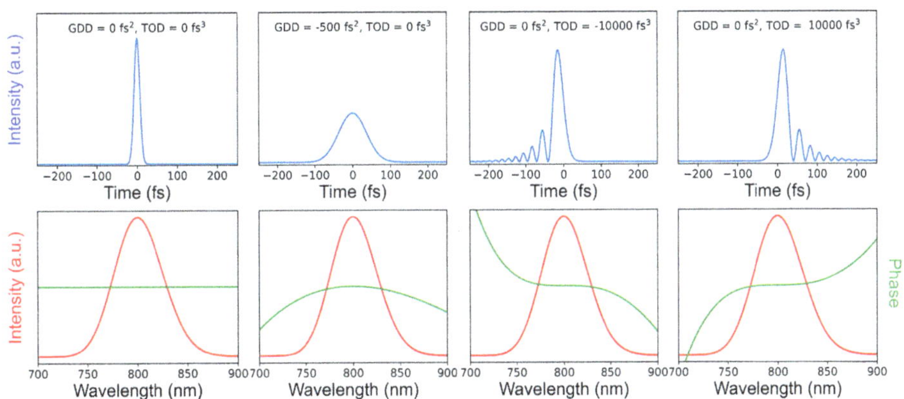

Figure 3.46. Fourier synthesis and temporal pulse shape dependence on GDD and TOD.

hard to define since the concept of FWHM is not applicable on profiles that are far from Gaussian. However, some works propose to define the pulse duration at large TOD, of the order of $\pm 10^6$ fs^3, as follows [36]:

$$\tau_{\mathrm{P}}^{\mathrm{OUT}} \approx 4[\ln(2)]^{3/2}\,\frac{\mathrm{TOD}}{\tau_0^2} \tag{3.108}$$

A large negative TOD leads to series of pre-pulses as shown in the numerical example. These kinds of pre-pulses have been generated and exploited in laser material processing for enhancing the resolution of nanoholes obtained in dielectrics [36]. The pre-pulses generate free carriers on material just before arrival of the main pulse. Tailored pulses induce a transient free-electron density and 'pre-hitting' that optimize the photoionization process. On the other hand, for positive TOD or short pulses, the free-electron density increases abruptly, the mechanism of the laser ablation is changed, and the size of the ablated spots is enlarged. It has been shows that many laser-based interactions such as laser-induced modifications in glasses for fabrication of optical waveguides [27], laser surface nanopatterning [8] etc, are more efficient in the subpicosecond regime than for sub 100 fs pulses duration, and can be further optimized with synthesized temporal shapes. These studies demonstrated that in laser processing, the materials should be not irradiated with the shortest available femtosecond pulses, but it is much more efficient to deliver flux of photons with an optimized temporal shape that fits better the transient absorption and enhances the mechanism of photoionization. Dynamic change of pulse duration and pulse shape can be well controlled from GDD, TOD or from a spectral phase mask modification in a programmable way. Several types of setups can be used, depending on the magnitude of desired dispersion, such as setups that includes an AOPDF or spectral phase modulators in stretcher or compressor with an SLM.

Figure 3.47(a) presents a sketch of a setup for temporal pulse shaping that uses SLM in 4-f configuration with reflective optics. SLM is an array of individuality addressed liquid crystal pixels that introduce a retardance in the range of a few radiands, typically up to 4π when an electric potential is applied on the electrodes. It can be a 2D type SLM with matrix of pixels for a 2D phase mask as for computer generated holograms and for spatial beam shaping, or 1D array SLM for generation of a spectral phase mask. Outside the laser amplification channel, the SLM can be placed at the region of a

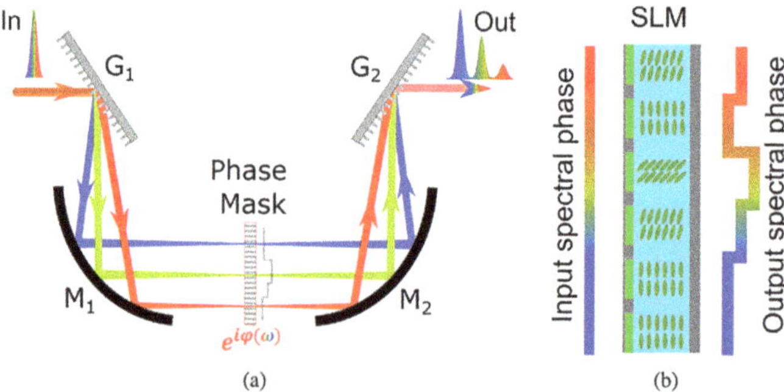

Figure 3.47. (a) Pulse shaping setup with SLM as phase mask. (b) Structure of SLM array with liquids crystal in transmission mode.

collimated spatially dispersed beam of a zero dispersion optical device built with two pairs of gratings. Each spectral mode passes through the phase mask (figure 3.47(b)). Retardance larger than 2π can be obtained by phase warping method. This means that the real phase profile that can be applied to the mask is composed by segments of modulo 2π. By unwrapping the real profile that can be applied to the mask, an approximation of the desired phase shift is obtained. However, the wrapping technique fails for phases with very steep slopes producing results that are far from the expected pulse shape. Other relevant parameters of SLM are: spectral range and maximum laser operation power that limits the laser beam parameters that can be used; number of pixels, pixels size and pitch, related to the spectral resolution and spectral range, bit depth and phase noise that set the phase resolution of the mask. For 2D arrays, the pixel sizes and number of pixels are similar in most cases to those of the video sensors with HD resolutions or higher. Instead, 1D arrays are optimized for spectral applications. In one direction, along a spatially dispersed beam, the array has a larger number of pixels in order to have a good spectral resolution; in the other direction, perpendicular to the dispersion direction, the pixels are long with length of the order of a few mm in order to fit the size of the beam.

Figure 3.48 shows several computer simulated temporal shapes from Fourier synthesis method. The blue curves represents the expected output temporal pulse

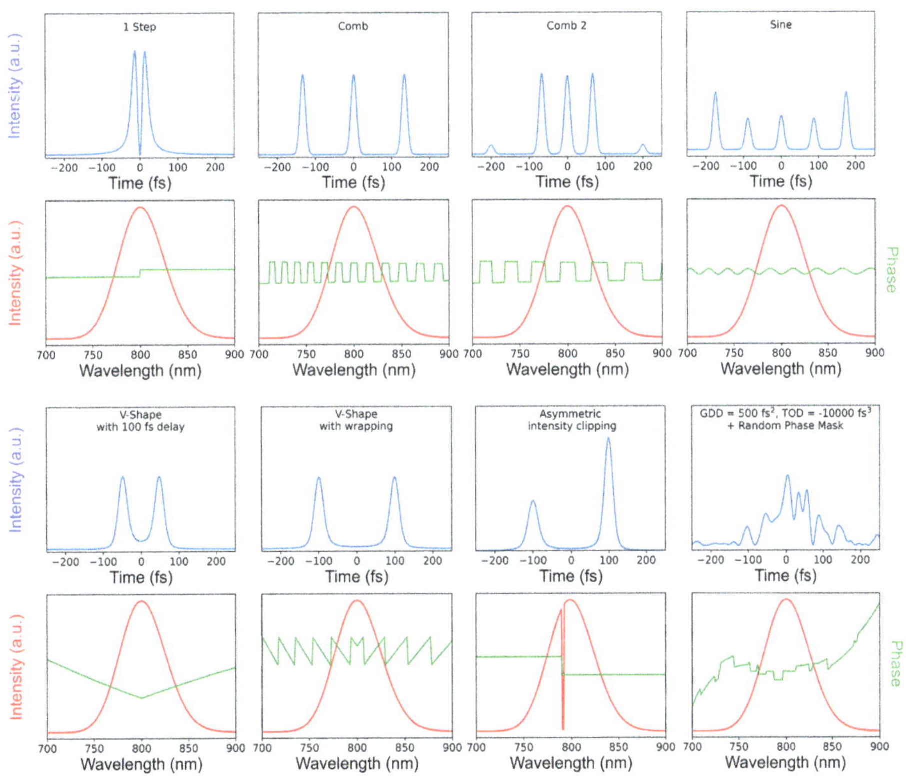

Figure 3.48. Multiple pulses generation by Fourier synthesis.

shape. On the bottom part of the temporal profiles, the red curves represent the spectral intensity, and the green curves represent the phase shift to be addressed on the SLM array in order to obtain the temporal shape. For all simulations an initial Fourier limited pulse with 10 fs pulse duration is used. An SLM is able to generate, up to some limits imposed by spectral and phase resolution, almost any phase mask described mathematically, such as phase combs, V-shapes, sines etc. These functions can be used to generate double and multiple pulses with variable delays and intensity ratios. The V-shape phases are present in both versions, with non wrapped retardance, and wrapped as it is in fact applied to the pixels arrays in the range of available retardance of $\pm 2\pi$ that can be physically applied on the SLM. The slope of the V-shape phase can be used to tune the delay between the two generated pulses. The two V-shape phases presented in the figure, show delays of 100 fs and 200 fs, respectively. The wrapping or non-wrapping has no effect on pulse duration or shape, since wrapping represents only a jump of 2π on the wave's phase. For clarity, the other images do not show the wrapped masks, but the full phase functions. Various type of masks are used to synthesize different pulse profiles, pulse trains with different number of pulses, pulse delays, and even intensity ratios between pulses. In advanced laser processing with optimized pulse shapes, the method of Fourier synthesis with SLM is compatible with optimization algorithms that dynamically modify the pulse shape, and from the experimental feedback adjust the phase mask in order to maximize a fitness parameter, linked to the desired physical parameter. Surprisingly, the optimal temporal profile is not a pulse with specific short or long pulse duration, but a quasi-random shape that in fact delivers the photon energy to the materials lattice in a flux that optimizes the absorption and photoionization mechanisms [26, 38].

An interesting method to obtain double pulses, is based on spectral phase jump generated without using SLM [1, 33]. The double pulses are obtained by clipping the intensity and introducing a phase step between the two spectral slices (figure 3.49). This can be achieved by selecting a part of the spatially dispersed spectrum in a non-dispersive device with gratings, like the one shown before. Each spectral part is reflected by separated delay lines that can tune the pulse delay. The spectral slices are

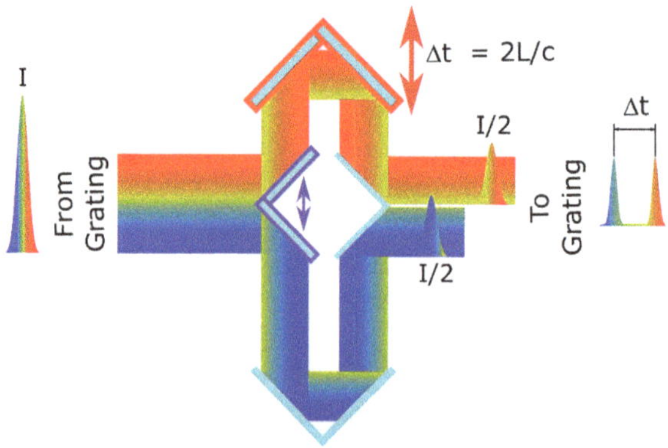

Figure 3.49. Double pulses generation with spectral phase jump.

recombined back on the same optical path by the second grating of the setup. The ratio of the selected spectral region is given by an asymmetric clipping, and can tune the ratio of temporal intensities of the two pulses. The phase step, the clipped spectral intensity and the double pulses temporal profiles are also exemplified in the Fourier synthesis examples (figure 3.48). This method can be safely used outside the amplification chain in a separately built two-grating setup, and with many precautions in the stretcher on the laser, since such an operation can lead to undesired temporal spikes that can permanently damage the optical components in the amplification stage and even the gratings in the compressor.

Another method to generate multiple pulses uses the effect of multiple reflections between a mirror with high reflectivity at the laser wavelength and thin film BS (TFBS), known also as pellicle BS (figure 3.50(a)). The commercially available pelicles have a transmission to reflectance split ratio of the order of 90:10, 50:50, 33:67 etc for a large spectral band (polarization dependent). A part of the beam is directly reflected by the pellicle (P), that could be 10%, 50% or 67% of the main input as given by the BS split ratio. The rest of the beam energy is transmitted to the mirror (M) placed behind the pellicle, at variable distance that can set the delay Δt between pulses. The mirror reflects the beam that is sequentially slpit in a series of secondary beams with lower and lower intensities. Usually the beam diameter is much larger (a few mm) than the distance between the pellicle and mirror (below 1 mm) and the spatial displacement of the secondary beams should not be an issue for the final beam profile. With proper optical mounts and alignment procedure, this device can provide multiple pulses with delays from hundreds of femtosecond to a few picoseconds. Figure 3.50(b) shows the autocorrelation traces recorded on the oscilloscope for a series of delays: 0.91 ps, 2.18 ps and 3.02 ps, obtained by changing the distances between pellicle and mirror. Oscilloscope traces are symmetric due to the well-known limitation of the second-order autocorrelators that cannot distinguish between symmetric and asymmetric temporal shapes. However, from traces we can easily identify the

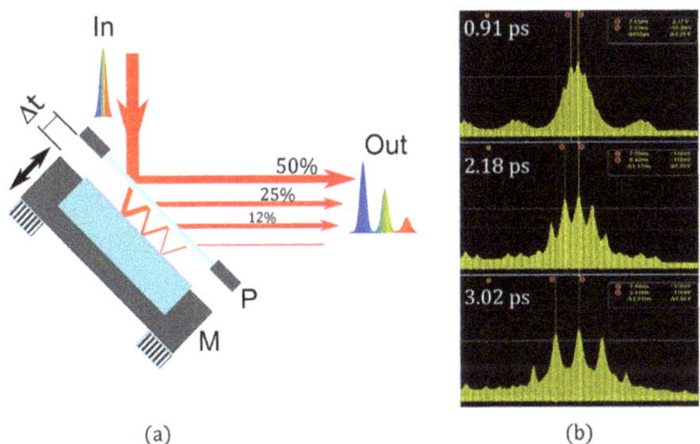

Figure 3.50. (a) Multiple pulses generation with pellicle BS (P) and a Mirror (M) with variable spacing between the two optical components. (b) Autocorrelation function of multiple pulses generated with pellicle BS, recorded on an oscilloscope at various pulse delays.

main pulse and the symmetrically shown secondary pulses, and then the delays can be measured. It has to be mentioned that this method can generate post-pulses, and not pre-pulses, so in certain applications this can be a possible limitation. The methods that use TFBS have been successfully implemented in experiments to generate multiple pulses, such as in plasma x-rays amplification with one long pulse and two short pulses [5], and in laser filamentation for generation of intense terahertz beams [32].

3.5.2 Spatial beam shaping

Laser machining with ultrashort pulses has demonstrated impressive performances of nanostructuring below diffraction limits. Laser processing, although it has the advantages of simplicity of equipment and protocols, relative to standard lithography, has the main drawback of the time consuming process caused by usually sequential steps involved in the programmable controlled workstations. As an example, processing a surface of $10\,\text{cm} \times 10\,\text{cm}$, with a pattern of parallel lines at $5\,\mu\text{m}$ pitch, with a typical laser scanning speed of $1\,\text{mm s}^{-1}$, the total processing time is around 6 h, which is unacceptable for an industrial product. The low throughput can be compensated by increasing the lasers repetition rate and the laser scanning speed. These industrial requirements pushed the technologies toward lasers working at repetition rates of hundreds of kHz up to MHz, and high speed positioning opto-mecanical devices in the range of hundreds of mm s^{-1} at reasonable positioning accuracy of the order of μm. However, working with lasers at MHz repetition rates changes the laser pulse energy regime and the thermal effects on irradiated area that dramatically modifies the laser interaction regime.

The main direction explored in solving the technological problem related to the need of a faster and more productive fabrication method, was parallel processing with multiple laser beams. The main idea is to shape an intense laser beam in order to cover larger surfaces with a spatial intensity distribution as close as possible to the desired pattern. Then, the pattern is imprinted by laser irradiation effects such as photochemical process or ablation, on the material surface as a maskless technology. Another approach is to generate a multiple beam that can simultaneously scan and cover the entire surface in a parallel way. Such multiple beams or spatial beam profiles can be obtained by spatial phase manipulation, that uses the 2D SLM already mentioned in the previous section. Other methods involve inferential generation of periodical pattern from interference of two or three spatially and temporally superposed beams, or near-field optical enhancement on self-organized micro-spheres used as beam focalization micro-optics.

Fourier optics. Tailoring of the spatial phase is based on Fourier optics [13] and involves formalism and algorithms as in computer generated holography, spatial Fourier transform, spatial filtering, optical convolution etc. The optical field of the *image*: $E_{\text{img}}(u, v)$, propagating through an optical system depends on the initial 2D optical field of the *object*: $E_{\text{obj}}(x, y)$ and on *filtering function*: $H(x, y)$ that can alter the phase and intensity. In an optical Fourier system that propagates the field

High Resolution Laser Microprocessing

without filtering ($H = 1$) the initial and final field distribution can be linked by Fourier relations as follows:

$$E_{\text{img}}(u,\,v) = \mathcal{F}[E_{\text{obj}}(x,\,y)] = \iint_{-\infty}^{\infty} E_{\text{obj}}(x,\,y)\mathrm{e}^{-\mathrm{i}2\pi(xu+yv)}\mathrm{d}x\mathrm{d}y \qquad (3.109)$$

The inverse Fourier transform is written as:

$$E_{\text{obj}}(x,\,y) = \mathcal{F}^{-}[E_{\text{img}}(u,\,v)] = \iint_{-\infty}^{\infty} E_{\text{img}}(u,\,v)\mathrm{e}^{\mathrm{i}2\pi(xu+yv)}\mathrm{d}u\mathrm{d}v \qquad (3.110)$$

where x and y are the spatial coordinates in the initial plane of the object field, u and v are named spatial frequency and are defined in the Fourier plane. With a lens, the Fourier image is obtained at $z \to \infty$, or in practical conditions at distances of the order of meters from the object, much larger compared with image details.

A lens is a well-known Fourier transformer. If a lens is used to form the image, the initial optical field distribution $E_{\text{obj}}(x,\,y)$ at $z = -f$ is Fourier transformed by the lens. The Fourier image $E_{\text{img}}(u,\,v)$ is obtained at the image focus plan F', at $z = f$, with lens positioned at $z = 0$.

Figure 3.51 shows an intuitive example of a beam with Gaussian spatial distribution with noisy transversal intensity profile, described at the object focus plan F. After focusing by a lens with focal distance f, the Fourier transform is obtained in the image focus plan F'. As expected, the Fourier transform of a Gaussian function with the width $2w_L$ is another Gaussian function with the width $2w_0$. In a particular case of collimated beam with diameter $2w_L$, the beam waist at focus is $2w_0$, given by the relation for propagation of an ideal Gaussian beam: $w_0 = \dfrac{\lambda f}{w_L \pi}$.

It is interesting to note that the noise from the object plane is transferred in the Fourier plan as high spatial frequencies which appear on the image with a speckle-like appearance. The relation between the spatial coordinates $(x,\,y)$ and spatial frequencies $(u,\,v)$ are given by:

$$\begin{aligned}
\delta u &= \frac{1}{\Delta x}, & \delta v &= \frac{1}{\Delta y} \\[2mm]
\Delta u &= \frac{1}{\delta x}, & \Delta v &= \frac{1}{\delta y}
\end{aligned} \qquad (3.111)$$

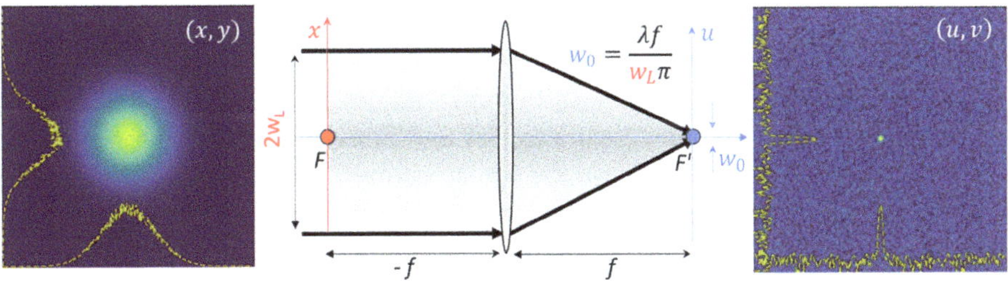

Figure 3.51. Fourier transform of a Gaussian beam profile generated by a lens with focal distance f.

The significance of these expressions is the following: small details δx on the object plane (as from noisy intensity) are transferred to the Fourier plane as large spatial frequencies Δu. The same principle applies in image processing, where the size Δx of the image gives the resolution δu of the 2D-FFT, and the pixel resolution δx of the image gives the size Δu of 2D-FFT image.

Spatial filtering. A very practical example of spatial beam shaping is *spatial filtering*. This is applied in an optical system known as $4f$, with two lenses placed in an *afocal* optical arrangement, or at distance $f_1 + f_2$ from each other. To exemplify the spatial filtering method, we consider an initial quasi-Gaussian beam profile with noisy intensity, and plane wavefront $\varphi(x, y)$, the optical field can be expressed as:

$$E_{\text{in}}(x, y) = g(x, y) \cdot \text{rand}(x, y) \cdot e^{-i\varphi(x, y)} \tag{3.112}$$

where $\text{rand}(x, y)$ denotes the random white noise, $\varphi(x, y)$ is the spatial phase that we consider negligible in this example ($\varphi = $ const.), and $g(x, y)$ is the 2D Gaussian transversal intensity distribution with:

$$g(x, y) = E_0 \cdot \left[e^{-\frac{(x-x_c)^2}{w_L}} + e^{-\frac{(y-y_c)^2}{w_L}} \right] \tag{3.113}$$

(x_c, y_c) is the beam centroid and w_L is the beam radius of a circular beam.

If we place a pinhole at the focus plane, the spatial frequencies Δu and Δv, larger that the pinhole diameter, are blocked by the diaphragm and only the rays with low spatial frequencies can pass the filter to form the new image after the second lens.

The spatial filtering is described in the focus plane as Fourier transform of the object $\mathcal{F}[E_{\text{obj}}(x, y)]$ convoluted with a filtering function that describes a circular aperture:

$$H(u, v) = \begin{cases} 1, & \text{if} \sqrt{u^2 + v^2} \leqslant r \\ 0, & \text{if} \sqrt{u^2 + v^2} > r \end{cases} \tag{3.114}$$

Then, the reconstructed Gaussian beam after the second lens will be:

$$\begin{aligned} E_{\text{filtered}}(x, y) &= \mathcal{F}^{-}[E_{\text{img}}(u, v) \times H(u, v)] \\ &= \iint_{-\infty}^{\infty} E_{\text{img}}(u, v) \times H(u, v) e^{i2\pi(xu+yv)} du dv \end{aligned} \tag{3.115}$$

In practice, the radius r of the aperture is chosen as three times larger than the beam waist ($r \approx 3w_0$) in order to prevent loss of energy by diffraction. The optical alignment of a spatial filtering device has to include optomechanics with precise XY positioning stage and Z stage with micrometer or differential adjuster drives (figure 3.52). The XY stage is used for positioning of the pinhole at the focus plane of the first lens, and Z stage is used for focusing the lens, usually a microscope objective, to the pinhole plane. The beam spatial filtering is implemented in experiments where the quality of the beam is critical, such as in holography, or to improve the beam spot at focus in ultra-precise laser processing. Also, an advanced beam shaping experiment requires a good profile

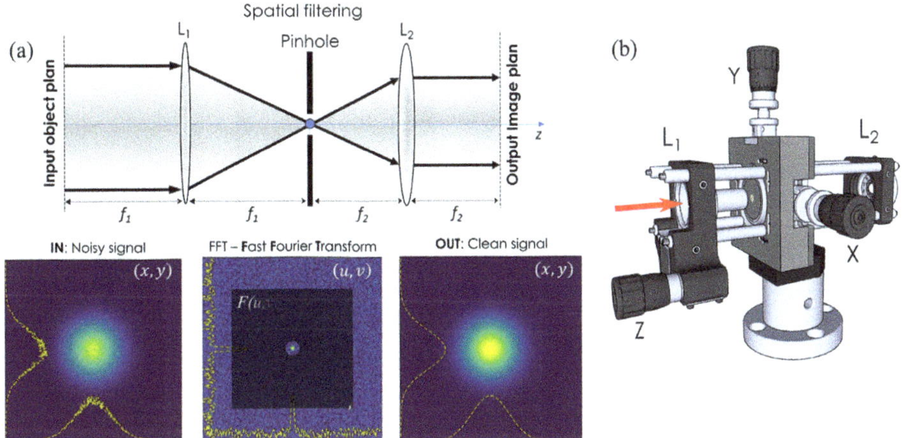

Figure 3.52. Spatial filtering. (a) A pinhole with radius comparable with the beam waist blocks the high spatial frequencies and cleans the beam profile. (b) Sketch of a commercially available opto-mechanical setup with stages for positioning of pinhole (XY) and focusing lens L₁ (Z). Created with SketchUp

of the initial optical field. Then, the spatial filtering of the laser beam can be a good option as the first stage of the experiment. In some spatial beam shaping schemes, the filtering is used also to block the undesired zero order produced by the diffraction patterns from the light modulator.

Spatial beam shaping with SLM. The most common beam manipulation method includes SLM or digital micromirror device (DMD). Although DMS is faster, with refresh rates of the order on tens of kHz, the SLM is preferred because it has a good control on wavefront phase modulation and allows for more complex optical patterns, even in 3D. SLM is used for both temporal or spatial shaping. The temporal shaping with SLM has already been disused in the previous section. In a similar way, the spatial modification is done via a 2D filtering function $H(x, y)$ that can alter the initial phase $\varphi_{in}(x, y)$, intensity $I_{in}(x, y)$, or both.

The most used commercial systems are '*phase only*' devices working in reflection or transmission. The computed phase is applied to the arrays of liquid crystal pixels. The optical system performs the Fourier transform of the initial optical field that includes the modified phase and produces on the Fourier plan the desired intensity profile. Figure 3.53 shows several examples of phases and beams profiles simulated by 2D-FFT reconstruction. The final image can be measure in the focus of a lens or by propagation far from the SLM plane. In a real experiment, an expanded laser beam with Gaussian profile irradiates the phase mask on SLM. Then, the field in the SLM plane is expressed as:

$$E_{in}(x, y) = g(x, y) \cdot H(x, y) \tag{3.116}$$

where $g(x, y)$ is the initial expended Gaussian beam and $H_{SML}(x, y)$ is the filtering function induced by SLM. Modulation can be in phase or amplitude, depending on the type of device, SLM or DMD. The wavefront modulation factor is express as:

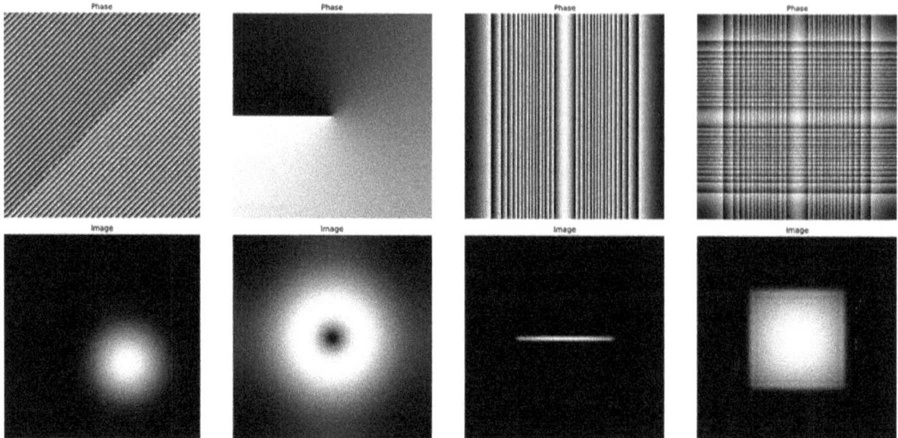

Figure 3.53. Image reconstructions (bottom) from phase (top) using 2D-FFT for: tilted beam, beam with optical angular momentum, line shape, rectangle shape.

$$H_{\mathrm{SML}}(x, y) = A_{\mathrm{SML}}(x, y) \cdot e^{-i\varphi_{\mathrm{SML}}(x, y)} \tag{3.117}$$

For a *phase only* SLM, the amplitude modulation is $A_{\mathrm{SML}} = 1$, then only the phase is accounted for in the optical field reconstruction at the image plan:

$$E_{\mathrm{image}}(x, y) = 2D - FFT[g(x, y) \cdot e^{-i\varphi_{\mathrm{SML}}(x, y)}] \tag{3.118}$$

In programming scripts such as in Python language, FFT functions are implemented in libraries (for example `scipy.fft`). The image reconstruction is made with `fft2()` function, then shifted by `fftshift()` to place the zero spatial frequency at the center of the image.

The simplest setup of beam shaping can be composed by an SLM and one lens that generate the Fourier image in the focus plane. However, a more advanced laser processing setup involves also a beam expander, a *high-pass* spatial filter to block the zero order diffraction, folding mirrors, relay imaging, focusing optics etc. The setup has to take into account all the propagation paths up to the sample in order to achieve the desired resolution and patterns on the sample. Figure 3.54 presents an example of a spatial beam shaping setup in 4f configuration. The SLM is placed at the conjugate plane of the focus plane of L_1. The first Fourier image is formed at the focus plane of the same lens. A blocker is used to spatially filter and remove the zero order diffraction. The second lens L_2 produces the inverse Fourier transform of the optical field. The ratio of the focal lenses f_2/f_1 is chosen in order reduce and to fit the beam diameter to the entrance aperture size of the focusing optics, that can be a microscope objective or an f-theta lens. The last lens is the focusing optics that produces the image pattern on the sample.

In experiments, one can require a certain wavefront profile on the sample. The SML offers a flexible means to generate multiple beams for parallel processing, flat-top rectangular beams to cover a large processed area, non-diffractive Bessel beams

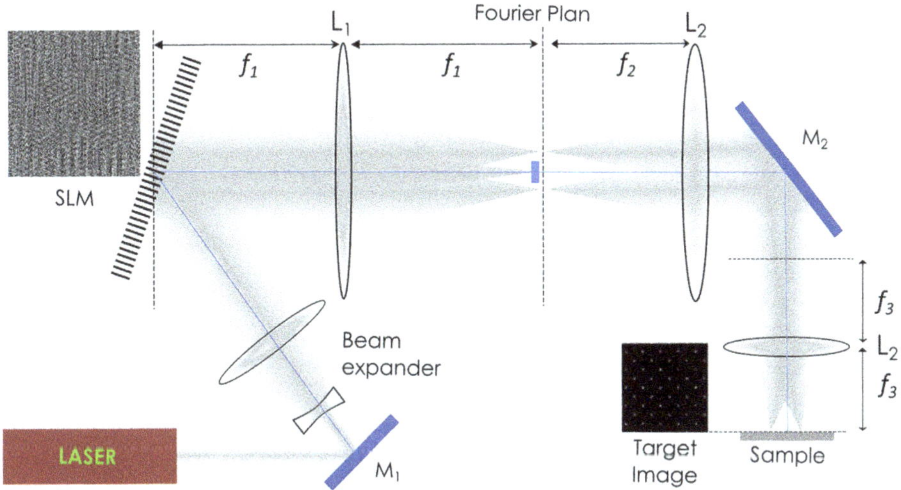

Figure 3.54. Experimental setup for beam shaping with SLM in 4f configuration.

to drill micro-holes with high aspect-ratio, or beams with angular orbital momentum. The mask to be applied to SLM can be computed from the desired target profile. Mathematically, the inverse Fourier transform $\mathcal{F}^{-1}$ cannot retrieve the initial phase from the final intensity profile. However, there are some approximations based on the Gerchberg–Saxton algorithm that iteratively retrieve the phase. As an example, the inset images from figure 3.54 show the target image on the sample plane. This method has been successfully implemented not only in laser processing [18, 37], but also in microscopy and optical tweezers [12].

References

[1] Banici R A, Cojocaru G V, Ungureanu R G, Dabu R, Ursescu D and Stiel H 2012 Pump energy reduction for a high gain Ag X-ray laser using one long and two short pump pulses *Opt. Lett.* **37** 5130

[2] Chichkov B N, Momma C, Nolte S, Alvensleben F V and Tunnermann A 1996 Femtosecond, picosecond and nanosecond laser ablation of solids *Appl. Phys.* **63** 109–15

[3] Born M, Wolf E, Bhatia A B, Clemmow P C, Gabor D, Stokes A R, Taylor A M, Wayman P A and Wilcock W L 1999 *Principles of Optics: Electromagnetic Theory of Propagation, Interference and Diffraction of Light* 7th edn (Cambridge: Cambridge University Press)

[4] Boyd R W 2008 *Nonlinear Optics* 3rd edn (Cambridge, MA: Academic Press)

[5] Cojocaru G V *et al* 2016 One long and two short pumping pulses control for plasma x-ray amplifier optimization *Opt. Express* **24** 14260

[6] Dems M, Wnuk P, Wasylczyk P, Zinkiewicz L, Wójcik-Jedlińska A, Regiński K, Hejduk K and Jasik A 2016 Optimization of broadband semiconductor chirped mirrors with genetic algorithm *Appl. Phys.* B **122** 266

[7] Eimerl D, Davis L, Velsko S, Graham E K and Zalkin A 1987 Optical, mechanical, and thermal properties of barium borate *J. Appl. Phys.* **62** 1968–83

[8] Englert L, Wollenhaupt M, Haag L, Sarpe-Tudoran C, Rethfeld B and Baumert T 2008 Material processing of dielectrics with temporally asymmetric shaped femtosecond laser pulses on the nanometer scale *Appl. Phys.* A **92** 749–53

[9] Fork R L, Martinez O E and Gordon J P 1984 Negative dispersion using pairs of prisms *Opt. Lett.* **9** 150

[10] Frantz L M and Nodvik J S 1963 Theory of pulse propagation in a laser amplifier *J. Appl. Phys.* **34** 2346–9

[11] Ghosh G 1997 Sellmeier coefficients and dispersion of thermo-optic coefficients for some optical glasses *Appl. Opt.* **36** 1540

[12] Gieseler J *et al* 2021 Optical tweezers–from calibration to applications: a tutorial *Adv. Optics Photon.* **13** 74–241

[13] Goodman J W 2005 *Introduction to Fourier Optics* 3rd edn (Greenwood Village, CO: Roberts & Company)

[14] Gordon J P and Fork R L 1984 Optical resonator with negative dispersion *Opt. Lett.* **9** 153–5

[15] ISO 11146-1:2021(en) 2021 Lasers and Laser-Related Equipment - Test Methods for Laser Beam Widths, Divergence Angles and Beam Propagation Ratios - Part 1: Stigmatic and Simple Astigmatic Beams

[16] Kaplan D and Tournois P 2002 Theory and performance of the acousto optic programmable dispersive filter used for femtosecond laser pulse shaping *J. Physique IV* **12** 69–75

[17] Koechner W 2006 *Solid-State Laser Engineering* **Vol 1** (Berlin: Springer)

[18] Lutkenhaus J, George D, Moazzezi M, Philipose U and Lin Y 2013 Digitally tunable holographic lithography using a spatial light modulator as a programmable phase mask *Opt. Express* **21** 26227

[19] Malitson I H 1965 Interspecimen comparison of the refractive index of fused silica *J. Opt. Soc. Am.* **55** 1205

[20] Nikogosyan D N 2005 *Nonlinear Optical Crystals: A Complete Survey* 1st edn (Berlin: Springer)

[21] Oksenhendler T and Forget N 2010 Pulse-shaping techniques theory and experimental implementations for femtosecond pulses *Advances in Solid State Lasers Development and Applications* ed M Grishin (London: InTech)

[22] Polyanskiy M N 2024 Refractiveindex.info database of optical constants *Sci. Data* **11** 94

[23] Sean Ross T 2013 *Laser Beam Quality Metrics* (Bellingham, WA: SPIE Optical Engineering Press)

[24] SCHOTT 2025 *Optical Glass* (SCHOTT Advanced Optics)

[25] Smith D and Baumeister P 1979 Refractive index of some oxide and fluoride coating materials *Appl. Opt.* **18** 111

[26] Srisungsitthisunti P, Zamfirescu M, Neagu L P, Faure N and Stoian R 2014 Real-time adaptive optimization of laser induced nano ripples by laser pulse shaping *Proc. SPIE 8967, Laser Applications in Microelectronic and Optoelectronic Manufacturing (LAMOM) XIX* 896704

[27] Stoian R 2020 Volume photoinscription of glasses: three-dimensional micro- and nano-structuring with ultrashort laser pulses *Appl. Phys.* A **126** 438

[28] Strickland D and Mourou G 1985 Compression of amplified chirped optical pulses *Opt. Commun.* **55** 447–9

[29] Stuart B C, Feit M D, Rubenchik A M, Shore B W and Perry M D 1995 Laser-induced damage in dielectrics with nanosecond to subpicosecond pulses *Phys. Rev. Lett.* **74** 2248–51

[30] Szipöcs R, Spielmann C, Krausz F and Ferencz K 1994 Chirped multilayer coatings for broadband dispersion control in femtosecond lasers *Opt. Lett.* **19** 201

[31] Treacy E 1969 Optical pulse compression with diffraction gratings IEEE *J. Quantum Electron.* **5** 454–8

[32] Ungureanu R G, Grigore O V, Dinca M P, Cojocaru G V, Ursescu D and Dascalu T 2015 Multiple THz pulse generation with variable energy ratio and delay *Laser Phys. Lett.* **12** 045301

[33] Ungureanu R G, Cojocaru G V, Banici R A and Ursescu D 2014 Phase measurement in long chirped pulses with spectral phase jumps *Opt. Express* **22** 15918

[34] Wang J, Li J, Huang J, Du W, Zhao M, He J and Zhu T 2024 Linear dispersion (GDD) design using grating group *Appl. Phys. Lett.* **124** 261101

[35] Weber M J 2003 *Handbook of Optical Materials (Laser and Optical Science and Technology Series)* (Boca Raton, FL: CRC Press) 90

[36] Wollenhaupt M 2009 Control of ionization processes in high band gap materials *J. Laser Micro/Nanoeng.* **4** 144–51

[37] Ye L, Perrie W, Allegre O J, Jin Y, Kuang Z, Scully P J, Fearon E, Eckford D, Edwardson S P and Dearden G 2013 NUV femtosecond laser inscription of volume Bragg gratings in poly (methyl)methacrylate with linear and circular polarizations *Laser Phys.* **23** 126004

[38] Zhang Z, Kosareva O, Zhang N, Lin L and Liu W 2020 Genetic algorithm for the location control of femtosecond laser filament *Sci. Rep.* **10** 12878

IOP Publishing

High Resolution Laser Microprocessing
Implementation and techniques
Bogdan Ştefăniţă Călin, Marian Zamfirescu and Niculae Puşcaş

Chapter 4

Optomechanical systems for laser microprocessing

All laser processing techniques involve the use of optomechanical components, systems and devices, for beam manipulation. This chapter contains information regarding the main characteristics of such systems, as well as tutorials on how to build your own devices. The nature of this chapter is of a 'how-to' guide, rather than theoretical. As such, we first establish a hypothetical situation where you have a laser system on an optical table and need to build everything in order to realize laser processing experiments.

The chapter starts at the output of the laser system and will follow the laser beam all the way to the surface of the sample. Although we do not go into detail regarding the physics involved (such as reflection, refraction, polarization, etc), we do provide reminders, schematics and references for further study. Therefore, we will discuss beam alignment procedures, beam expanders, spatial and spectral filters, delay lines for multi-pulse laser processing, laser spot size and numerical aperture, imaging, sample positioning and some complementary notions. The chapter will end with centralized information on how to build your own laser processing microscope.

A simplified schematic of a laser processing setup, including alignment, is presented in figure 4.1. We will start with the foremost step one should follow when preparing to perform an experiment involving lasers: beam alignment. For this scope, we provided step-by-step guides for two alignment procedures for the **primary laser** (i.e. the laser beam used for processing). In some situations, however, an experiment can greatly benefit from using a **secondary laser** (often called an alignment laser) with a beam that overlaps the optical path of the primary laser. The secondary laser is usually a low power He–Ne laser. After the general beam transport and alignment, we will move to the laser processing microscope. For a laser processing microscope, there are three parts that must be considered: beam focusing, imaging setup and sample illumination. Apart from the alignment of the three, laser power and optical components must be appropriately configured.

doi:10.1088/978-0-7503-3239-2ch4　　　　4-1　　　　© IOP Publishing Ltd 2025. All rights,

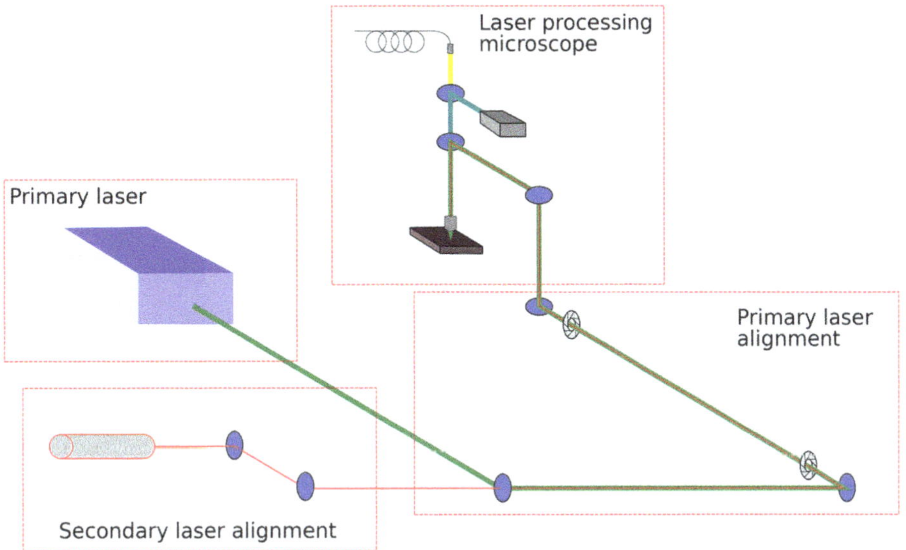

Figure 4.1. Schematic representation of a laser processing setup that includes beam transport, beam alignment and a laser processing microscope. Some parts, such as beam expanders, are omitted in this schematic, but will be covered in further analysis.

4.1 Focusing optics and image formation—short recap

Lasers and optics, in general, can be a formidable tool for analysis and materials processing. But in order to use light towards our desired purpose, we need to be able to control at least some of its properties, the principal one being its propagation. In the context of laser processing, light focusing and image formation take on an important role, but it is under a slightly different approach. Even so, fundamental notions and concepts are the same. In this section we will briefly discuss lenses, optical focusing and image formation before introducing several options regarding the optomechanical systems that can be used in laser processing.

In broad terms, lenses are optical components generally made of dielectric materials, transparent for a specific wavelength interval, that use refraction to change the propagation of an incident beam. Lenses can be convex or concave. In the case of a beam of light, incident on a convex lens and parallel to its optical axis (see figure 4.2), all rays of light are refracted towards a single point found at a specific distance from the lens, in the direction of propagation. On the other hand, a concave lens refracts the rays of light of the incident beam away from the optical axis, but in a controlled manner, i.e. if we extend backwards the refracted rays of light, they all converge to a single point (focal point). There are other types of lenses as well that modify the propagation of incident light in other ways, in order to achieve specific objectives (aspheric, F-theta, telecentric, axicon, etc), some of which will be touched upon in the following sections.

The way lenses form images is important for laser processing, as it is helpful for precise focusing on the surface of a sample before processing, as well as real-time

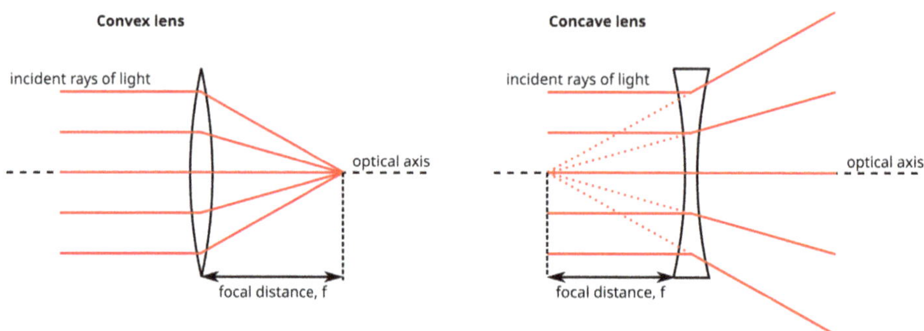

Figure 4.2. Focusing properties of convex and concave lenses.

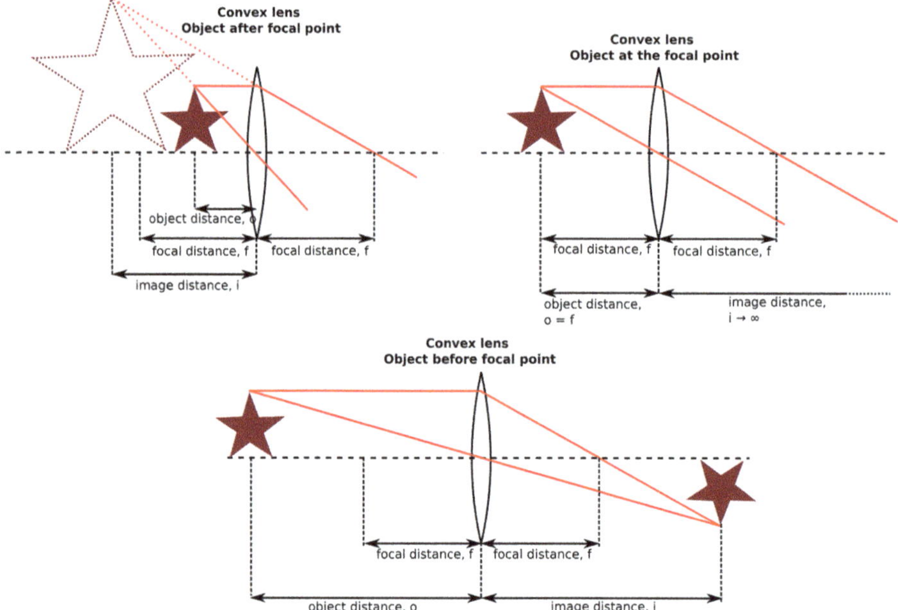

Figure 4.3. Image formation of convex lenses.

monitoring. Image formation is correlated with the position of the object relative to the focal point of the lens, and therefore the lens itself. Image formation via a convex lens is shown in figure 4.3. The lens-maker formula states:

$$\frac{1}{o} + \frac{1}{i} = \frac{1}{f} \tag{4.1}$$

where o represents the distance between the object and the lens, i is the distance between the lens and the formed image, and f is the focal distance of the lens itself.

We can see the three cases of image formation, shown in figure 4.3. In the first case, the object distance, o, is greater than the focal distance, f. The image is formed after the lens, and is called a 'real image', because physical rays of light intersect at the point of image formation. In the second case, however, the object distance, o, is equal to the focal distance, f, and the refracted rays become parallel, therefore we can say that the image is forming at infinity (or, more practically, it does not form at all). Note how this relates to the lens-maker formula as well: if $o = f$, then i must be ∞ so that $\frac{1}{i} = 0$. In the last case, the object is very close to the lens, so that the object distance, o, is smaller than the focal distance, f. In this case, light rays refracted by the lens will not converge towards a single point. On the contrary, the rays spread out. However, an image is still formed, if we extend the refracted rays backwards. In this case, we say that we obtain a 'virtual image'.

4.2 Beam alignment procedures

The first and foremost activity when building a laser processing setup is beam alignment. This is important to obtain a good quality laser spot at the point of interaction with the sample.

There are several configurations that can be used to do the alignment, depending on your layout, available space and requirements (see figure 4.7). Even so, there are a couple of fundamental properties of each configuration. Firstly, you need at least two mirrors with *kinematic mounts* (i.e. fine-adjustment screws) to point a laser beam wherever you need it (not accounting for the mirrors that are part of the processing system). Secondly, you need two reference points, one **near the mirrors**, and the other **near the target** or processing system. Thirdly, the alignment procedure is iterative, i.e. you will repeat some steps until the alignment is satisfactory or successive position changes are indistinguishable.

Although the schematic shown in figure 4.4 is a top-down view, the same principles apply for alignment in each and every direction. The reference points can be mechanical irises, a ruler, a card, or anything that can show differences in spot position. The most efficient way to align a laser is to use two irises, but it can be done with one item as well, just as precisely, as we will show in this chapter.

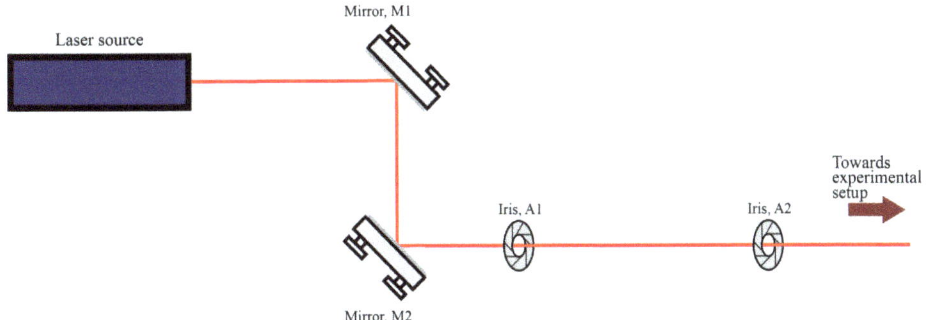

Figure 4.4. Z-type beam alignment schematic.

4.2.1 Alignment procedure with two mirrors and two irises

The alignment procedure is as follows:

1. Place mirrors and irises on the optical table, at their approximate position. You can use screw threads on the optical table for guidance (see Figure 4.4).
2. Position mirrors by hand so that the laser spot is close to the opening of the first iris.
3. Close irises until only a small opening remains, **slightly smaller than the laser spot**.
4. Fine-tune the **M1 mirror** (the one closer to the laser source) until the laser spot hits the center of the **A1 iris** (the one closer to the laser source, see Figure 4.5).
5. Fine-tune the **M2 mirror** (the one farther from the laser source) until the laser spot is centered on the **A2 iris** (the one farther from the laser source, see Figure 4.6)

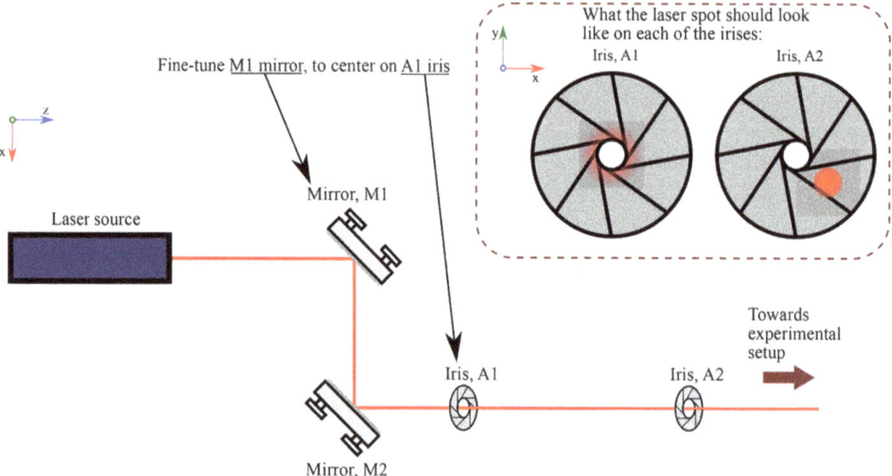

Figure 4.5. First fine-tuned alignment step.

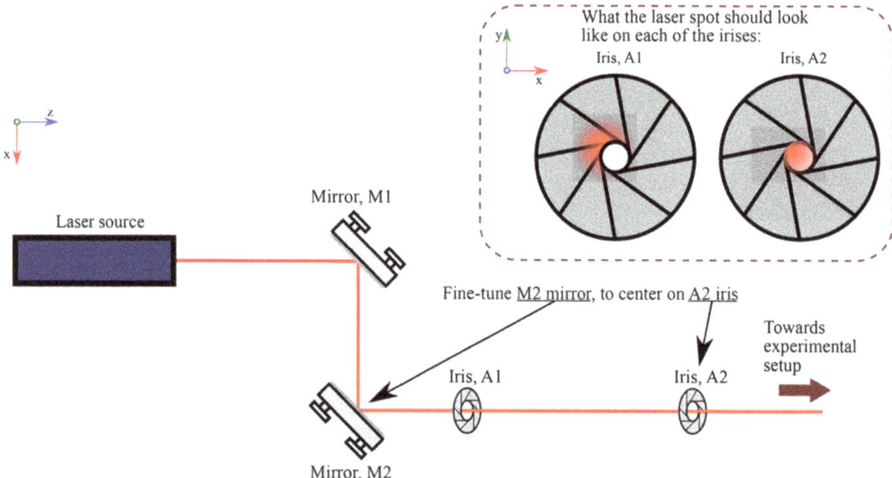

Figure 4.6. Second fine-tuned alignment step.

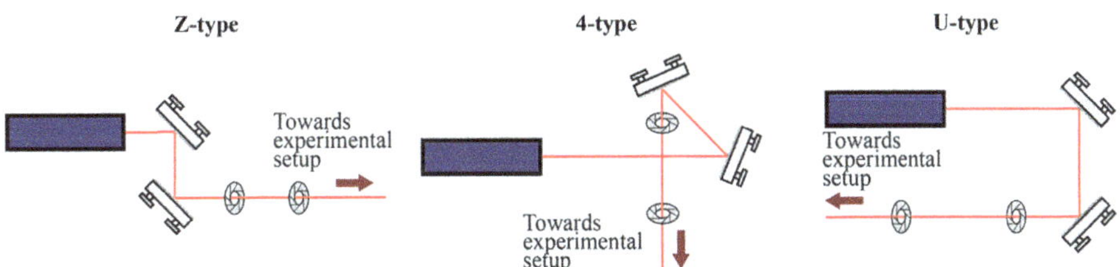

Figure 4.7. Different types of laser beam packing and alignment configurations.

6. You will notice that centering the laser on the farthest iris (A2) means the laser beam will not be centered on the first iris (A1) anymore. Therefore, you **repeat steps 4 and 5 until the laser beam hits the center of both irises**.

The steps shown above can be used for various two-mirror configurations (see figure 4.7).

4.2.2 Alignment procedure with two mirrors and one iris

A similar procedure can be approached when there's only one iris available (see Figure 4.8). To be more precise, any target can be used in this case, as long as it can be repositioned on the optical table along a straight line, and also has features that can be used as reference. For example, you can affix a cutout of graph paper to an optomechanical post and use it to align a low power beam. The alignment steps are almost identical with the previously discussed method, with the main difference that you need to move the single iris (or target) between two positions: one near the

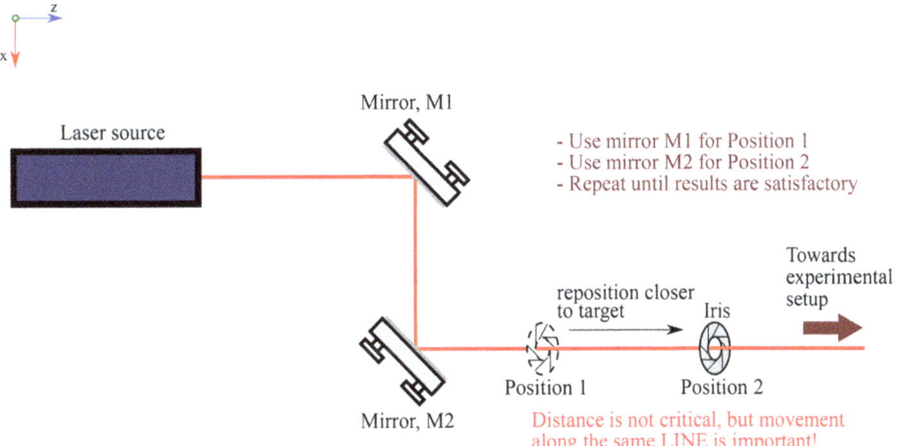

Figure 4.8. Alignment using a single iris or target.

mirrors, and one near the target. The longer the distance between these two positions, the more precise the alignment is going to be.

1. Place mirrors and iris on the optical table, at their approximate position. You can use screw threads on the optical table for guidance.
2. Position mirrors by hand so that the laser spot is close to the opening of the iris at the first position (**close to mirrors**).
3. Close the iris until only a small opening remains, **slightly smaller than the laser spot** (not necessary if using a different device other than an iris).
4. Fine-tune the **M1 mirror** (the one closer to the laser source) until the laser spot hits the center of the iris.
5. Move the iris closer to the target (away from the mirrors, in the direction of intended alignment).
6. Fine-tune the **M2 mirror** (the one farther from the laser source) until the laser spot is centered on the iris, when it is found at the new position.
7. Similarly, **repeat steps 4–6 until the laser beam hits the center of the iris at both positions**.
8. **Observation:** it is imperative for the iris (or target) to be moved back and forth along the exact same line or direction. The actual distance between the two positions for multiple steps is not critical, but its repositioning along the intended alignment direction is important.

4.2.3 Alignment of an invisible laser beam

What happens when the laser beam is not visible? We can try using a device or material that can detect and/or measure the laser spot, but that is a rather inefficient approach which makes the alignment procedure difficult to perform. In this case, the most commonly encountered method is using another low power alignment laser (usually a He–Ne laser). The basic idea is to overlap the main laser beam with an alignment beam, and then use the latter to perform the alignment.

Important: Make sure you take all necessary precautions when working with invisible laser beams. It is important to use appropriate protection when working with any laser sources, and it is especially important when working with invisible laser beams.

Beam overlap (main laser and alignment laser) is schematically presented in figure 4.9. In order to perform this alignment procedure we require an additional optical element: a dichroic mirror, or beam-splitter (sometimes named 'dichroic beam-splitters'). It is important for the dichroic mirror to have a polished back as well. Most mirrors, be they metallic or dielectric, have a translucent back as a safety measure: it is meant to diffuse the incident laser radiation, in case the reflective film is ablated. The dichroic mirror must have appropriate parameters, i.e. to be transparent for the main laser beam and reflective for the alignment beam.

The overlapping procedure of the two laser beams (main laser and the alignment laser) is as follows:

1. Place the **dichroic mirror D1** in front of the main laser beam, with the reflective surface *away* from the source. A 45° angle relative to the intended

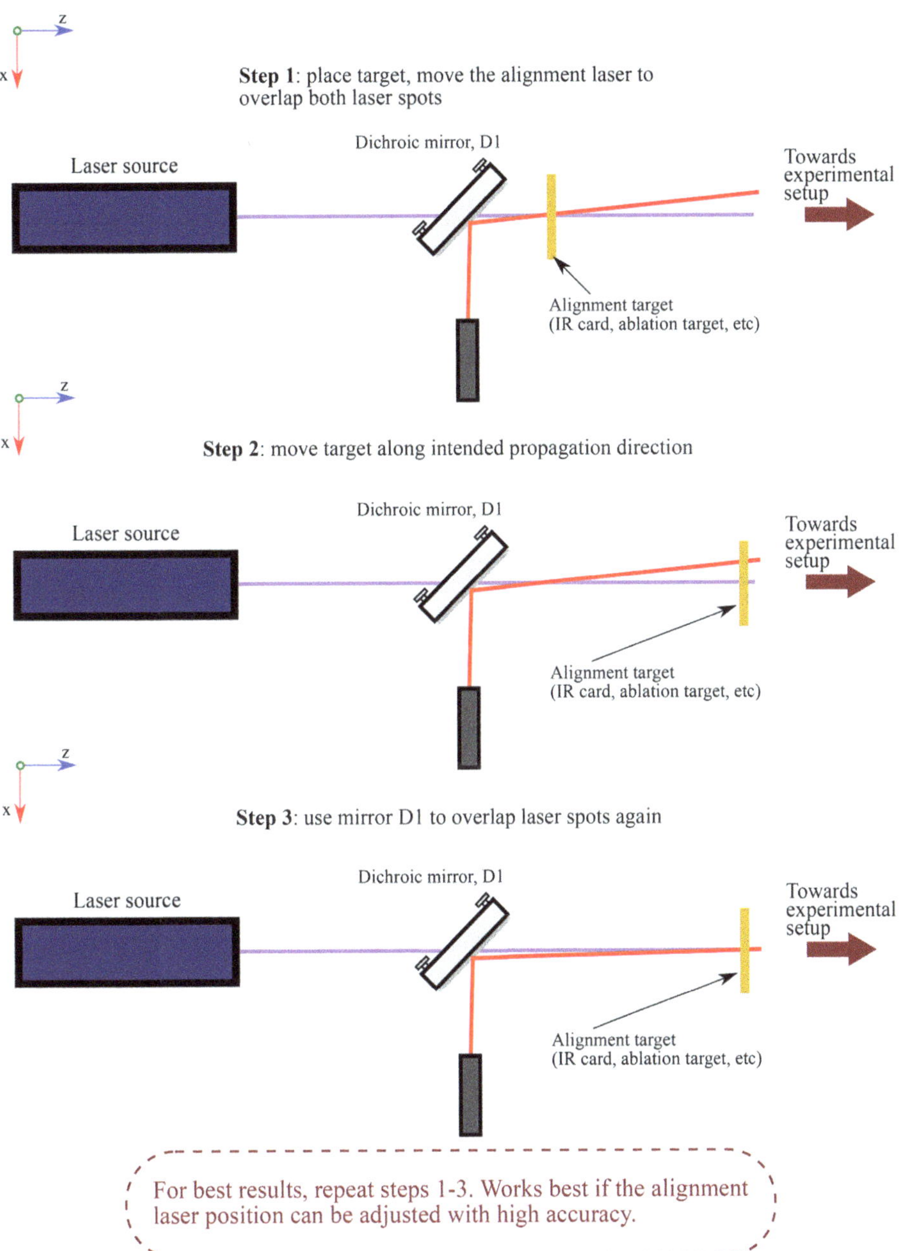

Figure 4.9. Schematic of the alignment procedure for an invisible laser beam.

propagation direction is usually the best choice, so that the shape of both laser beams is maintained (i.e. the transverse intensity profile).

2. Position the **alignment laser** so that the beam is reflected off of the dichroic mirror in the intended direction of propagation (see figure 4.9).

3. Position the **alignment target** close to the dichroic mirror, and adjust the position of the **alignment laser** so that it overlaps the main laser beam. A fluorescent type card or an ablation target can be chosen. The position of the alignment target is not necessary to have high accuracy, because, in this case, we are interested in the relative position of the two laser spots.
4. Reposition the **alignment target** away from the dichroic mirror, in the direction of propagation of the main laser beam. The two laser spots will move away from each other, as the distance increases. The longer the distance, the higher the accuracy of the overlap.
5. Adjust **the dichroic mirror D1** so that the spot overlaps the main laser beam on the alignment target, at the new position.
6. Repeat steps 3–5 to increase the accuracy.

The procedure above shows characteristics similar to alignment procedures previously described. In short: when the target is close, adjust the alignment laser, and when the target is far, adjust the dichroic mirror. After the two laser beams are overlapped, the alignment laser beam can be used for alignment, using any procedure described above.

4.3 Numerical aperture

In optics, the description of the numerical aperture varies a bit, depending on the application, but it is generally used to describe the angular acceptance for incident light, i.e. the angular interval that the incident light can have and interact with the optical component according to its specifications (see Figure 4.10). This parameter is a dimensionless number, usually defined as:

$$\mathrm{NA} = n \sin(\theta) \tag{4.2}$$

where NA is the numerical aperture, n is the refractive index of the medium in which the lens or microscope objective is placed, and θ is the maximum angle with respect to the optical axis, that incident light can have and pass through the optics (be it lens or objective).

In some cases, increasing the numerical aperture offers higher resolving power of the microscope objective, i.e. better resolution and smaller spot size. This is useful for both imaging and laser processing. We cannot modify the construction of the lens or objective itself, in order to increase the innate angular acceptance. What we can do, however, is to change the refractive index of the medium in which the focusing device is found, i.e. n from equation (4.2). Most often, a drop of immersion oil is placed on the sample, and the microscope objective is immersed in it, therefore increasing the refractive index of the medium in which light is focused (see figure 4.11).

4.4 Laser micro- and nanoprocessing workstation

The general schematic of a *laser micro- and nanoprocessing workstation* does not differ much from an industrial laser processing system used for laser cutting, drilling or welding. However, the type of laser used, the performances required in terms of

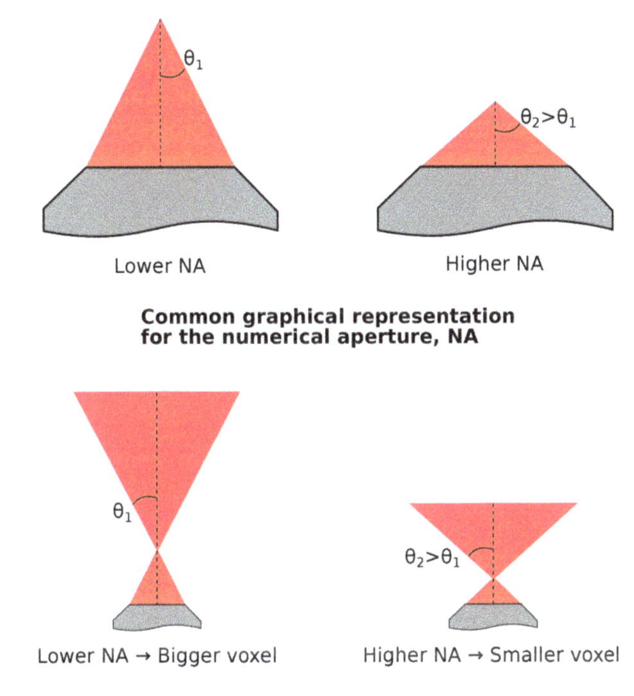

Figure 4.10. Schematic representation of the numerical aperture, in the context of imaging and laser processing.

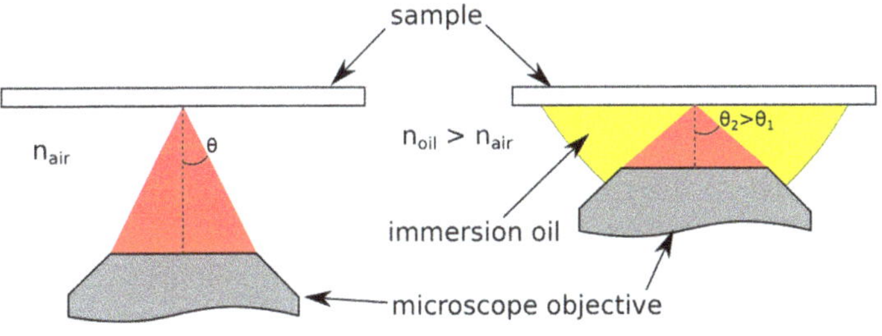

Figure 4.11. Schematic representation of increasing NA using immersion oil.

processing accuracy, and even the physical and/or photochemical effects involved in laser–material interactions, impose several particularities to take into account when a laser processing workstation is designed and built. For example, the laser cutting and drilling are processes almost always assisted by high-pressure gas, oxygen, nitrogen or argon, while in micro- and nanoprocessing the flux of high-pressure gases cannot be used in proximity of the sample because the huge amplitude of generated vibrations that could induce poor quality of micro-structures.

A laser micro- and nanoprocessing workstation has almost the structure of an optical microscope, and most of the time uses optics, such as microscope objectives or short focal lenses, for tight focusing of laser beam on the sample surfaces or inside the transparent materials. For these reasons, this laser processing device is commonly known as *microscope for laser processing*. Indeed, the device includes an optical microscope for real-time visualization of the processed area. Apart fro the microscope, other components are included in laser processing workstations, such as the optics for laser guidance and focusing, mechanical positioning systems for sample positioning or beam steering etc. The building blocks of a laser processing workstation are:

- Laser source and beam delivery optics.
- Beam manipulation components.
- Beam monitoring.
- Optical systems for beam focusing.
- Translation stages and/or galvano scanners for controlled displacement of laser beam relative to sample.
- Visualization system.
- Command and control systems, synchronization electronics and software.
- Vibration attenuation, temperature control and safety.

Proper design of a laser micro- and nanoprocessing workstation has to consider several aspects that we will discus in the following subsections.

4.4.1 Laser source and beam delivery optics

Laser source and alignment of beam delivery components have already been discussed in the previous sections. These parts are mechanically decoupled from the main laser processing microscope. The laser source and laser workstation are linked to each other by the laser beam itself and its optical parameter: laser wavelength, laser power and pulse energy, pulse duration and beam diameters, that will impose the type of optical coating and size of mirrors, lenses and their mounts (see Figure 4.12).

Beam delivery can be done by free-space optics or by optical fibers. A free-space beam delivery system is in general imposed by the case of ultrashort laser pulses, below ps. Spectral dispersion induced by the optical components in the beam path has to be controlled in order to avoid temporal beam distortions at the sample. Mirrors with low group delay dispersion (GDD) are specially designed for femtosecond laser pulses and preferred in experiments with such laser systems. In the path of a femtosecond laser beam it is recommended to use reflective optics as much as possible to the detriment of transmission optics, in order to avoid nonlinear optical effects that could be induced by laser beam in transparent glasses.

High pulse energy can damage the optical coatings, even for longer pulse duration. Laser-induced damage threshold (LIDT) of optical coatings is an important parameter that characterizes the coatings of laser mirrors. A metallic mirror has the LIDT of the order of 0.1 J cm^{-2}, while a dielectric mirror can be optimized to work up to 2–4 J cm^{-2}. From this parameter it is obvious that dielectric mirrors are net superior to the metallic mirrors, also because their GDD can be engineered and optimized.

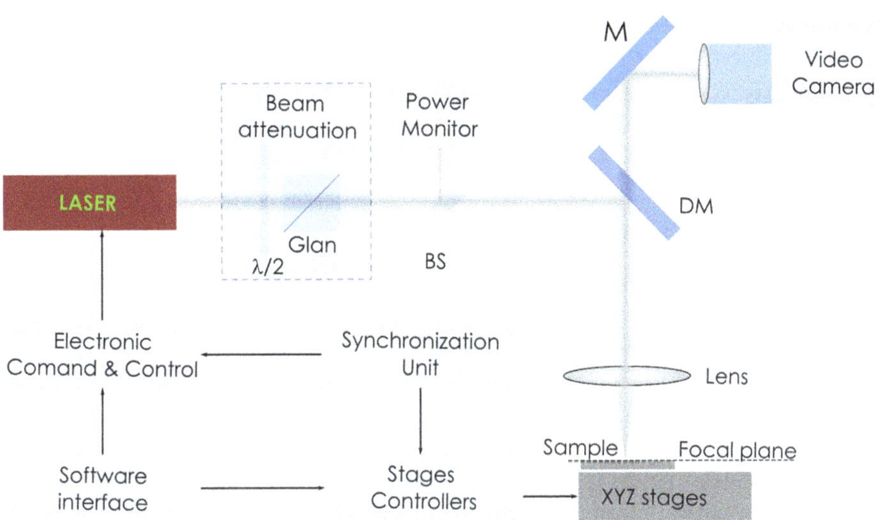

Figure 4.12. Schematic of the laser processing workstation.

If necessary, the laser intensity can be reduced by increasing the beam diameter using a telescope in front of the laser source. It has to keep in mind that a beam should properly fit the size of the clear apertures of optical components respecting the rule of 1/3 as ratio between the diameter of a Gaussian beam and diameter of active area of the optical components, in order to avoid any beam clipping.

4.4.2 Beam manipulation components

The processing microscope could include in the propagation path various optical components for beam attenuation, spatial beam shaping, or even for temporal and spectral beam manipulation. Some beam shaping devices are quite complex and need dedicated space close to the main body of the processing workstation. Beam manipulation devices can include a telescope for beam size adjustment, 4-f lenses arrangements for spatial filtering, axicon lens for Bessel beam generation, SLM for dynamic control of spatial, spectral and temporal shaping of the pulse. An example of beam shaping arrangement is presented in figure 3.54 form section 3.5 *Pulse shaping*.

4.4.3 Beam monitoring

Some laser parameters should be measured in real time, at least the laser power or pulse energy for reproducible results. Pulse duration, spectral distribution, beam size, M2 quality factor are not expected to vary much during a processing session. However, in complex systems with temporal pulse shaping included in ultrashort laser pulses experiments, spectral distributions and, if possible, temporal profile should be monitored in order to prevent any drift from the initial values. On time basis, there are two types of laser parameters fluctuation, long time drift, that is mainly due to variation of temperature of the environment around the laser, and the short variations that are closely linked to the amplification processes in the gain medium of the laser: thermal lens of the laser crystal, stability curve of the cavity, pump energy fluctuation. The long thermal drifts can be minimized by stabilization of the temperature in the laser room. The short term variations are in general related to a bed alignment of the laser cavity, and cannot be decreased below a value that depends on laser design.

4.4.4 Optical systems for beam focusing

Microprocessing involves tight focusing of the laser beam, or in some special cases, such as for spatial beam shaping, the imaging of an optical mask in the sample plane. The simplest beam focusing system is a lens with short focal length. However, better processing performances impose corrections for spherical aberrations. Complex optics, such as aspherical lenses, best-form lenses, f-theta lenses or microscope objectives should be used, depending on expected processing resolution. In the case of spatial beam shaping, the focusing optical system will be correlated with the beam manipulation components, such as SLM and 4-f optical filtering system. The smallest feature that can be obtained by laser interaction is related to the how tightly a laser beam is focused on a sample and on the nonlinear optical effects that induce reversible modifications on materials. In laser processing, as in microscopy, the smallest spot diameter depends on focal distance of the focusing optics, f, laser wavelength, λ, and beam diameter, D_L, at the entrance pupil of the lens:

$$d_0 = \frac{4f\lambda M^2}{\pi D_L} \tag{4.3}$$

where M^2 is the quality factor of the beam, with $M^2 = 1$ for a ideal Gaussian beam and $M^2 > 1$ for a real beam. The above equation is the same as for calculation of the beam waist $w0$ for a quasi Gaussian beam. A microscope objective has not directly defined the focal length, but the numerical aperture NA. In this case, the smallest spot size is given by the diameters of the Airy disk:

$$d_0 = 1.22\frac{\lambda}{\text{NA}} \tag{4.4}$$

This equation is valid only for a laser beam covering the entire entrance aperture. Otherwise, the numerical aperture NA has to be adjusted according to the effective size of the input beam.

However, the size of the ablated spot is not the size of the focused spot. It depends on the level of the laser beam intensity and on the photophysical or photochemical modification threshold of the material. More details are presented in subsection 6.1.1 *Laser ablation threshold.*

It has to be mentioned that the construction of a laser focusing microscope objective for laser processing differs from the design of the usual imaging objectives used in optical microscopy. The first requirement for this component is to be compatible with high intensity laser, especially if the microscope objective has to be used for laser ablation. Optical cements are avoided inside objectives with multiple lenses, as found in any aberration corrections optical system. Antireflective coatings are specially designed for high laser damage threshold (LIDT). The design of a laser processing objective has to avoid any internal focusing and *hot-spots* that could induce damage on coatings and optical glasses. The coatings are optimized for one or several processing lasers wavelengths, at 1064nm, 800 nm, 532 nm, or 355 nm. Even objectives for UV spectral range at 266 nm can be designed using fused silica glass and proper coatings.

Optical characteristics considered for objective selection criteria are briefly descriptive below.

- *Spectral transmission.* Typically, the maximum transmission of objectives is around 80%–90% at the working laser wavelength. It depends on the number of lenses used inside the optical system and antireflexive coatings.
- *Numerical aperture (NA)* was already discussed in the previous section. It gives the resolving power of the objective. NA >1 can be obtained only for immersion oil objectives, but can be used only for low laser intensity in applications such as two-photon photopolymerization (TPP).
- *Magnification.* The value of magnification (10 × , 20 × , 100 × , ...) is defined only in conjunction with the tube lens used for image formation. Typically, the compatible focal length for tube lenses is 200 mm, 170 mm, or 160 mm, depending on the standard used by microscopes providers.
- *Optical corrections.* Spherical or chromatic corrections give the type of objectives (APO, Plan APO, EF, Fluor, etc). In laser processing, if only one laser wavelength is used, the chromatic correction is not critical. However, if multiple laser wavelengths are used, an APO corrected objective is more suitable. PLAN refers to flat-field optical correction (field curvature) and is used for obtaining a sharp image for the entire field of view.

Mechanical characteristics cannot be neglected when a laser processing microscope is constructed. The mechanical parameters are: diameter of the entrance aperture; thread diameter; parfocal distance; length and diameter of the objective body; the working distance (see Figure 4.13).

The thread of the optical mounts or of the microscope objective have to be considered in order to have proper compatibility of the mechanical parts. For microscope objectives several standards exist on the market. The most common thread type is the **RMS**, that stands for *Royal Microscopy Society*. It is used for most of the objectives for optical microscopy. However, each provider of optomechanical components imposes their own standard. The typical threads are listed in table 4.1.

It is very common in optical setups to use the SM1 thread (1.035–40). The nominal diameter of this optical mount is very close to the M26 thread of some optical microscope objectives. However, these two type of components cannot be directly interconnected. Even if their connectivity is apparently possible, doing this could result in a permanent damage of the objective thread. An adapter can be used between optical mounts with different threads. However, using a unique standard for all the optical setups gives more stability and is ergonomic for the assembly.

The working distance, w_d, of an objective microscope has to be large enough, in order to accommodate the variety of sample thicknesses, especially when processing

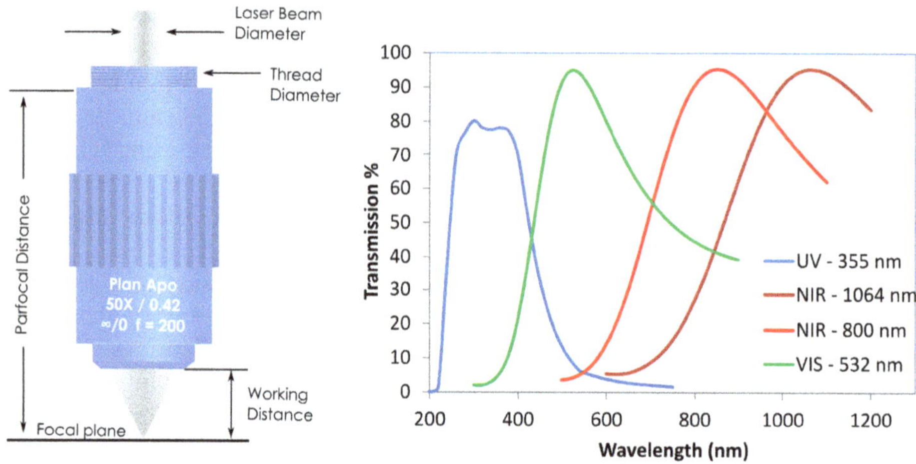

Figure 4.13. Relevant mechanical and optical parameters of a microscope objective.

Table 4.1. Mounting threads standards for microscope objectives (according to ISO 9345:2019).

Type	Nominal diameter (mm)	Pitch (mm)
RMS	20.32	0.706
M25	25	0.75
M26	26	0.706
M27	27	0.75
M32	32	0.75
SM1	26.29	0.75

inside the volume of transparent materials. In the case of ablation, the working distance should be of the order of more than 10 mm in order to avoid deposition on optics of any debris from ablation plasma. Infinity-corrected long working distance objectives are commercially available with simultaneously high numerical aperture, magnification and working distance.

4.4.5 Translation stages and galvanometric scanners

At this level, the main performance characteristics of the processing workstation are determined. Two configurations are possible. In the first configuration, the laser beam is fixed and the sample is moved with very high precision, depending on the performance of the motorized translation stages. However, the processing speed and productivity are limited by the characteristics of the stages. In the second config-uration, galvanometric scanners are used: the sample remains fixed while the beam is moved at very high speed. This configuration provides high throughput, while sacrificing processing resolution or the size of the processed area. Mechanical

Table 4.2. Micro-positioning systems and their characteristics.

Type	Speed	Precision of positioning	Range of movement
Stepper motors	Low (few mm s^{-1})	Average (few μm)	10^2 mm
DC motors	High (tens of mm s^{-1})	Low (tens μm)	10^2 mm
Air bearing stages	High (tens of mm s^{-1})	Average (few μm)	10^2 mm
Direct Drive stages	High (tens of mm s^{-1})	Average (few μm)	10^2 mm
Piezo stages	Low (few mm s^{-1})	High (tens of nm)	10^{-1} mm
Galvano scanners	Very High (few m s^{-1})	Low (tens μm)	10^2 mm

positioning systems also include a Z-axis for placing the sample in the focal plane. In some systems, high-speed focus control is achieved by using appropriate optical configurations in the beam manipulation block.

The highest resolutions are obtained with a fixed focused beam and a mobile sample. In this configuration, the laser can be focused by a microscope objective close to the diffraction limit. However, the overall performance in microstructuring is determined by the micro-positioning system. There are a wide variety of translation stage configurations with different characteristics in terms of positioning precision, travel range, and translation speed. Table 4.2 summarizes the main types of micro-positioning devices and their performance.

For laser power adjustment or for certain designs involving circular traces, rotational stages are used. These stages are, in fact, similar to linear translation stages, with mechanical components that convert linear motion into rotation.

4.4.6 Visualization system

In any processing workstation, proper positioning and alignment of the sample are essential. Sample visualization is achieved using an optical microscope integrated into the laser workstation.

The main components of the microscope include: a camera with sensor size and pixel dimensions optimized for the required optical resolution; an imaging lens, referred to in microscopy as the tube lens, typically with a focal length of 200 mm according to standard microscope design, operating in conjunction with the focusing optics, often in a confocal configuration; a visible-range illumination device adapted to the camera specifications; and optical filters at the laser wavelength, which protect the camera sensor from overexposure or permanent damage.

For practical reasons, a right-angle folding mirror can be included in front of the visualization camera to facilitate the alignment of the image field of view with the illumination and the laser spot. Figure 4.14(b) illustrates the confocal configuration commonly used in laser processing systems. The main advantage of this arrangement is that the distance l_{12} between the focusing optics L_1 and the imaging lens L_2 is not fixed. Within this interval, the beam is collimated, allowing the insertion of additional optical components, such as filters or beam-splitters, without degrading the image resolution. However, due to the limited pupil diameters of the two lenses, vignetting effect occurs. Vignetting represents the clipping of rays with spatial

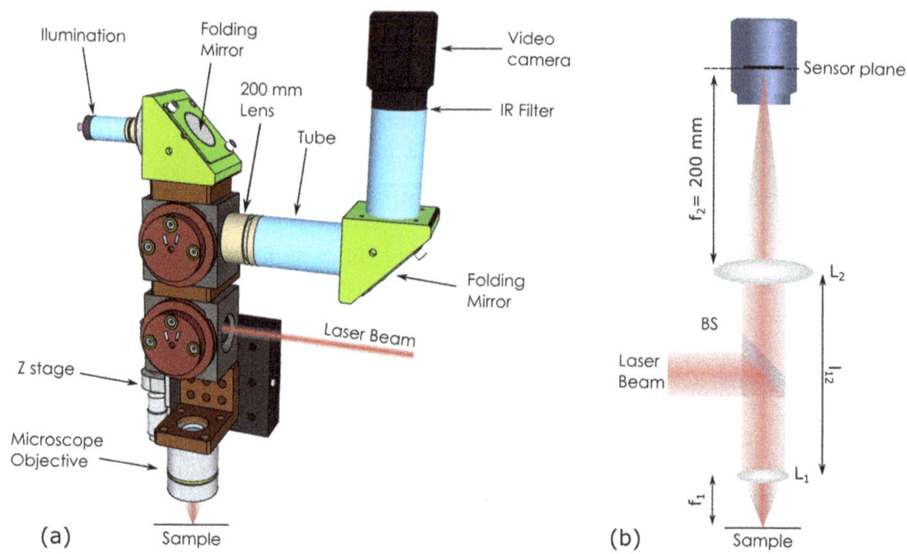

Figure 4.14. Example of visualization system included in a microscope for microprocessing. (a) Sketch of a processing microscope; (b) confocal configuration of the microscope. Created with SketchUp.

frequencies that cannot be transmitted through the optical system's apertures. As a result, pixels of the sensor located farther from the center of the image appear darker, reducing the effective field of view. The condition for the onset of vignetting is satisfied when the following relation holds:

$$\phi_2 < \frac{l_{12}\phi_s}{f_2} + \phi_1 \tag{4.5}$$

where ϕ_2 is the entrance pupil of the imaging lens and f_2 is its focal length, ϕ_s is the diagonal of the imaging sensor, and ϕ_1 is the entrance pupil of the focusing optics, which can be expressed as $\phi_1 = 2f_1$ NA, with f_1 the focal length and NA the numerical aperture of the focusing lens.

For a given configuration of optical components, typically based on $1''$ diameter optics, the vignetting effect can be avoided by limiting the separation l_{12} to a maximum value:

$$l_{12}^{\max} = \frac{\phi_2 f_2 - 2f_1 \, \text{NA}}{\phi_s} \tag{4.6}$$

As a numerical example, we consider the following parameters: NA $= 0.5$, $f_1 = 2$ mm, $f_2 = 200$ mm, $\phi_s = 17.46$ mm (corresponding to a $1.1''$ standard sensor size), and $\phi_2 = 20$ mm as the effective clear aperture of a $1''$ mounted lens. The resulting maximal separation between L_1 and L_2 is:

$$l_{12}^{\max} = 229 \text{ mm},$$

which is sufficiently large to accommodate the necessary optical components, while still remaining within a reasonable z-traveling distance of the focusing stage.

4.4.7 Command and control system

In a laser processing workstation, the laser pulses must be synchronized with other equipment such as translation stages, beam attenuators, and shutters. Although each sub-assembly is connected to a computer and operated through a graphical user interface, the computer itself cannot directly control external devices in real time due to the limitations of the operating system. Typically, each piece of equipment is equipped with an electronic control unit that manages the direct operation of the device, provides interfaces for communication with other controllers, and ensures connectivity with the computer.

The fastest way to control an electronic system is through external triggers. Complex control units, such as those used for translation stages, provide ports for input and output triggers, digital inputs/outputs, and analog interfaces. Multiple devices can be synchronized via their trigger ports. In such an arrangement, one device must provide the **Master** trigger signal, while the other devices act as **Slaves** (see Figure 4.15). The computer software determines only the logic and timing sequence of operations for each device and sends the command to the master controller to start the operation. Once initiated, the rest of the chain of operations automatically follows from the trigger sequence.

At certain time intervals, on the order of milliseconds, which is feasible for an operating system, the computer can interrogate the controllers to check their current status (e.g. position, speed, limit-switch state, shutter state, laser power, etc). Based on the responses, the software algorithm decides the next operations and their execution order. In complex and fast-processing systems, a significant portion of these algorithms is offloaded to modern devices such as FPGAs (field-programmable gate arrays), which are integrated circuits with programmable internal logic that operate much faster than a standard computer.

In some experiments, programmable delays must be introduced between system sequences, for example, to control pulse bursts or to insert a temporal delay between a laser pulse and the start of an acquisition sequence. Dedicated delay generators can

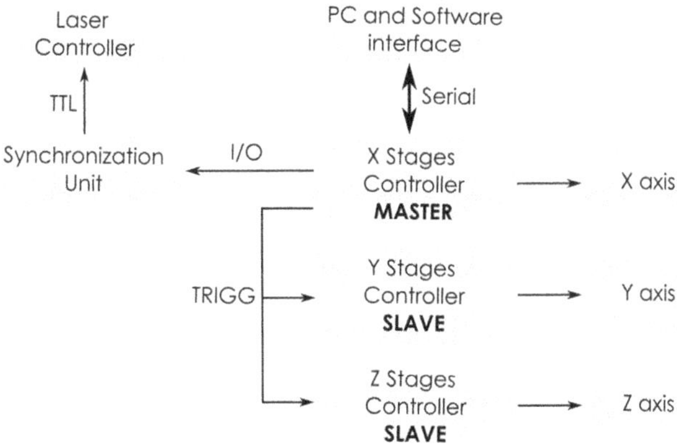

Figure 4.15. Typical configuration for electronic command and control of a laser processing workstation.

be included in the electronic path to fine-tune timing sequences with picosecond accuracy, limited only by the electronic time jitter of the specific configuration.

In certain laser systems, the laser operates independently of any external signal and cannot be triggered. In such cases, all sequences must be configured around the laser controller as the **Master**, with the laser providing the reference time t_0 as the origin of the entire timing sequence.

4.4.7.1 Programming software

The users interact directly with a graphical user interface (GUI) that provides common controls such as Push Buttons, Radio Buttons, Text Inputs, and Text Areas. Behind any intuitive interface stand thousands of lines of code for decision blocks, and algorithms.

Any complex programming script for laser processing should include, as a core library, at least a few essential commands for basic movements:
- `move_abs()` for absolute movements;
- `move_rel()` for relative movements;
- `home()` for homing the stage axes;
- `get_pos()` for interrogating the current position of the axes;
- `set()` for setting different stage parameters such as velocity, acceleration, triggers, and digital I/O states;
- `laser_on()` to start the laser;
- `laser_off()` to stop the laser;
- `stop()` for emergency stopping of the stage movement, probably the most important command, and the first one that should be implemented in any new script.

A minimal example of a control script might look like this:

```
# Initialize and home stages
home()

# Move to an absolute position
move_abs(x=10.0, y=5.0, z=0.5)

# Interrogate current position
pos = get_pos()
print("Current position:", pos)

# Set motion parameters
set(velocity=2.0, acceleration=1.0)

# Perform a relative movement with laser on
# to generate a line of 1 mm length in x diretion
laser_on()
move_rel(x=1.0, y=0.0)
laser_off()

# Emergency stop (if needed)
stop()
```

To properly work, the functions mentioned above should be defined according to the programming manuals and software libraries provided by the producer of the stage controllers.

4.4.8 Safe processing environment

The processing environment is equally important when the highest performance is expected. The main environmental factors that impact processing quality are temperature, vibrations, dust, and humidity.

4.4.8.1 Temperature and humidity control

It is well known that a temperature variation of 1 °C can induce a positioning drift of about 1 μm. Then, such temperature variation during a long processing session will compromise the results when sub-micrometer resolution is expected. Temperature, humidity and dust are well controlled in a clean-room environment. ISO Class (ISO 14644-1 standard) or Class 100 000 is generally sufficient for laser stability and for the reliable operation of translation stages. Typical conditions include a temperature set point of about 20–22 °C ($\pm$0.5 °C) and a relative humidity of about 45% ($\pm$5%).

4.4.8.2 Vibrations control

Vibrations have the most direct impact on processing quality. Active optical tables with honeycomb internal structures are commonly used as support for the laser processing workstation. In most cases, the laser system and the processing cabinet are placed on the same table to avoid beam-position drift. The optical table suppresses vibration frequencies in the range of 1 Hz to a few tens of Hz, which are typical for laboratory vibration sources such as vacuum pumps, valves, elevators, or foot traffic in the building.

However, the best performance of complex moving platforms cannot be ensured by optical tables alone. Over large areas, metallic optical tables are subject to bending effects. To optimize flatness below $\pm$0.5 μm for large XY translation stages, the main processing microscope body is mounted on very stiff materials such as graphite blocks.

4.4.8.3 Laser safety

Last but not least, appropriate safety measures must be implemented in accordance with local regulations. Lasers used in processing applications fall within at least **Class 3B**, and most often **Class 4**, according to the international laser safety classification (IEC 60825 and ANSI Z136). This implies that direct exposure can cause permanent damage to the eyes and skin. Additional indirect hazards are associated with high voltages from power supplies, the release of harmful nanoparticles and gases from processed samples, as well as fire risks due to flammable materials that may come into contact with the laser beam.

To mitigate these risks, specific safety measures are adopted: interlocks and shutters for laser beams, dust collectors and smoke extractors in the processing area, proper enclosures for the laser and workstation, laser safety goggles, protective windows, filters, and curtains designed for the laser wavelength and intensity, updated working procedures. When the laser and processing workstation are fully enclosed within a cabinet equipped with interlocks and observation windows incorporating certified laser filters, the system can be classified as a **Class 1 laser product**, which is considered safe for both trained operators and untrained personnel. Equally important are regular safety training for staff and periodic risk assessments, which should also be conducted whenever significant modifications are made to the experimental setup.

IOP Publishing

High Resolution Laser Microprocessing
Implementation and techniques
Bogdan Ştefăniţă Călin, Marian Zamfirescu and Niculae Puşcaş

Chapter 5

Laser–matter interaction for microprocessing

5.1 Laser source

The general parts are similar for any form of laser processing, and they can be split into three categories: **laser source**, **beam delivery** and **laser–matter interaction** (or material properties). Lasers have achieved a development stage where they represent complex, yet optimized and easily controllable systems. There are exceptions when it comes to ultrashort-pulsed and ultrahigh-power laser systems, where the user has the ability to control and fine-tune beam and pulse characteristics for stability and automation. This chapter is written in the context of experiments performed by the authors. As such, when appropriate, we will offer specific technical information regarding the systems we used, or the experimental conditions.

Laser characteristics directly influence both the beam delivery and the experimental results after irradiation.

When the laser–matter interaction is mostly linear, as is the case for some laser processing types such as ablation, cutting, drilling, welding and others, a good way to describe what happens at the target surface is with respect to time. Moreover, in this case, the laser spot (beam size at target surface) is usually different than the affected area. It is usually smaller, as heat is dispersed through the target. This does not happen, however, in the case of ultrashort pulses (or, rather, it is negligible).

Multiphoton processing can be split into two categories: stochastic interaction (statistical phenomena) and volume processing. Stochastic multiphoton processing is usually based resonant phenomena. For volume multiphoton processing, pulse duration directly affects the size of the **volume pixel** (voxel), as well as the energy tolerance per pulse (or peak power). Energy and power tolerance are important parameters as they can impose restrictions when it comes to processing various materials. If the power/energy tolerance is low, i.e. the material is sensitive to variations of the mentioned parameters, processing might prove difficult, as every laser source possesses some degree of fluctuation of said parameters [2, 3].

There is a peak power fluctuation when the laser is triggered as well (laser spiking), which can produce unwanted effects. The shorter the pulse, the greater the

doi:10.1088/978-0-7503-3239-2ch5   5-1

effects, due to peak power. Pulse duration also determines the type of ionization that takes place inside the material. For example, in experiments we conducted, when processing photosensitive glass using picosecond pulses, the dominant ionization mechanism is determined by two- or three-photon absorption. As we approach the femtosecond regime, tunneling ionization becomes predominant.

Repetition rate is another important parameter. Higher repetition rate is usually advantageous. This is because the repetition rate can be directly linked to the processing time. The term 'energy dose' is used in such premises. For example, if the material permits, we may use higher average powers coupled with a higher writing velocity to obtain similar processing in a shorter period of time. High repetition rates also determine a different kind of laser–matter interaction. Pulses strongly overlap at the surface (or volume) of the target, which means effects are statistically averaged over time (not to be confused with stochastic multiphoton processing, where it doesn't matter if there are overlapping pulses, but how a pulse interacts with the target). Low repetition rate can be useful for single-pulse processing or pump–probe experiments. Laser power needs to be stable for high-resolution laser processing. In most cases, for volume multiphoton processing, high values of the average power are not necessary. In our experimental activity, we mostly used average powers of tens of mW, in some cases going as high as hundreds of mW. This is determined by the type of material we want to process, as well as the focusing optics. We will go through each of the topics discussed above (type of interaction, processing technologies, optics and others).

For the experimental activity, we used several laser systems, each for a specific set of applications. The first system is a processing station specialized for laser-based 3D printing at micron and sub-micron scale (Nanoscribe Photonic Professional [1]). The first laser source is a high-frequency femtosecond fiber laser. It is used primarily for two-photon absorption processing. The active medium of the oscillator is an Er-doped fiber, mode-locked using a saturable absorber mirror. The oscillator generates laser pulses with a wavelength, $\lambda = 1560$ nm, 80 MHz frequency and pulse duration of below 100 fs. Energy extracted from the oscillator is used to seed an amplifier, which is another Er-doped fiber. For polymerization, we need the second harmonic, with $\lambda = 780$ nm. This is achieved using a periodically poled $Li : NbO_3$ (PPLN) crystal. The laser system then delivers 120 fs pulses at a frequency of 80 MHz, with a maximum average power of 140 mW, centered on $\lambda = 780$ nm.

The second laser processing station is using a $Nd : YVO_4$ active medium. It delivers 7–15 ps pulses with a frequency of 500 kHz. Pulse duration depends on the wavelength, i.e. the fundamental wavelength, $\lambda_1 = 1064$ nm, is 7 ps, the second harmonic, $\lambda_2 = 532$ nm, 10 ps and the third harmonic, $\lambda_3 = 355$ nm, below 15 ps pulse duration. The maximum average power is 50 W for λ_1, 25 W for λ_2 and 15 W for λ_3.

5.2 Beam delivery

High resolution laser processing is usually realized using a stationary beam and a mobile sample. This is because we require a high quality incident laser beam when working close to the diffraction limit (or below!) and/or when the processed sample is sensitive to laser parameter variations. This kind of system employs the use of strongly focused light as well, using high magnification microscope objectives. Depending on the desired laser spot diameter, we can also use oil immersion (or polymer immersion for two-photon polymerization), in order to increase the numerical aperture of the microscope objective.

Let us look at an example, the most appropriate being the setup used for two-photon polymerization. As mentioned above, the first laser system we used is specialized for two-photon polymerization. Beam delivery is done through an inverted microscope from Zeiss. It is equipped with a telescope and spatial filtering. Focusing is done through a series of microscope objectives, each for a different processing method. The simplest method is using a 63× objective (LD Plan-Neofluar). The 1.7 mm working distance is helpful for processing through thick transparent substrates, albeit with lower resolution as structure height increases.

For IP photoresists (two-photon absorption optimized photoresist formula, from Nanoscribe [1]), focusing with the 63× objective usually results in a voxel with a 2 μm traverse diameter and 4 μm height. The second objective has a 100× magnification and NA = 1.4. It is used for processing while immersed in photoresist, a method known as 'dip-in laser lithography' (DiLL) [1]. Samples are usually processed with an inverted Z-axis movement. This method is used for obtaining a smaller voxel (usually 1 μm traverse diameter and 2 μm height but can obtain smaller features) and tall structures. The third available processing method is employing a different 100× objective, with a numerical aperture, NA = 1.3. This objective is used for oil immersion processing. It offers the highest resolution available (90 nm lateral feature size, 300 nm height voxel), although structures are limited in height to 100 μm in normal conditions (170 μm thick substrate, Zeiss Immersol, IP-L photoresist), due to its short working distance and increased NA. Notice resolutions are in nanometers, even below 100 nm while we mentioned the incident radiation is $\lambda = 780$ nm, i.e. processing is done below the diffraction limit.

The second laser processing station is similar to the first, as it is configured to have a stationary laser beam and mobile sample. It is, however, constricted to only 2D automatic processing, the third dimension being added manually. This setup is coupled with the *picosecond laser source*, therefore it is used for multiphoton processing of larger volumes compared to the previously described system, for photopolymerization. While it is more difficult to assess the dimensions of a voxel, because this type of processing is terminated with a chemical etching step, we have measured that we can obtain channel widths of up to 12 μm. These dimensions allow for simpler layer-by-layer fabrication. This setup is more versatile and is practically independent of the laser source, provided appropriate optics are mounted.

Considering the above information, we can represent a general schematic for multiphoton laser processing stations for both linear and multiphoton processing. Such a schematic is presented in figure 5.1, adapted for the experimental setup we have used for our research. It is worth noting that, while the left schematic in figure 5.1 represents the photopolymerization station, the right schematic can be, as previously mentioned, adapted to work with virtually any laser.

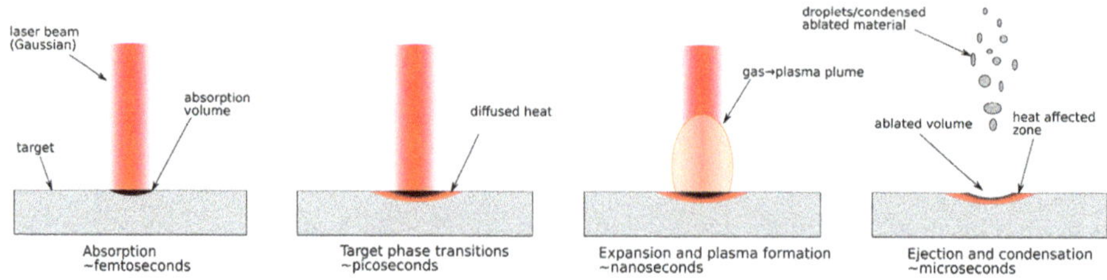

Figure 5.1. General schematic of the high-resolution laser processing stations used in the experimental activity. Left setup is generally used for photopolymerization, while the right setup can be used for linear, stochastic and volume multiphoton processing.

5.3 Laser–matter interaction

Incident laser radiation parameters and target material characteristics determine the type of laser–matter interaction and its results. Let us begin with the most common type of laser processing: ablation. Laser ablation is, simply put, removal of material from a target, using a laser beam. It can be done with most existing lasers that can be focused tightly enough to obtain an intensity above a threshold value, which depends on the material and environmental conditions. We previously put ablation in the 'linear' category of laser–matter interaction. The reason for this is that the main absorption mechanism is based on linear interaction. It is not meant to describe laser ablation as a simple process. It is a highly complex process, which is not completely characterized and understood, even to the date of writing this text. There are, however, several theoretical approaches that can be used to calculate the results of ablation, within very good approximations.

When laser light is incident on a target, it interacts with that material, regardless of intensity. Even if the material is transparent, there is still a weak interaction between the electromagnetic field and the atoms or molecules of the target material.

Let us consider an ideal target material, that absorbs all the incident radiation. First, electrons transition to superior energy levels. If the energy is high enough, electrons will be able to leave the atoms or molecules. The next process is usually referred to as 'electron thermalization'. This process describes the interaction between the ejected electrons and the quantum systems of the target material (atoms or molecules). We have several processes taking place, including collisions, excitations, ionizations, and others, but the main effect is local increase in temperature. As the incident radiation keeps interacting with the target, the electron density grows, and, therefore, the temperature grows. The target material begins to undergo phase changes, from solid to liquid, to gas and finally to plasma.

We have two situations at this point. If the laser pulse is short enough, then plasma is formed after the whole pulse arrives on the target. This is usually the case for cold ablation as well, i.e. phase changes occur fast enough to not have additional thermal effects around the laser-processed area. If, however, the pulse is longer, then plasma is formed while it is still propagating towards the target. This means that it will interact with the plasma before the target (i.e. absorption, phase changes, harmonics, etc).

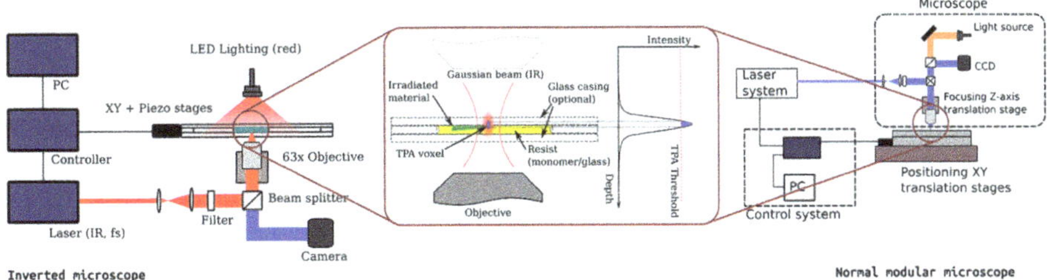

Figure 5.2. Simplified schematic describing the main dynamics of the ablation process.

The ejected material will move away from the target surface, with a general direction perpendicular to the target surface, independent of the incidence angle of the laser pulse. As it moves farther, it will expand and undergo phase changes in reverse order (i.e. from plasma to gas, to liquid (and condensate), to solid).

A simplified schematic describing the ablation process is presented in figure 5.2.

Multiphoton interaction is usually used to refer to processing methods where the main interaction mechanism relies on multiphoton absorption. In the case of stochastic multiphoton processing, there are many processing techniques, each with different objectives and applicability. They do, however, share many characteristics of the experimental setup and activity. Stochastic multiphoton processing usually does not involve tight focusing optics, small laser spots, ultra-precise positioning stages. They still represent high-resolution processing technologies, because they strongly depend on very precise incident laser parameters, such as pulse duration, repetition rate and especially intensity.

In this category, surface nanostructuring techniques are often encountered. They are used to increase functionality of target materials by creating nanostructures with various shapes and sizes, at the surface of the material. These structures can be either disordered, such as nanopatterning and spikes/cones, or periodic, such as laser induced periodic surface structures (commonly known as 'ripples').

Volume multiphoton processing is a different category altogether, and is usually associated with micrometer-level 3D printing techniques. In this case, two-photon and three-photon absorption represent, in most cases, the main physical phenomena that is used to initiate a chemical reaction. This can be done indirectly, as in the case of two-photon polymerization, or indirectly, as is the case of laser-assisted etching. In the case of two-photon polymerization, the material is irradiated with infrared radiation. A substance called a photoinitiator has high absorption for the second harmonic of the incident laser beam. When two-photon absorption occurs, photoinitator molecules either break or get ionized, and they immediately start chemical reactions inside the target material, therefore being considered direct processing. In the case of laser-assisted etching, the two-photon absorption is used to ionize certain elements inside the target. The released electrons are captured by other ions, which change their properties. Further processing steps provide the necessary chemical reactions in order to obtain the etching, and therefore it is considered indirect processing. These topics are discussed further in chapter 6.4.

References

[1] Nanoscribe GmbH *https://www.nanoscribe.de* (last accessed 5 January 2019)
[2] Rüdiger P 2009 Noise in laser technology *Optik Photonik* **4** 48–50
[3] Rüdiger P 2009 Noise in laser technology *Optik Photonik* **4** 45–7

IOP Publishing

High Resolution Laser Microprocessing
Implementation and techniques
Bogdan Ştefăniţă Călin, Marian Zamfirescu and Niculae Puşcaş

Chapter 6

Applications of high-resolution laser microprocessing

Emergent microprocessing technologies

The fabrication of micro-devices using lasers is made possible by the controlled modification of materials through irradiation with coherent light. When an ultrafast laser beam (with pulse durations on the order of picoseconds or femtoseconds) is focused on the surface or inside a transparent material, and the laser fluence exceeds a certain threshold, a small volume of the material is modified through *photophysical* or *photochemical* processes, resulting in structures with dimensions on the order of micrometers, sub-micrometers, or, in some special cases, even below 100 nm.

Due to the relatively long heat diffusion time in most materials compared to the ultrashort duration of sub-picosecond laser pulses, the material volume adjacent to the processed area remains unaffected. This characteristic makes ultrafast laser processing a suitable technique for the fabrication of micro- and nanostructures on metal films, ceramics, polymers, dielectrics, etc, and it has been implemented in applications such as microelectronic circuits, microsensors, microfluidics, photonic devices, and other micro-opto-electro-mechanical systems (MOEMS) [12, 29].

With the development of new laser systems featuring ultrashort pulses and high repetition rates, many of the technological limitations of lasers have been overcome —such as low throughput, processing resolution constrained by thermal effects from long pulses, or the physical limits imposed by optical diffraction. As a result, lasers have been widely adopted in many industrial applications as a **maskless technology**, more commonly known as **direct laser writing** (DLW) (see Figure 6.1). Laser processing methods can be classified based on the nature of the laser–material interaction. Two main categories and their sub-classes are listed below:

1. **Photophysical processes**
   - **Micro- and nanoprocessing by laser ablation**. These techniques are used to produce micro- and nanostructures by removing material through an ablative process. It is a subtractive method and shares many

doi:10.1088/978-0-7503-3239-2ch6   6-1

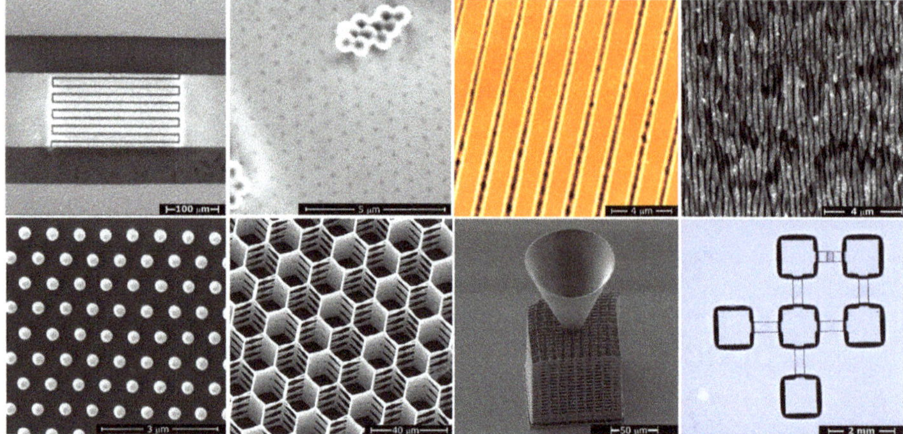

Figure 6.1. Direct laser writing of micro- and nanostructures: (a) interdigital capacitors fabricated by laser ablation of gold films on Si substrate; (b) periodical pattern on nanoholes obtained by near-field laser ablation; (c) nanogratings fabricated by near-field laser ablation; (d) surface nanotexturing by LIPSS effect; (e) 2D photonic crystal in photopolymers; (f) 3D scaffold for tissue engineering; (g) cone-shape microtarget for ultraintense laser experiments; (h) microfluidic chip fabricated by picosecond laser-assisted etching in Foturan glass (image courtesy of Felix SIMA) [24].

technological characteristics with large-scale laser processing techniques, such as laser cutting or drilling. The main differences compared to laser macroprocessing are that the microstructures are created by tightly focusing the laser beam using short focal length optics, the positioning mechanics offer submicrometer precision, and the lasers deliver ultrashort pulses to avoid heating the surface. Laser microprocessing by material ablation is used for fabrication of structures for micro-electronics and photonic devices [4, 22, 28].

- **Laser-induced periodical surface structures—LIPSS.** At low laser fluence, below the ablation threshold, surfaces can be nanotextured without significant material removal. This method is fast and versatile; it can be used to generate nanopatterns on almost any type of surface, without the need for vacuum conditions or the laborious fabrication steps required in photolithography [10].
- **Near-field laser ablation.** The processing resolution can be pushed beyond the diffraction limit by utilizing the evanescent optical field in the proximity of the substrate, enabled by micro-optical focusing elements such as silica or polystyrene microbeads [27]. Optical masks can be engineered for ablation within a near-field optical enhancement regime, employing so-called *photonic nanojets*. This parallel processing technique enables large-area nanostructuring with a single laser shot [11].
- **Refractive index modifications of glasses.** Laser densification of transparent materials induces localized modifications of the refractive index and the formation of micro- and nanostructures within the bulk.

This effect is exploited for the fabrication of optical waveguides and other 3D photonic structures in glass [19, 23].

- **Laser ablation in liquids**. Nanoparticles can be produced through liquid-assisted ablation. When combined with microprocessing techniques, laser-induced periodic surface structures (LIPSS), or near-field ablation, this approach enables the formation of novel surface textures such as microcones or enhanced patterns generated within a liquid environment [2]. In certain cases, the liquid may interact with the laser radiation, inducing photochemical reactions rather than direct laser ablation [20, 26].

2. **Photochemical processes**

- **Single-photon photopolymerization**. In *2D laser lithography*, UV-curable photoresists are processed in a maskless configuration by directly focusing a UV laser onto positive or negative photosensitive polymers. This approach enables the direct transfer of computer-designed structures, eliminating several technological steps such as mask fabrication and alignment.

- **Two-photon photopolymerization (TPP)**. Three-dimensional structures are obtained using femtosecond laser pulses in negative photoresists through a nonlinear two-photon absorption process and photopolymerization. The nonlinear absorption enables the fabrication of polymer structures with 3D features below the diffraction limit, down to 100 nm. This has led to the development of a technique known as *3D laser lithography*, which is used in the fabrication of various optomechanical structures such as microfluidic devices, micromechanical components, 3D scaffolds for tissue engineering, micro-optics and waveguides, 3D targets for ultraintense laser experiments [7, 18].

- **Photochemical modification of glasses**. A focused laser beam can induce photochemical reactions within the volume of photosensitive glasses such as *Foturan*. Through subsequent thermal treatment and chemical etching, the irradiated material can be selectively removed, enabling the creation of complex three-dimensional structures, such as microtubes and microreservoirs for microfluidic applications, within the glass substrate [24].

6.1 Microstructuring by laser ablation

Complex and reproducible geometries can be obtained through laser ablation, with feature sizes reaching the diffraction limit, or even below 500 nm under certain conditions. The key input parameters in laser microprocessing include: the beam spot size at focus ($2w_0$, measured in μm); laser pulse energy (E_{pulse}, in μJ); laser repetition rate (f, in kHz); the sample position relative to the focal plane (Δz, in μm); the laser fluence threshold for the laser–material interaction effect ($F = 2E/A$, expressed in mJ cm^{-2}); the scanning speed (v, in mm s^{-1}); and the scanning path strategy. For a Gaussian beam, the laser fluence is twice that of a top-hat beam profile, so a factor of 2 must be included in the calculation formula. It is important to note that the beam spot size at focus (or beam waist, $2w_0$) is not necessarily the same as the size of the laser-modified area, Δs. The beam diameter $2w$ is defined at $1/e^2$ from the maximum irradiance profile $I(x, y)$. However, the interaction threshold can be below or above this irradiance value. Therefore, the effective processed area depends on the laser fluence and the threshold for the modification effect.

The main drawback of laser-based fabrication of microstructures using the direct writing method is the long processing time required for large surfaces. Therefore, an essential step is to design and optimize the laser scanning strategy. Usually, the design of microstructures is created using CAD software and then converted into a computer numerical control (CNC) programming language such as G-code (Geometric Code), or into another programming language (e.g. Python, LabVIEW, C#, etc) capable of controlling the sample positioning system or beam steering equipment with the required speed and positioning accuracy. Any design is composed of basic elements such as lines, circles, arcs, and dots, which are produced as the laser beam moves at a certain speed relative to the sample's coordinate system. The scanning speed directly influences two key output parameters: the total processing time, which should ideally be as short as possible, and the exposure time, which must be carefully adjusted to ensure the desired quality and reproducibility of the microstructures.

The exposure time can be expressed as the number of laser pulses irradiating the same area, and it is directly linked to the laser repetition rate. There are two exposure regimes: a *static regime*, in which the beam remains stationary during exposure, and a *dynamic exposure regime*, where the beam moves relative to the sample.

Laser irradiation in static regime. When the laser beam is not moving relative to the sample's reference coordinate system, the laser beam produces dots on the sample's surface. The number of pulses contributing to the irradiation of the sample depends solely on the laser repetition rate, f, and the exposure time, Δt:

$$N = f\Delta t \tag{6.1}$$

This type of exposure is generally used only to create dots on the surface with a specific number of pulses and pulse energy, such as in the case of experiments for determining the ablation threshold.

Laser irradiation in dynamic regime. Traces on sample surfaces are obtained by moving the laser relative to the sample (figure 6.2). The number of pulses contributing to the exposure is estimated based on the overlap of pulses as the laser beam moves across the sample surface at a certain scanning speed. In this case, the number of overlapping pulses is given by:

$$N = f\frac{\Delta s}{v} \tag{6.2}$$

In the relation above we consider the diameter of the laser-modified area, Δs, and not the beam diameter, because the photons at the irradiance below the threshold effect do not contribute with their energy at ablation process (figure 6.2(b)).

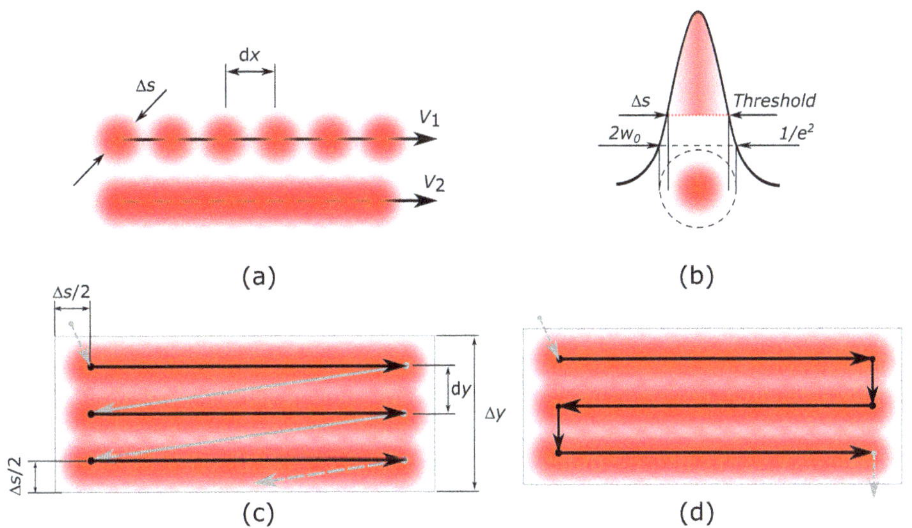

Figure 6.2. Scanning path strategies. (a) The number of overlapping pulses is influenced by the laser repetition rate and scanning speed; (b) The effective spot size of the laser-modified area is give by the laser threshold level of the modification effect. (c) and (d) Scanning using a meander path to optimize processing time or structure quality.

The distance between two pulses measured from center to center is $dx = v f^{-1}$. It is obvious that as the scanning speed increases, the number of overlapping pulses decreases.

At speeds $v > f \Delta s$, the distance between two pulses dx is larger than the spot size Δs. Therefore, the sample is practically irradiated by isolated single pulses, and the lines produced by laser ablation will exhibit discontinuities (figure 6.2(a)). Increasing the scanning speed in an attempt to reduce processing time may negatively affect the quality of the resulting structures.

In some designs, it is necessary to generate ablated strips with a width Δy larger than that of a single trace, which corresponds to the size of the laser-modified spot, Δs. Therefore, the strip must be covered by multiple laser passes to achieve the desired width (figures 6.2(c) and (d)). One of the scanning strategies presented must be selected as a compromise between processing efficiency and microstructure quality. The single direction hatching illustrated in figure 6.2(c) is recommended when high structural fidelity is required. Mechanical translation systems often exhibit a positioning inaccuracy known as backlash. To mitigate its effects, negative-stage movements with the laser active should be avoided. In such cases, the system can compensate for backlash following a reverse movement, and material processing is carried out only during forward (positive) translations. If the backlash error is negligible or has been properly compensated, the bidirectional meander strategy shown in figure 6.2(d) can be implemented. This approach effectively reduces the overall processing time by approximately 50%, while maintaining acceptable structural quality.

The scanning pitch, dy, between successive single traces of the hatch, should be smaller than Δs in order to maintain a reasonable overlap between traces. Consequently, in the programming software that controls the laser irradiation process, the number of passes must be determined based on the parameter Δs, which is defined by the laser–material interaction. The total width of the strip generated by N_{lines} is:

$$\Delta y = (N_{\text{lines}} - 1) \cdot dy + \Delta s \tag{6.3}$$

with $dy \leqslant \Delta s$. The total length of the line generated by laser is:

$$\Delta x = L_x + \Delta s \tag{6.4}$$

where L_x represents the stage movement programmed in the software. In practice, the effective area covered by laser irradiated spot is larger that the rectangle described by stages movements. The design must take into account that the final rectangle (Δx, Δy) will have dimensions larger by $\Delta s/2$ on all sides, due to the finite size of the laser beam, which cannot be neglected.

At this stage, the only remaining unknown parameter is Δs, the size of the ablated spot. This can be evaluated by directly inspecting the laser-ablated area under the working laser parameters. However, it must be noted that any change in irradiation conditions, such as laser fluence, exposure time, scanning speed, or beam focusing, will directly affect the size of the ablated spot. A more rigorous, albeit laborious, approach is to measure the ablation threshold. From this value, it is in principle possible to estimate the size of the ablated spot under varying irradiation conditions.

6.1.1 Laser ablation threshold

The ablation threshold of a material depends on several factors, including the nature of the material, the pulse duration and temporal profile of the ultrashort laser pulse, the contrast between the main peak intensity and pre-pulses, the level of amplified spontaneous emission (ASE) or other temporal background components, and the number of laser pulses. From the perspective of fabricating structures with specific resolution, it is experimentally important to determine both the laser ablation threshold and the dimensions of the structures produced at a given laser fluence. In this context, we define the ablation threshold as the minimum laser fluence at which permanent modifications of the irradiated surface are observed. It is important to note that the ablation threshold is not equivalent to the laser-induced damage threshold (LIDT), which is typically used to characterize optical coatings exposed to high-intensity laser beams. LIDT refers to any form of laser-induced modification— whether permanent or temporary, including topographical changes or alterations in optical properties such as the local refractive index. In contrast, ablation specifically refers to the permanent removal of material as a result of laser irradiation.

Two types of ablation tests can be employed. The first method provides a rough estimation of the structure size for given values of laser pulse energy and writing speed, which are directly used as parameters in the sample processing protocol. This approach allows for a rapid, though approximate, determination of the minimum laser fluence required to induce surface modification. The second method provides a more precise determination of the threshold laser fluence, taking into account the number of focused laser pulses delivered to the material, the pulse energy, and the beam size. Examples of ablation tests performed using the two methods mentioned above are presented below.

The first laser ablation test is shown in figure 6.3. These measurements consist of producing series of parallel lines with a pitch of 20 μm, using various sample scanning speeds and laser powers. For this example, the laser power was varied from 0.11 mW to 0.9 mW The material used was a film of bimetallic silver–chromium (AgCr) alloy deposited on a ceramic substrate. The test was performed using a femtosecond laser beam with a wavelength $\lambda = 775$ nm, pulse duration $\tau = 200$ fs, and repetition rate $\nu = 2$ kHz. The focusing optics consisted of a microscope objective with a focal length of $f = 20$ mm; the beam diameter at the input aperture of the objective was 4 mm; the beam quality factor was $M^2 = 1.5$, and the estimated laser beam waist was $2w_0 = 7.4$ μm. Laser fluence, defined as the laser energy density per unit area, is measured in J cm^{-2} and accounts for the Gaussian beam profile. From the scanning map, the width of the structures obtained under specific irradiation conditions can be determined. In this particular case, the surface of the ceramic substrate is highly inhomogeneous and exhibits significant roughness, which compromises the laser processing accuracy at low fluences and high scanning speeds. For this reason, each line was repeated multiple times using the same laser parameters to obtain an averaged response of the laser-ablated surface. In figure 6.3, the lowest laser power used was $P = 0.11$ mW, corresponding to a pulse energy of $E = 55$ nJ and a laser fluence of $F = 0.25$ J cm^{-2}. Under these conditions, the laser-ablated lines are discontinuous and barely visible. At twice the laser power ($P = 0.23$ mW), the lines are more clearly defined. The width of the ablated lines increases as the laser scanning

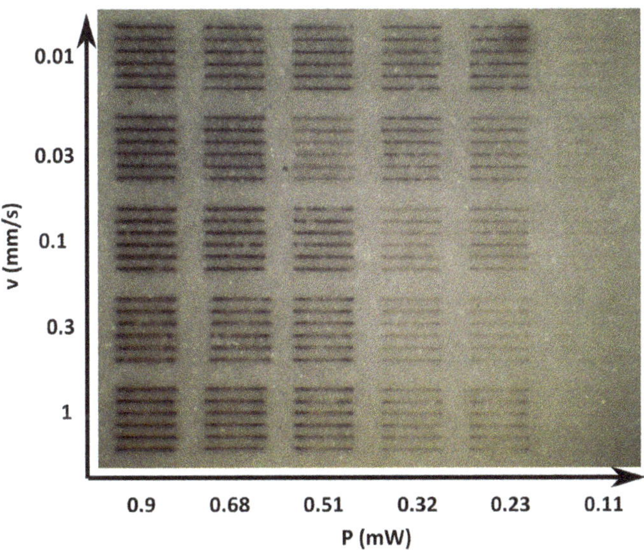

Figure 6.3. Map of ablation test on film of bimetallic silver–chromium (AgCr) alloy deposited on ceramic substrate.

speed decreases, from 1 mm s^{-1} down to 0.01 mm s^{-1}. The resulting line widths fall within the range of 3 μm to 10 μm. For scanning speed of 1 mm s^{-1} the equivalent number of overlapping pulses is $N = 5$. Although this method allows for a quick estimation of the minimum power required for ablation, it is not highly accurate and cannot reliably predict the ablated line width for other laser powers.

A method for the precise determination of the ablation threshold assumes that the key experimental parameters are the number of laser pulses and the average laser power. In the example from figure 6.4, a map of ablated points is created by laser ablation, with the laser power varied from 0.93 mW to 20 mW. The number of pulses is varied from 1 to 15. For a constant laser power and number of pulses, a series of equidistant points with a pitch of 20 μm is generated to statistically measure the spot diameters produced by laser ablation on the surface. In this map, the minimum fluence applied is not necessarily the exact value at which surface modification becomes observable but is chosen to be as close as possible to the ablation threshold. The diameter of the resulting crater depends on the laser fluence according to the relation described in [15]:

$$d(F) = \frac{d_0}{\sqrt{2}} \sqrt{\ln(F/F_{\text{th}})} \tag{6.5}$$

where $F = \frac{8E}{\pi d^2}$, and d_0 represents the diameter of the focused laser spot, which is distinct from the diameter of the crater formed by laser ablation. From this relationship, the threshold energy for laser ablation can be determined graphically.

The experimental data are fitted with a linear relationship of the form:

$$d^2(E) = \frac{d_0^2}{2} \ln(E/E_{\text{th}}) = \frac{d_0^2}{2} \ln(E) - \frac{d_0^2}{2} \ln(E_{\text{th}}) \tag{6.6}$$

From the relation above, assuming the linear dependence $f(\ln(E)) = a \ln(E) + b$, we can extract the spot size and ablation threshold as follows:

$$d_0 = \sqrt{2a}$$

$$E_{\text{th}} = \exp\left\{\left(-\frac{2b}{d_0^2}\right)\right\} \tag{6.7}$$

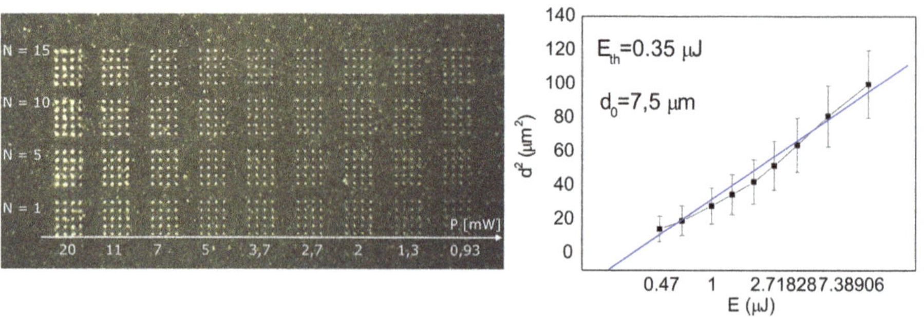

Figure 6.4. Laser ablation map with series of pulses at varying average power.

In the example shown in figure 6.4, the same film of AgCr deposited on ceramic substrate was processed. The fitting parameters for ablation with a single laser pulse ($N = 1$) are $a = 29.78$ and $b = 28.47$. The threshold pulse energy is determined to be $E_{th} = 0.38\ \mu J$, corresponding to a fluence $F_{th} = 1.64\ J\ cm^{-2}$. The beam spot diameter obtained from the fitting parameters is $d_0 = 7.7\ \mu m$, which is close to the theoretically estimated beam waist diameter of $2w_0 = 7.4\ \mu m$.

It is worth noting that the ablation threshold determined for single-pulse processing differs from the ablation threshold observed during surface scanning. In the previous example, at a scanning speed of 1 mm s^{-1} and pulse repetition rate of 2 kHz, the ablation threshold was $E_{th} = 0.25\ J\ cm^{-2}$. This difference arises from the cumulative effect of multiple laser pulses contributing to the ablation process.

The lateral size of the ablated spot depends on both the laser beam size at the sample surface and the laser pulse energy. Together, these parameters determine the laser fluence. By comparing the ablation threshold fluence with the applied laser fluence or irradiance, the diameter of the ablated spot on the surface can be estimated. In the case of a flat-top transverse irradiance profile, the ablated spot size corresponds closely to the laser beam size. However, this is not the case for a Gaussian beam profile, where the irradiance decreases gradually from the center outward. Figure 6.5(a) shows the longitudinal irradiance profile of a focused Gaussian beam. A dotted curve marks the region of constant irradiance corresponding to the laser ablation threshold. If the sample is positioned at a distance Δz from the focal plane, an area with a diameter s will be ablated, as this is the region where the irradiance exceeds the ablation threshold. This effect suggests that the ablated spot size can be controlled as a function of the focusing position Δz. Figures 6.5(b) and (c) illustrate the radial distribution of the transverse irradiance profile as the beam is defocused relative to the sample position. If the beam intensity is superior but close to the ablation threshold ($P > P_{th}$), only the central part of the Gaussian profile contributes to the ablation process. This situation typically occurs when the beam is tightly focused at low average laser power to obtain the finest laser patterns on the sample surface. As the beam is slightly defocused, the peak irradiance drops below the ablation threshold, and the ablation condition is no longer met, resulting in no surface modification.

A counterintuitive effect occurs under defocused conditions when the laser power is only a few times greater than the ablation threshold (e.g. $P = 3P_{th}$). At large Δz

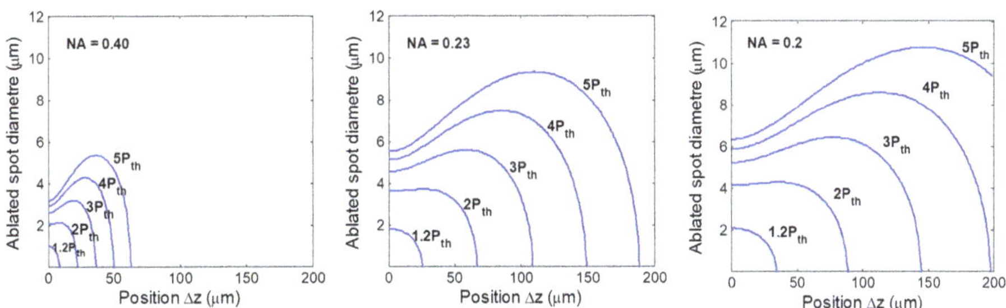

Figure 6.5. Ablation spot size as function of Δz position

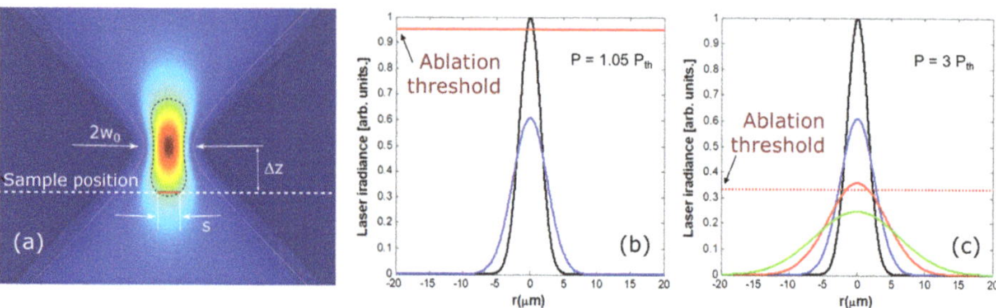

Figure 6.6. Ablation spot size as function of focus position.

distances between the sample and the focal plane, the beam size increases, and one might expect the ablated spot to become larger. In fact, the opposite occurs: the ablated spot becomes smaller as the defocusing increases, eventually disappearing when the laser fluence drops below the threshold (figure 6.6(c)). This effect is illustrated more clearly in figure 6.5. For three different focusing optics, microscope objectives with varying numerical apertures (NAs), the ablated spot size is estimated as a function of the Δz position. The spot size corresponds to the area where the irradiance of the defocused beam exceeds the ablation threshold.

The size of the laser spot focused by a microscope objective depends on the NA of the focusing optics and the laser wavelength. However, the beam size can be defined according to different criteria, as follows:

$$\text{Airy disc diameter:} \qquad d_{\text{Airy}} = 1.22\frac{\lambda}{\text{NA}}$$

$$\text{Full Width Half Maximum:} \quad d_{\text{FWHM}} = 0.51\frac{\lambda}{\text{NA}} \qquad (6.8)$$

$$\text{Intensity at } 1/e^2: \qquad d_{1/e^2} = 0.82\frac{\lambda}{\text{NA}}$$

Typically, the beam size is smaller than the entrance aperture, so the laser beam is expanded by a telescope to achieve the smallest possible beam waist. However, expanding the beam diameter to reduce the spot size comes at the cost of sacrificing laser power due to beam clipping at the entrance aperture and increased diffraction. The laser power is preserved if the beam diameter remains smaller than the aperture size—typically about one-third of the aperture diameter for a Gaussian beam. In this case, the effective numerical aperture NA_{eff} of the objective must be reevaluated as:

$$\text{NA}_{\text{eff}} = \text{NA}\frac{d_{\text{laser}}}{d_{\text{obj}}} \qquad (6.9)$$

where d_{laser} is the diameter of the laser beam at the input of the microscope objective, and d_{obj} represents the diameter of the entrance aperture. The laser beam size at the focal plane must be carefully estimated using one of the relations from equation (6.8), considering the effective NA.

6.1.2 Micro-structures fabricated by laser ablation

Laser microprocessing is a convenient alternative to classical lithography for fabricating small series of devices such as sensors, microstrips for metallic electrodes, microfluidic structures, and others. The main advantage of laser processing over lithography is that no physical mask is required to produce a computer-generated design. Furthermore, the chemical etching step is eliminated, avoiding the use of chemicals harmful for the environment such as solvents and corrosive solutions. A typical surface structured geometry consists of metallic and non-metallic layers deposited on a silica or ceramic substrate. Usually, on a silica substrate, a thin native layer or a thermally grown think layer of SiO_2 is present. Additionally, a thin buffer layer of chromium (Cr) or titanium (Ti), with typical thickness of several tens of nanometers, is deposited to improve the adhesion of the top layer to the substrate (figure 6.7). Most of the time, the top layer is a metallic material needed for configuration of interdigital electrodes. Metals have in general ablation threshold one order of magnitude lower compared with dielectrics. This difference is a great advantage in producing selective ablation. Then, by setting the right laser intensity, only the top layer is removed and the substrate remains unaffected.

An example of a laser-processed device based on interdigital configuration is presented below. Structured devices operating in the microwave frequency range, from hundreds of MHz to a few GHz, are typically fabricated using microstrip techniques or in a coplanar configuration (CPW—CoPlanar Waveguides). The basic structures may include interdigital capacitors and rectandular strips as inductances. For the frequency ranges mentioned above, these structures typically have dimensions of a few millimeters, with fine structural details on the order of hundreds of micrometers. To shift the operating frequency into the tens or even hundreds of GHz, a significant reduction in the physical dimensions of the structures is required, down to tens or hundreds of micrometers, with submicrometer processing resolution. When fabricating these structures by direct laser writing, the CPW configuration is preferred due to its simpler implementation compared to other fabrication methods.

The example shown in figure 6.8 illustrates structures fabricated by combining classical lithography (figure 6.8(a)) with laser ablation (figures 6.8(b)–(d)). Standard lithography was employed to remove metallic material from large areas, while laser ablation was used to generate the fine patterns of the interdigital capacitors in a coplanar design. The resulting structure is a millimeter-wave bandpass filter consisting of a series of interdigital capacitors and shunt inductances arranged in a composite right/left-handed (CRLH) configuration. This structure exhibits

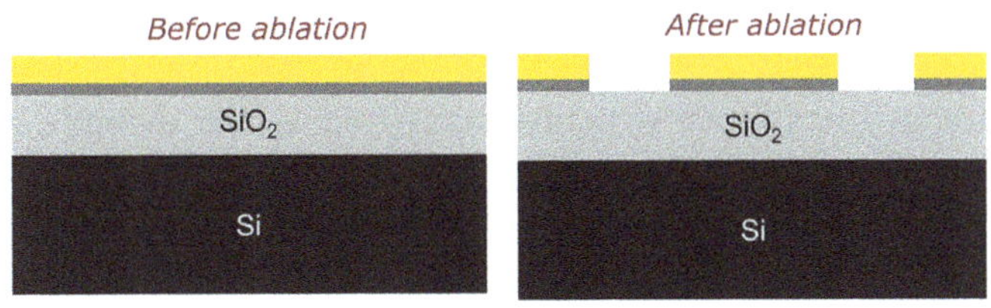

Figure 6.7. Laser ablation process on thin films.

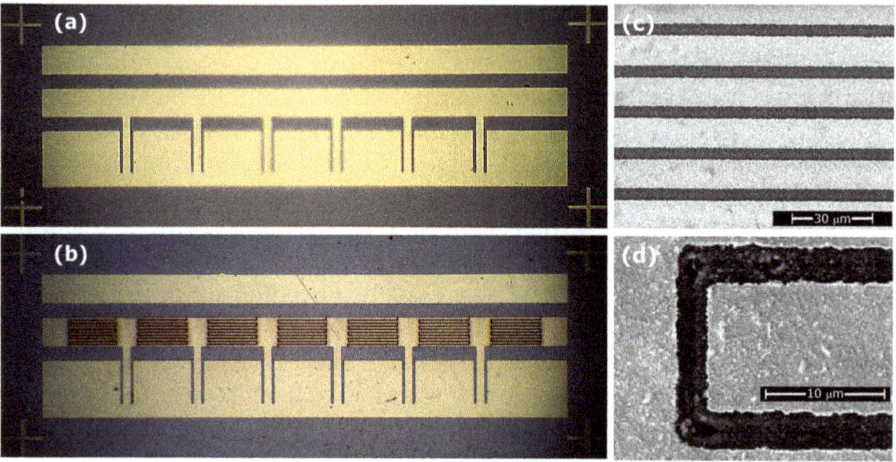

Figure 6.8. Millimeter wave CRLH bandpass filter fabricated by combination of standard lithography and femtosecond laser ablation.

metamaterial behavior in the electromagnetic spectral domain at frequencies of several tens of GHz. Its primary characteristic arises from the low attenuation of this electromagnetic device over a broad spectral range [22].

To fabricate these structures, a 100 nm thick gold film deposited on a silicon substrate was processed using laser ablation. The interdigital capacitor consists of five pairs of digits, each with a width of 10 μ m and interspaces of 5 μ m (figure 6.8 (d)). Due to the relatively large volume of material removed from the metallic film during ablation, some debris may redeposit on the processed surface, potentially causing short circuits within the electrode structures. These redeposited particles can be effectively removed using an ultrasonic bath. Figure 6.8(d) presents a scanning electron microscopy (SEM) image of a detail from the interdigital structure, illustrating the typical ablation pattern of the metallic film. The processing was performed using a femtosecond CPA laser system with the following parameters: pulse duration of 200 fs, pulse repetition rate of 2 kHz, and a central wavelength of 775 nm. The sample was translated using computer-controlled XYZ stages equipped with stepper motors, providing scanning speeds in the range of a few mm s^{-1} and a bidirectional positioning repeatability of less than 1 μm.

In the SEM image, the lighter gray areas correspond to the metallic regions of the structure, while the darker regions represent the substrate where the metallic film was removed by laser ablation. It is evident that the horizontal and vertical interspaces produced by ablation do not exhibit identical widths. This asymmetry may pose a potential issue, as it can slightly affect the expected transmission and reflection bands of the CRLH structure compared to the theoretical design. The observed width variation is attributed to slight astigmatism of the focused laser beam, originating from the resonant cavity of the laser system. An astigmatic laser beam exhibits an elliptical focal spot near the nominal focal plane. In practice, the beam behaves as if it has two distinct focal planes: one for the vertical direction and another for the horizontal direction. At each focal position, the beam profile is elliptical, with the major axis oriented vertically or horizontally, respectively.

Another drawback of laser processing is also evident in the SEM image. The edges of the ablated metallic film exhibit roughness on the order of fractions of a micrometer. This effect is primarily caused by non-uniformities in the thin film, which locally induce variations in the ablation threshold. Combined with the inherent pulse-to-pulse laser energy fluctuations (typically around 5%), this leads to imperfections in the laser-ablated structures.

The corners of the interdigital structure appear rounded, as expected, due to the circular shape of the focused laser beam. At these corners, deeper ablation of the silica substrate is observed. This occurs because of the speed profile of the translation stages, which is typical of any mechanical positioning system. Whenever the stage accelerates or decelerates, the system requires several milliseconds to reach the nominal scanning speed. As a result, for a constant laser intensity, the total exposure time along the scanning path increases locally, producing deeper ablated regions or wider ablation lines. This drawback is common in laser processing methods that rely on mechanical translation stages. It can be mitigated by implementing an adaptive laser power control strategy to compensate for instantaneous variations in translation speed.

6.2 Laser-induced periodic surface structures

Many applications require surface texturing at the nanoscale over large areas. Despite the excellent performance of modern lithographic techniques, these methods cannot be easily applied to large objects, in ambient air, for small production series, or within short processing times. Under certain conditions, laser surface processing offers an interesting and practical approach for large-area nanotexturing. Laser-induced periodic surface structures (LIPSS) have been frequently observed on irradiated material surfaces near the ablation threshold since the earliest studies of laser ablation more than five decades ago [5, 17, 25]. This phenomenon has been reported on nearly all types of materials, including metals, dielectrics, semiconductors, and polymers (figure 6.9).

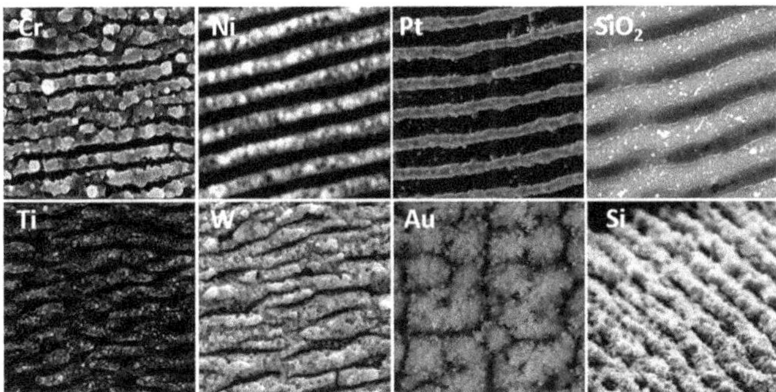

Figure 6.9. LIPSS on various materials. The periodic structures are oriented perpendicular to the laser polarization axis. Scale bar is 1 μm.

Beyond their significance for fundamental research, LIPSS structures have recently been extensively investigated for practical applications, including the fabrication of hydrophobic surfaces, bioinspired surfaces for cell growth, and surfaces with modified tribological properties.

6.2.1 Mechanisms of LIPSS formation

Over time, various hypotheses have been proposed regarding the formation mechanisms of laser-induced periodic surface structures (commonly referred to as ripples). Initially, the generation of these periodic patterns was explained by an interference model involving the incident laser radiation and optical waves scattered by surface roughness. Within the well-known model of LIPSS formation developed by Sipe *et al* [25], the fringe patterns can be predicted by electromagnetic field structures generated under the interference of the surface-scattered waves. This model successfully explained the polarization dependence of the LIPSS orientation, and the LIPSS period dependence on the incident laser wavelength, λ_0, angle of incidence, θ, and refractive index of the material, n:

$$\Lambda_{\text{LIPSS}} = \frac{\lambda_0}{n(1 + \sin \theta)} \tag{6.10}$$

However, with the advent of ultrashort-pulse lasers in the femtosecond range, these classical theories were challenged by new experimental evidence. The observed structures exhibited dimensions well below the diffraction limit, in some cases reaching scales of only a few tens of nanometers. Table 6.1 resents the experimental result of LIPSS periods for several materials, metals and dielectrics. The samples were irradiated by ultrashort focused pulses emitted by a CPA Ti:sapphire laser, at 200 fs pulse duration, 2 kHz repetition rates, at 775 nm laser wavelength end laser fluence just above the ablation threshold.

For each material, several types of LIPSS were generated with periods ranging from 100 nm up to 750 nm. It is obvious that in femtosecond pulses irradiation regime the inferential model no longer corresponds with the experimental observations.

In general, LIPSS can be classified into two main categories: low spatial frequency LIPSS (LSFL), which has a period comparable to the laser wavelength λ and is oriented perpendicular to the laser polarization axis, and high spatial frequency LIPSS (HSFL) which exhibits a spatial period ranging from half that of LSFL down to approximately $\lambda/6$, with an orientation parallel to the laser polarization axis. New theoretical models have been proposed, taking into account mechanisms such as self-organization processes and plasmonic effects occurring at the material surface [6, 21]. However, none of the existing theories fully explain the experimental observations in a unified manner. Challenges remain in accounting for effects such as:

- coexistence of different types of structures with distinct periods and orientations,
- dependence of structure orientation on the laser polarization direction,

Table 6.1. Periods Λ for low spatial frequency LIPSS and high spatial frequency LIPSS for various materials and their comparison with excitation wavelength and surface plasmon wavelength.

Material	Period Λ_{LIPSS}	Orientation	Refraction index @ $\lambda = 775$ nm	λ_0/n	Plasmon wavelength Λ_{SP}	Plasmon propagation distance
Ni	HSFL: 100–190 nm HSFL: 300–330 nm LSFL: 600–660 nm	$\parallel \perp \perp$	2.37–i 4.28	327 nm	744 nm	0.86 mm
Ti	HSFL: 90–120 nm LSFL: 460–530 nm	$\parallel \perp$	2.43–i 3.45	319 nm	707 nm	0.2 mm
Pt	HSFL: 190–220 nm LSFL: 550–700 nm	$\parallel \perp$	2.77–i 4.86	279 nm	750 nm	1.05 mm
Cr	HSFL: 120–140 nm LSFL: 550–690 nm	$\perp \perp$	4.07–i 4.32	190 nm	557 nm	0.005 mm
W	HSFL: 170–300 nm	$\perp$	3.68–i 2.77	210 nm	838 nm	0.26 mm
ZnO	HSFL: 150–220 nm	$\perp$	1.962	395 nm	869 nm	-
SiO2	HSFL: 300–350 nm LSFL: 600–750 nm	$\perp \perp$	1.548	500 nm	923 nm	-
Au	-	-	0.232–i 4.316	3.340 mm	754 nm	19.49 mm
Ag	-	-	0.166–i 4.846	4.668 mm	758 nm	38.99 mm

- variation in LIPSS period with material properties,
- influence of laser parameters such as pulse energy and number of pulses.

The presence and even the coexistence of both LSFL and HSFL structures adds an additional layer of complexity to the development of a comprehensive theory describing the formation of periodic nanostructures. One of the most recent hypotheses is based on the self-organization of the material under the influence of thermocapillary forces, driven by the Marangoni effect. In this model, material redistribution occurs in the molten phase on a timescale of several hundred picoseconds. The periodicity of the structures is explained by the excitation of surface plasmon polaritons (SPPs), which generate a periodic thermal gradient. This gradient induces thermocapillary forces that drive the self-organization of the molten material, resulting in periodic structures with a period comparable to the wavelength of the plasmons.

Figure 6.10 shows the SEM images and, respectively, cross-sectional SEM images of LIPSS structures generated on a thin film of platinum deposited on silica wafer. The SEM image reveals LSFL with period of the order on 710 nm. On cross-sectional SEM the droplets formation of Pt is observed, suggesting the movement of material in liquid phase, as predicted by the self-organization model under the Marangoni effect.

Within the plasmonic theory of the formation of LIPSS structures, their period (Λ) depends on the dielectric permittivity of the material ε_m, on the transparent dielectric medium above with the dielectric constant ε_d, but also on the wavelength λ_0 of the incident radiation:

$$\Lambda_{\mathrm{LIPSS}} \approx \lambda_{\mathrm{SP}} = \lambda_0 \cdot \mathrm{RE}\left[\left(\frac{\varepsilon_\mathrm{d} + \varepsilon_m(t)}{\varepsilon_\mathrm{d}\varepsilon_\mathrm{m}(t)}\right)^{1/2}\right] \tag{6.11}$$

where $\varepsilon_\mathrm{m}(t)$ represent the transient dielectric permittivity due to ultraintense optical excitation of the order of TW cm^{-2} (Drude–Lorentz model). This parameter is a function of time, but also of the intensity of the incident radiation and depends on the density of the electron plasma created by the intense optical radiation. Thus, upon irradiation with an intense laser beam, for a very short time comparable to the

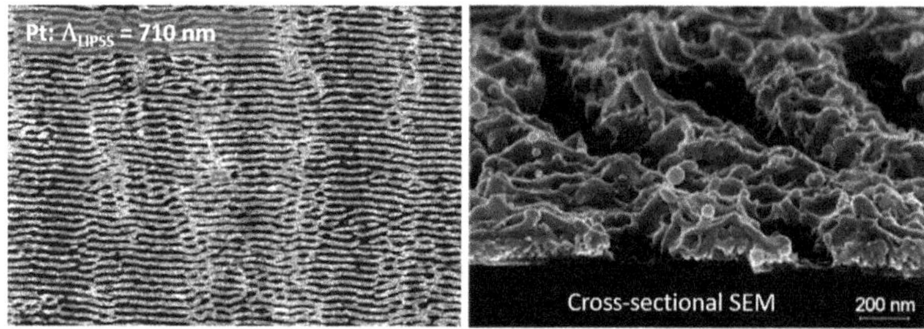

Figure 6.10. LIPSS formation on Pt film driven by self-organization of the molten material (Marangoni effect).

lifetime of the electron plasma, even a dielectric material will be considered in a 'metal-like' state, and material constants such as the dielectric constant undergo a transient change from the tabulated values determined under linear optical irradiation conditions. The variation of the laser intensity leads to a transient change in the dielectric permittivity $\varepsilon_m(t)$ and implicitly to the change in the wavelength of the surface plasmons that are assumed to give the period of the LIPSS structures. However, the variation of the laser intensity can only be done in a rather narrow range, around the ablation threshold. At much lower values of the laser intensity, the surface remains unchanged, or requires a longer irradiation time, while at values above the threshold there is a massive ablation of the material and the diminution until the destruction of the periodic structures. The same thing happens when varying the number of pulses, respectively, the scanning speed of the surface. In general, at least a few superimposed pulses are needed to induce the periodic structuring effect. For a few materials, it is possible to achieve LIPSS structures with a single laser pulse. Irradiation with a very large number of pulses, for a given energy, is likely to lead to the destruction of the periodic structures.

Although there are currently various physical models developed in the world to explain the mechanisms of generation of these nanostructures, the phenomena involved are still not fully understood. At present, many studies have as their main goal the removal of the limitations imposed by the characteristics of structured surfaces (the strong dependence of the morphology of the structures on the type of material and the presence of bifurcations) and to improve the contrast of quasi-periodic nanostructures.

6.2.2 Optimization of laser parameters for LIPSS nanotexturing

The formation of LIPSS is highly sensitive to laser irradiation parameters, such as pulse energy (or laser fluence) and scanning speed (which determines the number of pulses per spot). For practical applications, prior optimization of these parameters is essential. The first step in the process is the determination of the ablation threshold for the material to be irradiated. Around this value, the surface morphology can be tuned to achieve the desired nanotexturing. Figure 6.11 illustrates a typical setup for laser surface irradiation. The laser beam is first attenuated to the desired energy

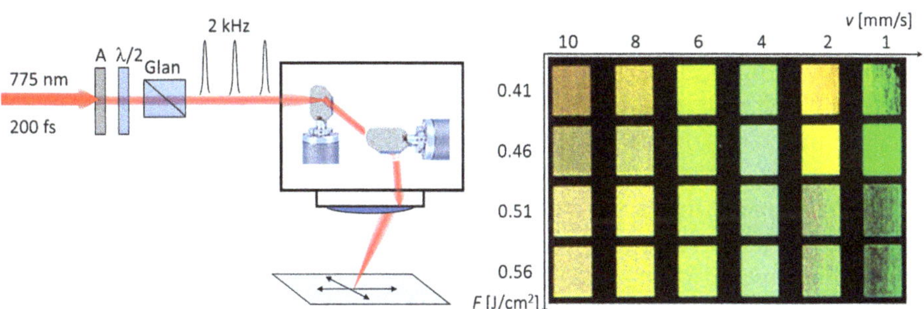

Figure 6.11. Setup for LIPSS generation with femtosecond laser pulses. Maps of scanning parameters are made in order to change the morphology of the nanostructured surface.

using neutral density filters in conjunction with a variable attenuator consisting of a Glan-laser polarizer and a half-wave plate. The half-wave plate can be mounted on a motorized rotational stage to enable computer-controlled adjustment of the pulse energy. An additional half-wave plate can be placed after the attenuator module to control the laser polarization and, consequently, the orientation of the LIPSS.

For large-area LIPSS formation, a practical configuration involves the use of galvanometric scanners combined with an f-theta focusing lens for beam focusing and steering, while the sample remains fixed. For industrial-scale applications, a hybrid scanning scheme can be implemented, in which the galvanometric scanner irradiates the surface line by line while the sample is sequentially repositioned using a motorized translation stage. This approach enables coverage of very large areas, limited only by the travel range of the translation stage.

A major technological challenge arises when irradiating complex 3D surfaces. Since LIPSS formation is highly sensitive to laser fluence, any deviation of the irradiated spot from the focal plane of the f-theta lens disrupts the formation regime, resulting in non-uniform nanotextures. This issue can be mitigated either by using non-diffractive Bessel beams or by employing advanced optomechanical systems capable of rapidly adjusting the beam focus position.

Bessel-like beams can be generated using an axicon lens or by employing a spatial light modulator (SLM). When focused, these beams exhibit an extended depth of focus (confocal parameter) on the order of a few millimeters along the z-axis, while maintaining an almost constant spot size. This is significantly longer compared to Gaussian beams, which typically have a confocal parameter of only a few hundred micrometers at a focal length of about 100 mm. The main drawback of using Bessel beams in laser processing is the considerable loss of laser intensity, as only a fraction of the optical power is concentrated in the central lobe of the beam. Nevertheless, if the initial laser power is sufficiently high, the remaining intensity can exceed the ablation threshold of the material.

An alternative approach for processing non-planar surfaces involves rapid lens positioning within the optical path to dynamically adjust the focal plane. However, this method introduces limitations in processing speed and accuracy, since precise synchronization is required among the galvanometric scanners, the focusing stage, and the surface topography.

The LIPSS formation regime can be switched from LSFL to HSFL primarily by adjusting the laser optical power and scanning speed. Figure 6.12 shows the scanning maps for speeds ranging from 1 mm s^{-1} to 8 mm s^{-1} and laser fluences between 0.115 J cm^{-2} and 0.179 J cm^{-2}. Low laser fluence combined with a low number of pulses (corresponding to higher scanning speeds) tends to produce LIPSS with fine periods on the order of about 100 nm. In contrast, irradiation regimes with higher laser fluences and larger numbers of pulses promote the formation of LIPSS with longer periods. The range of laser fluences that supports LIPSS formation is relatively narrow. For example, in the case of the Cr film, a variation in laser intensity of only about 50% is sufficient to shift from one formation regime to the other. A further increase in laser pulse energy drives the process into the ablation regime, resulting in the destruction of the previously formed LIPSS. Under

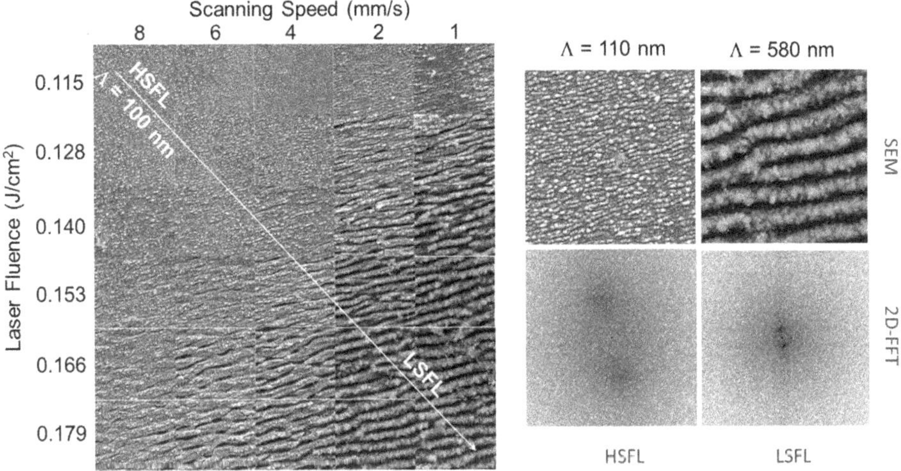

Figure 6.12. Scanning parameters for optimizing LIPSS on chromium films.

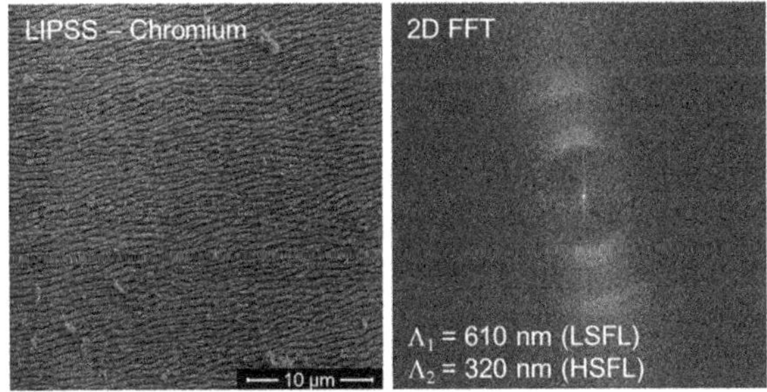

Figure 6.13. 2D FFT analysis of Cr LIPSS. Two types of LIPSS coexist on the same irradiated area.

intermediate conditions of laser fluence and scanning speed, both LSFL and HSFL can form simultaneously. Figure 6.13 shows an SEM image of LIPSS on a Cr film, along with the corresponding 2D FFT analysis, which clearly reveals the presence of two distinct spatial periods: $\Lambda_{LSFL} = 610$ nm and $\Lambda_{HSFL} = 320$ nm.

Direct measurement of the LIPSS period from SEM images is not accurate, as these structures are not perfectly periodic gratings but rather quasi-periodic features with bifurcations and small variations in periodicity and orientation. Therefore, it is more appropriate to define an average periodicity. The most reliable method to extract this value is through 2D fast Fourier transform (FFT) analysis of the SEM image. This approach can be implemented using various image processing tools, including widely used software such as *ImageJ*, or through custom scripts in programming environments such as *MATLAB* or *Python*. The image processing of SEM data involves the following steps:

- Image calibration: Use the scale bar provided by the SEM acquisition software to determine the calibration factor. The known distance (typically in micrometers) is correlated with its pixel length, yielding the calibration factor f in px μm^{-1}.

- 2D FFT computation: Apply a two-dimensional FFT to the SEM image to obtain its spatial frequency spectrum. Symmetric features observable in the Fourier image correspond to the spatial frequencies of the periodic structures. The distance from the center of the FFT image to the intensity maxima represents the spatial frequencies ν_{px}, expressed in px^{-1}.
- LIPSS period calculation: Convert the spatial frequencies from px^{-1} into the LIPSS period in micrometers using the following relation:

$$\Lambda \ (\mu m) = \frac{1}{f \cdot \nu_{px}} \tag{6.12}$$

6.2.3 Applications of nanotextured surfaces

There is a broad range of concepts and experimental efforts aimed at exploiting LIPSS in various applications, leveraging their key advantages: they can be generated on almost any material, under ambient conditions, and with relatively high throughput using modern laser systems operating at high repetition rates. One promising application is color marking of metallic surfaces, which can be used for purposes such as security marking of original products as an anti-counterfeiting method, or as a versatile tool in decorative design. Figure 6.14 presents a color map of LIPSS fabricated on a nickel foil. Since LIPSS behave as quasi-periodic gratings with defined periods and orientations, their diffractive properties produce colored effects on the irradiated surface. The perceived colors depend on both the laser irradiation conditions and the angle of incident light. The laser-processed Ni foils can serve as molds for pattern replication in roll-to-roll (R2R) manufacturing processes.

Other promising applications of LIPSS focus on the development of hydrophobic and superhydrophobic surfaces. A lotus-like effect can be achieved when surfaces are textured at the submicrometer scale, leading to water repellency. Such surfaces have potential uses as antibacterial and self-cleaning coatings, as well as in

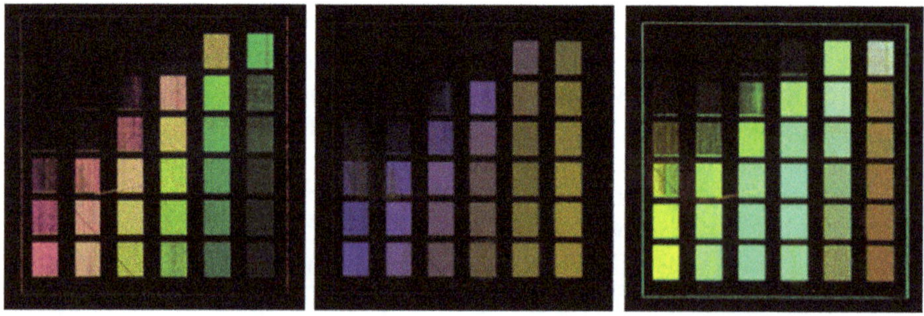

Figure 6.14. Colored appearance of a Ni foil surface at different illumination and viewing angles. The colorization effect arises from the diffraction of white light on LIPSS nanogratings.

applications requiring protection against water icing, corrosive fluids, or bacterial growth.

Surface-enhanced Raman spectroscopy (SERS) is an efficient analytical technique for detecting extremely low concentrations of chemical and biological species. The principle of SERS relies on the interaction between electromagnetic radiation and the conduction electrons of metallic nanostructures. Metallic nanoparticles such as gold, silver, or copper, with dimensions below 100 nm, exhibit strong optical field confinement well below the wavelength of the incident light. Commercial SERS substrates typically consist of random arrangements of gold nanoclusters or highly engineered geometries such as gold-coated pyramids fabricated on silicon chips. In practical detection procedures, the laser intensity is constrained by the damage threshold of the sample under analysis. Compared to these standard substrates, periodically self-organized nanostructures, such as LIPSS, can couple incident laser light more efficiently to surface plasmons, thereby enabling effective excitation at lower laser intensities. Consequently, considerable signal enhancement can be achieved with laser-induced periodic structures due to their strong interaction with polarized light and localized surface plasmons on metallic LIPSS.

This efficient coupling represents the key advantage of employing LIPSS for the fabrication of SERS substrates. As a result, SERS spectra exhibit sharp vibrational bands characteristic of specific molecular species or fluorescent labels, enabling reliable identification even within complex biological mixtures and facilitating multiplexed analysis. Furthermore, the use of LIPSS structures offers the additional benefit of easily and rapidly covering large areas with quasi-periodic nanostructures by scanning the surface with an ultrashort-pulse laser beam under controlled parameters.

6.3 Parallel near-field laser ablation

In typical laser processing configurations, the geometry of a structure is created either by keeping the sample fixed and moving the laser beam or by translating/ rotating the sample while maintaining a fixed beam position. In both cases, the process is inherently sequential, requiring point-by-point or line-by-line scanning of the surface, which results in relatively long processing times. The achievable resolution depends not only on the focusing optics but also on the precision and repeatability of the mechanical scanning systems.

A recently developed approach for parallel processing over large areas exploits the near-field enhancement effect of optical radiation at the interface of micro- and nanospheres made of transparent materials, or even metallic nanoparticles, deposited by self-assembly on the substrate surface. Upon irradiation, each microsphere acts as a microlens, focusing the incident laser beam into a sub-wavelength spot and enabling localized surface modification. However, this technique is limited by the inherent hexagonal symmetry of self-assembled microsphere arrays, restricting the achievable pattern geometries (figure 6.15).

In this section, an alternative method for large-area parallel surface structuring is presented. This overcomes the limitations of microsphere-based near-field processing. The approach involves the fabrication of microstructured polymer masks that

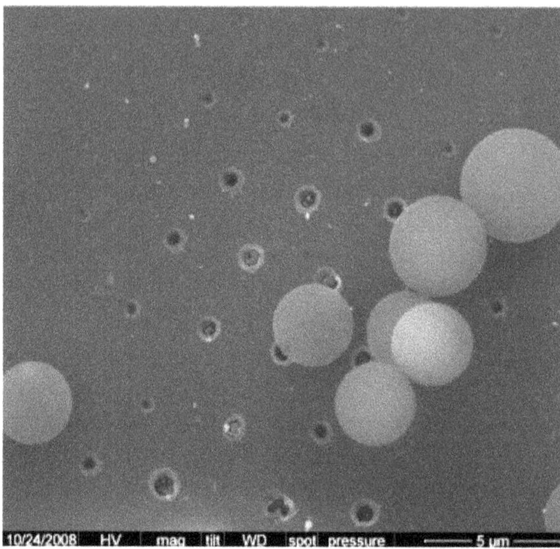

Figure 6.15. Parallel near-field laser ablation using polystyrene microspheres as focusing microlenses.

function as micro-optical elements for focusing laser radiation onto solid substrates (metal, semiconductor, dielectric). This technique eliminates the constraint of hexagonal symmetry inherent to microsphere-based methods and significantly reduces processing time compared to conventional sequential laser scanning [11].

The fabrication of microstructured masks is carried out via a photopolymerization process, using either conventional lithographic techniques or direct laser writing (DLW), according to a predetermined geometry. In the DLW approach, laser radiation is focused either on the surface of a photoresist or within the volume of a transparent photosensitive material. The effects induced by the laser strongly depend on the properties of the photoresist, as well as on the laser pulse duration and pulse energy.

Most photoresists are designed to polymerize under UV exposure and exhibit transparency in the near-infrared (NIR) range. However, when a femtosecond laser operating at 800 nm (e.g. Ti:sapphire source) is tightly focused within the volume of a photoresist, TPP occurs at the focal point. In this nonlinear process, two NIR photons combine to induce the same photochemical reaction as a single UV photon, enabling localized polymerization with submicrometer precision.

By computer-controlled scanning of the laser focal point within the photoresist volume, arbitrary 2D and 3D structures can be fabricated according to the predefined design. The achievable resolution and accuracy depend primarily on the optical focusing system, the performance of the laser scanning hardware, and the characteristics of the laser source. A detailed description of the TPP processing protocol is provided in the following section dedicated to 3D laser lithography. The fabricated masks, being transparent at the processing laser wavelength, function as micro-optical elements that locally enhance the optical field through near-field focusing at the mask–substrate interface. This near-field intensity enhancement

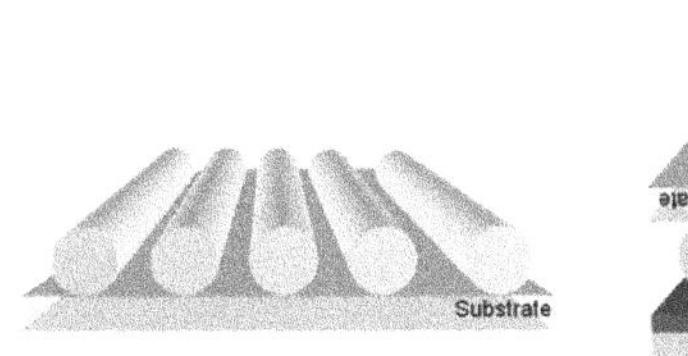
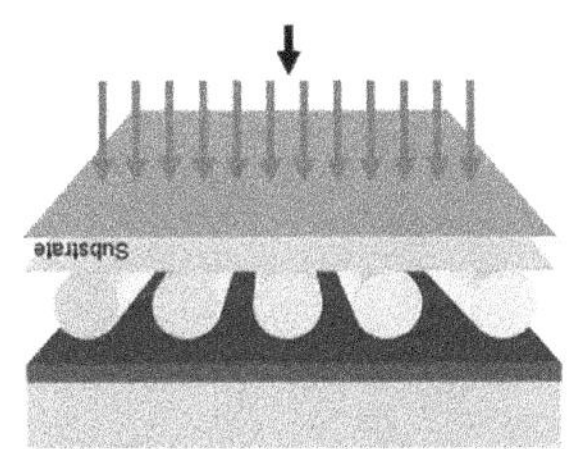
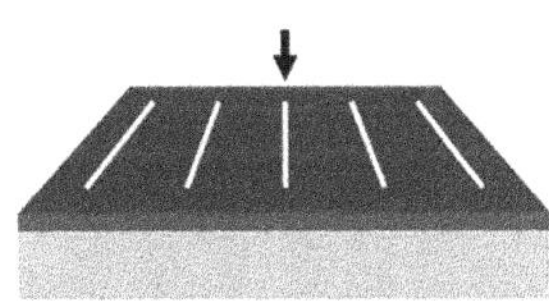

Figure 6.16. Parallel near-field laser ablation.

enables parallel laser processing, allowing the simultaneous structuring of large areas with improved efficiency compared to conventional sequential scanning techniques.

In figure 6.16, the mask consists of a transparent support layer and an array of micro-optical focusing elements fabricated on this support. The support can be made from glass, quartz, or another material transparent to the processing laser wavelength. The micro-optical elements are designed with various geometries—such as micro-spheres, cylindrical structures, or other regular and irregular shapes—which may be arranged in periodic or aperiodic patterns with micrometric or sub-micrometric dimensions. These micro-optical focusing elements are fabricated from a polymer material that is also transparent to the laser wavelength used for film ablation. A critical criterion for selecting the polymer is its optical damage threshold, which must exceed the ablation threshold of the target material to ensure mask durability during processing. Figure 6.17 presents atomic force microscopy images for a polymers mask and the structured silicon surfaced under femtosecond laser irradiation in the near-field enhancement regime. In this example, SU-8, a negative-tone photoresist, was employed as the polymer material. SU-8 exhibits an optical damage threshold of approximately 1.7 J cm^{-2}, significantly higher than the ablation thresholds of most metallic films, making it an excellent choice for this application. Other polymers such as PMMA can be used and adapted to the desired mask fabrication technique [8].

The material to be processed can be the surface of a solid material, or a thin film of material deposited on a substrate. The mask is placed on the surface of the material to be processed, so that the micro-optical focusing elements are in contact with or in proximity to the surface of the solid material. The mask and the material to be processed are irradiated over a large area, either with a single laser pulse or with a train of laser pulses, with a fluence below the optical damage threshold of the polymer material from which the focusing optical elements are manufactured. The average fluence of the laser beam is also lower than the damage threshold of the material to be processed. Through the near-field optical enhancement effect, the laser radiation is focused at the interface between the micro-optical elements and the material surface. Locally, the optical irradiance exceeds the optical interaction threshold, causing laser ablation only on the material surface, without affecting the mask.

As shown in figure 6.15, a mask generated by self-organized microspheres will produce hexagonal patterns, while cylindrical micro-optical focusing elements generate line-type microstructures by laser ablation. Combinations of cylindrical

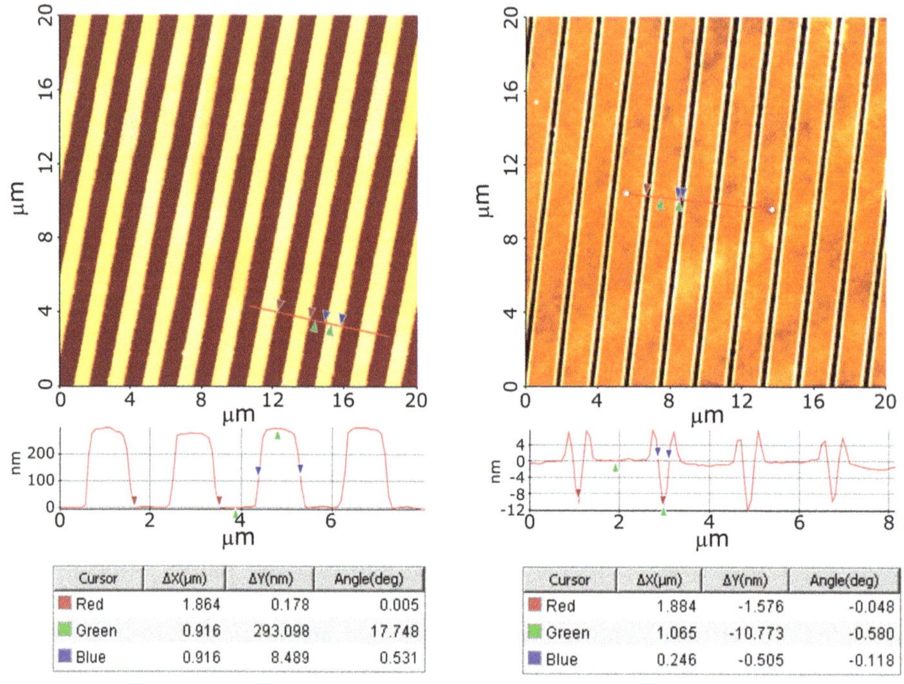

Figure 6.17. Parallel near-field laser ablation of periodical structures. (a) polymer mask; (b) laser-processed silicon surface.

structures arranged according to a certain predetermined arrangement produce by laser ablation in near-field enhancement regime, microstructures with a shape corresponding to the arrangement of the cylindrical structures. The arrangement on the surface of the structures obtained by laser ablation respects the arrangement of the cylinders in the component of the microstructured masks (figure 6.17).

6.4 Two-photon polymerization

In the case of laser-based additive manufacturing at macro-scales, there are various physical phenomena that either do not influence, or are negligible towards the experimental results. Some of these effects are surface charge accumulation, fluid dynamics, near-field optical phenomena and others. However, these physical phenomena become increasingly important as spatial features get smaller, especially at submicrometer scales. In this chapter we are going to briefly discuss a fast growing laser-based additive manufacturing technology that is used to obtain arbitrary structures with sub-micron resolution, below the diffraction limit. The method is based on two-photon absorption (TPA) processes and is usually called laser direct writing via TPP (LDW via TPP), or simply 3D Lithography (3DL). Photoresists used in 3DL experiments are generally liquid before and during processing. This imposes a different approach to the experimental setup and sample preparation. This section covers standard sample (or, rather, commonly encountered) setups, as well as examples of custom sample setups in order to reach specific fabrication parameters.

Standard sample setups are centered on drop-casting the photoresist on a substrate. As the name suggests, 'drop-casting' refers to a simple method of applying liquid photoresist on a substrate by putting a drop in the processing area. The drop-casting method can be used with complementary methods for increasing resolution and improve geometrical features. One such method is oil immersion. This method is used for increasing the resolving power of the microscope used for processing, through increasing the numerical aperture. Another method is immersing the objective directly in photoresist, in order to process taller structures (DiLL [1]). These methods are described in more detail below.

Non-standard sample setups were developed to solve application specific issues. One method is based on containing the photoresist between a substrate and a thin glass cover. The glass cover is placed at a precise distance from the substrate. This setup improves some geometrical aspects of structures while limiting fluid movements and voxel dimensions. It is usually employed for writing taller structures that benefit from a larger voxel, i.e. when DiLL method is not efficient. The glass cover method can be further modified to printing structures at substrate edges, connecting multiple substrates together or printing on non-planar substrates.

Photoresists used for additive manufacturing experiments presented in this work are mainly composed of photoinitiators (PI) and monomers (M), a solution which is also known as 'pre-polymer'. The basis for TPA additive manufacturing is photo-polymerization. The solutions we used are photoresists where the polymerization process is based on chain reaction. Polymerization of chain-reaction photoresists is composed of three main steps: initiation, propagation and termination. When exposed to radiation with an appropriate wavelength, the photoinitiator is ionized and produces chemically active species, such as free radicals (R) [16]. Resulting free radicals can attach to monomer molecules, which further results in a different molecule with active terminations. These new molecules determine the propagation step, in which other monomer molecules continue to attach to the active terminations, resulting in a polymeric chain. This process is known as 'chain polymerization'. The growth of a polymeric chain is stopped when a different active termination is connected to the first, be it a free radical or the termination of a different polymeric chain. This process determines the termination step. A more detailed description of photopolymerization processes can be found in [13]. This process is described by the following equations:

$$\text{Photons} + \text{PI} \rightarrow \text{PI}_{\text{excited}} + \text{R}_{\text{active}} \tag{6.13a}$$

$$\text{R}_{\text{active}} + \text{M} \rightarrow \text{RM}_{\text{active}} \tag{6.13b}$$

$$\text{RM}_{\text{active}} + \rightarrow \text{RM}_{2,\text{active}} \tag{6.13c}$$

$$\text{RM}_{2,\text{active}} + \text{M} + \cdots + \text{M} \rightarrow \text{RM}_{n,\text{active}} \tag{6.13d}$$

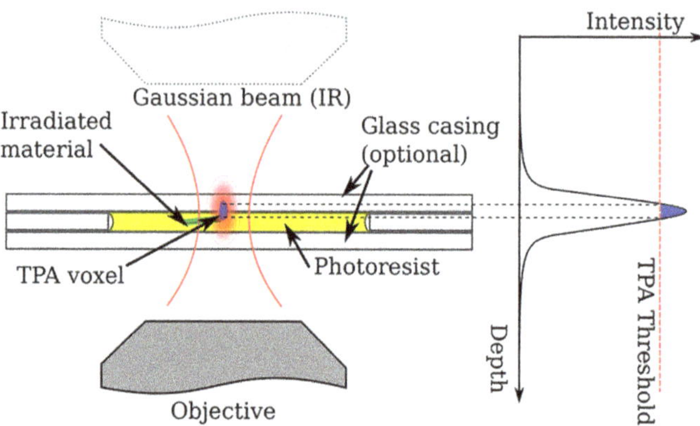

Figure 6.18. Formation of a voxel, presented in the context of TPP.

$$RM_{n,active} + R_{active} \rightarrow RM_n R_{terminated} \qquad (6.13e)$$

$$RM_{n,active} + RM_{k,active} \rightarrow RM_{n+k} R_{terminated} \qquad (6.13f)$$

where we recall that PI represents the photoinitiator, R represents free radicals and M represents monomer molecules.

Photopolymerization is used in various applications, both in industry as well as research and development. Among the applications, we can mention surface/material curing, stereo-lithography, imaging, printing, and others [3, 9]. For additive micro-manufacturing, however, photopolymerization must occur through TPA processes for several reasons. The two main reasons for this is that TPA allows for optical processing below the diffraction limit, as well as in a defined volume (voxel). There are several requirements for obtaining a voxel. Generally, ultrashort pulses with a Gaussian traverse beam profile and pulse shape are used. This allows good control over voxel formation and positioning. The working principle is presented in figure 6.18.

In order to apply TPA to additive micro-manufacturing, three main parameters of the laser pulses must be considered. The first parameter is the central wavelength. The central wavelength must be appropriate for specific materials, i.e. the photoresist must be transparent to this wavelength, as well as showing strong absorption to the second harmonic. The second parameter is pulse duration. In order for TPA to occur, high electric field intensities are required. However, this must be achieved with as low as possible thermal effects. This means ultrashort pulses are advantageous. The third main parameter is the beam profile. A Gaussian beam profile allows for processing below the diffraction limit by having only the peak of the profile above the intensity threshold for TPA.

6.4.1 Typical sample setup for two-photon polymerization in drop-casted photoresist

Commonly encountered sample setups are centered around drop-casting the photoresist on a substrate. The starting position is necessary to be on the surface of the substrate (or very close to it) so that the written structure adheres to the substrate, and does not change position during the exposure. In order to improve photoresist adherence on the substrate, surface treatment steps can be included. Substrate surface cleaning depends on the photoresist, and varies from simple washing with a solvent such as isopropyl alcohol (IPA) or ethanol, to surface treatment with a plasma cleaner, heating or applying complex chemical compounds. The working principles are presented in figure 6.19.

Figure 6.19(a) presents the general working principle. This is the simplest method of employing TPA-based processing in liquid photoresists. This method is the fastest and with highest reproducibility of microstructures. Laser pulses are focused using a microscope objective with appropriate working distance, where the focal point can reach above the substrate while maintaining a safe distance from the substrate. In this case, we have to take the optical path difference into account in order to obtain height precision. The polymer that results from the exposure has a slightly different refractive index than the unexposed photoresist. While this allows us to observe the fabrication in real time, it can also affect the traverse profile of the laser, as the height of the microstructure increases. This determines a drop in resolution and higher intensity requirements for polymerization. It is evident that the higher the magnification, the smaller the spatial features. This method is limited by the working distance of the objective, which is dependent on the magnification. In order to use higher magnifications, additional steps in the sample setup are required (presented in figures 6.19(b) and (c)).

The second method, presented in figure 6.19(b), comes as a solution to improve the resolving power of the microscope objective, therefore obtaining a smaller laser spot and, implicitly, smaller spatial features of the structures. The resolution improvements brought by the addition of immersol are great, lowering the minimum imprint from 1 μm up to 0.2 μm in diameter. This also allows for obtaining surface structures with features as low as 90 nm, albeit limited to 2D. This is done by focusing the laser deeper in the substrate, leaving only the peak of the voxel reaching the photoresist. However, the main drawback of this method is that it limits the height of the structures, depending on the working distance of the objective (a maximum of 120 μm in our case). In this case, the same issue appears for higher structures as for the case presented in the previous paragraph (figure 6.19(a)).

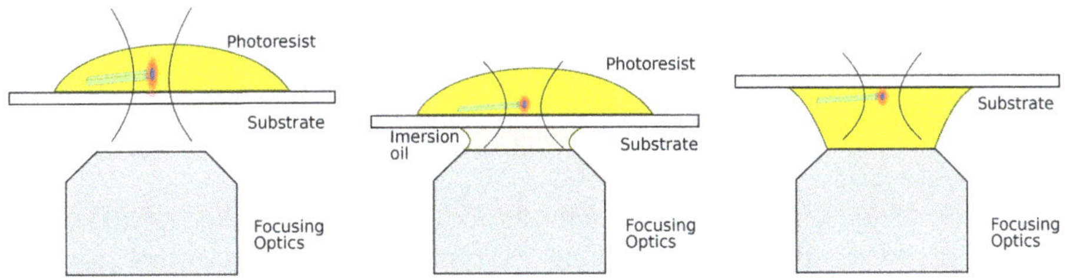

Figure 6.19. Typical sample setups for TPP in liquid photoresists: (a) normal writing, (b) oil-immersion writing, (c) dip-in laser lithography [1].

Figure 6.20. Typical sample preparation for normal writing (63× objective, photoresist on top of substrate).

In order to completely avoid the height-resolution issues, we can employ the polymer-immersion sample setup presented in figure 6.19(c).

If we want to obtain a microstructure taller than 120 μm (oil immersion case, figure 6.19(b)), or even taller than the droplet of photoresist (direct focusing case, figure 6.19(a)), we can use the polymer-immersion setup. In this case, we don't have the height limits imposed in the first two cases. We can increase a given microstructure height as long as the photoresist droplet remains attached to the microscope objective. The traverse intensity profile of the laser does not suffer alterations as structures gain height. It can prove to be time-inefficient for processing larger volumes due to the voxel size. This sample setup is presented in figure 6.19(c). The first two methods can be applied only in the case of substrates that are transparent to the laser wavelength. The third method, however, can be used for processing structures on any adhering substrate.

Preparing a sample for typical writing in drop-casted photoresist is fairly straightforward. We first position the substrate on a metallic holder. We then fix it with a rubber-based adhesive. We drop-cast the liquid photoresist on the adequate surface, and then insert it into the laser processing station. Basic steps of sample preparation for normal writing using a 63× objective, with the photoresist drop-casted on top of the substrate, are presented in figure 6.20. Similar steps are followed for the other two sample setups mentioned in this chapter, with small differences: for oil immersion we add the oil on the surface oriented towards the microscope objective, while for dip-in laser lithography we drop-cast the photoresist on the surface oriented towards the microscope objective.

6.4.1.1 Atypical sample setup for two-photon polymerization in drop-casted photoresist

Non-standard sample setups usually come as additional steps to the typical methods, in order to solve specific issues. One example is scaffold printing for tissue engineering. In this case fluid movements need to be limited, resolution and traverse intensity profile need to remain the same throughout the fabrication process, while

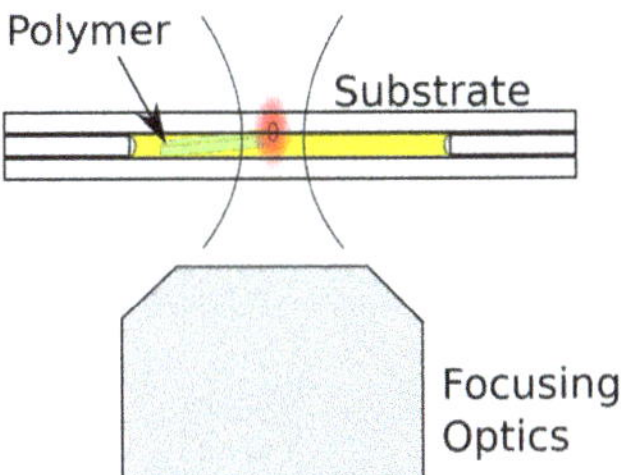

Figure 6.21. Glass window, sandwich-type sample setup, used for TPP of micron-level resolution structures on non-transparent substrate or scaffolds for tissue engineering.

also having a larger voxel for time efficiency (we used a voxel of 2 μm diameter and 4 μm height). One of the non-standard processing setup offers a solution for this issue by employing a glass cover. This glass cover, along with appropriate spacers, encases the liquid photoresist between itself and the substrate. This case is schematically presented in figure 6.21. The exposure becomes similar to cross-linking photoresists, where the sample is a thin film. This setup allows for faster and more precise printing of microstructures as tall as 120 μm. Using this setup, we can also write structures on non-transparent materials.

Another example we encountered comes as a solution to fabricate couplers for photonic integrated circuits (PICs) at the edge of the substrate. Polymer-based PICs are more efficiently used with end-coupling at the edge of the substrate. This is usually done by exposing a specific set of photoresists, mostly based on cross-linking. For example, photoresists such as polymethyl methacrylate (PMMA) or SU-8 are subjected to pre-processing steps. The pre-processing results in a thin solid film that is further subjected to UV or TPA-based exposure. In the case of PICs, this means that input and output waveguides can easily be constructed at the edge of the photonic chip, which allows for an effective and easily applicable coupling method, known as 'end-coupling'. For on-surface coupling of light, different versions of Bragg gratings are employed, yet these are only efficient in the case of high index contrast. Photopolymers with good optical properties possess a low refractive index, which means a low contrast and therefore a lower efficiency of Bragg gratings. Since we used liquid photoresists, writing structures at the edge is limited by several factors. One factor is determined by the surface tension of the photoresist drop, which limits the height of the structures. Moreover, having liquid photoresist at the physical limit of the substrate, while using an inverted microscope and a moving sample is not recommended, as the substance might flow or drop onto optical devices below. Standard sample setups don't allow reaching the edge of a substrate in an easy and reproducible manner. A solution we used is presented in figure 6.22. This sample setup can also be used to interconnect structures already existent on different substrates [14].

Generally, there are three main steps to prepare a glass window sandwich-type sample setup (figure 6.23). The first step is placing the substrate and photoresist (figure 6.23(a)). The substrate should reside on top of the metallic sample support. This is because we first have to assemble the whole sample before applying any adhesive to secure the sample for processing. The substrate can be either transparent

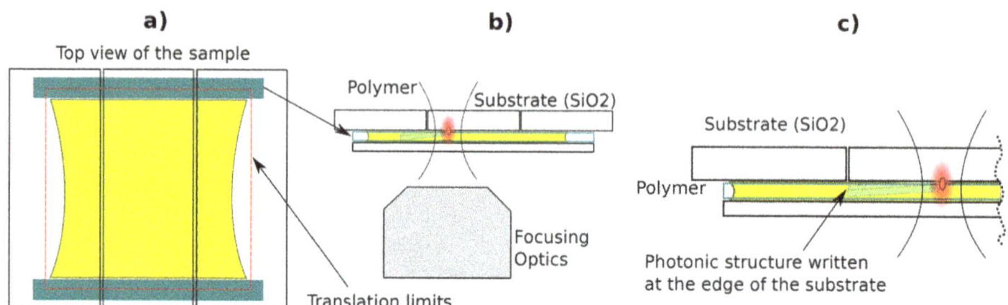

Figure 6.22. Glass window, sandwich-type sample setup, used for fabrication of structures on multiple substrates and/or interconnecting structures.

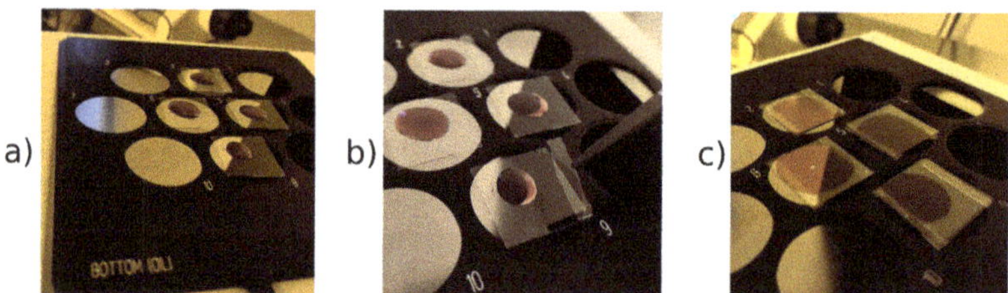

Figure 6.23. Steps to obtain a glass window, sandwich-type sample setup: (a) place substrate and drop-casted polymer; (b) place glass spacers; (c) place the glass window.

or opaque. The second step is placing and aligning the spacers (figure 6.23(b)). During this step, spacers can touch the drop-casted liquid photoresist and move them out of place due to surface tension pulling the spacer towards the higher volume of photoresist. Positioning of the spacers becomes stable after the photoresist spreads between spacers and substrate, which determines a more even pull of the photoresist on the glass spacers. The third step is applying the glass window (figure 6.23(c)). After applying it, we must wait for the arrangement to stabilize, as the surface tension moves it out of place similarly to spacer movements. The glass window can also require repositioning while waiting. The wait time depends on the viscosity of the photoresist. After the setup stabilizes, we apply a rubber-based adhesive to the corners to secure it in place. Adhesive application should be done at the end in order to keep all the elements aligned and parallel by avoiding the spread of the adhesive between any two elements. Unparallel surfaces can modify substrate flatness and affect light propagation and focusing.

6.4.2 Immersion methods for sample development

Additive micro-manufacturing through means of TPP is generally realized using high viscosity, liquid, negative photoresists, drop-casted or spin-coated on a glass substrate. In our case, the substrate is fixed on a thin metallic support, specifically designed for Nanoscribe Photonic Professional [1]. As described above, the exposed photoresist undergoes a series of chemical reactions, mediated by an intermediary molecule known as 'photoinitiator'. These chemical reactions bond monomer molecules, resulting in a solid microstructure. After exposure, the unexposed material must be removed, i.e. the sample undergoes a development phase. This generally means that sample development is done through immersion in a specific solvent, for a specific amount of time. The unexposed material is dissolved in the solvent, leaving behind the microstructure defined by the exposure path.

In our case, we set the metallic support with all the samples in an operating support, as shown in figure 6.24(a), (an elevated frame that holds the metallic support in place while it allows for ease of access to all samples, from both the top and the bottom). We carefully remove the adhesive holding the glass substrate in place so as not to induce bending forces that can result in breaking the substrate, figure 6.24(b).

As mentioned previously, when using 3D lithography, the exposed material can show lower tensile and flexural strength while still undeveloped (i.e. inside the unexposed material). In this phase, the shape of the microstructures can be affected for several reasons, among which fluid movement shows the strongest influence. As such, different immersion methods can provide better results for differently shaped microstructures. We can use various containers for the developing solvent that are better suited for specific microstructures. Some examples are shown in figure 6.25.

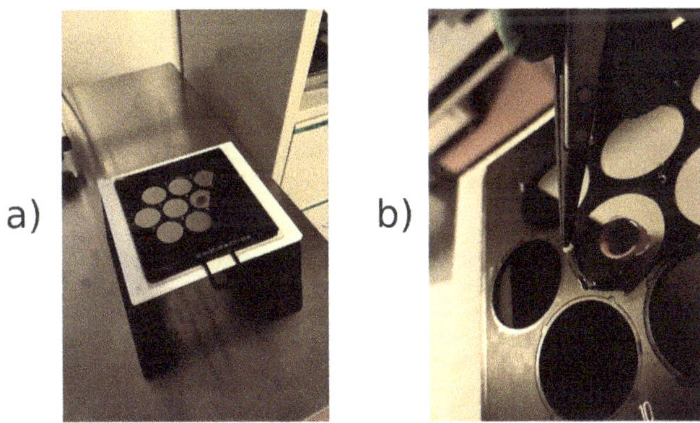

Figure 6.24. (a) The metallic sample support, positioned on the elevated operating support, before detachment and immersion in solvent for the development phase, (b) sample detachment.

Figure 6.25. Example of containers and tools used for development of 3D printed microstructures, through submersion in specific solvents.

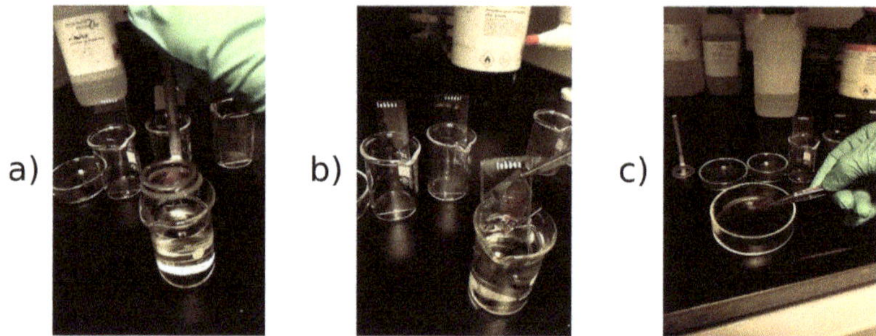

Figure 6.26. Different immersion types used to develop microstructures presented in this chapter, (a) flat positioning in tall container, (b) inclined positioning in tall container and (c) flat positioning in short/wide container

Development is often done using three types of immersion: flat positioning in a tall container, inclined positioning in a tall container and flat positioning in a short/ wide container. These methods of immersion are presented in figure 6.26.

Flat positioning in a tall container is the generally used method of immersion and provides good results for 'well rounded' 3D structures (the sizes are comparable on each of the *XYZ* axes), as well as for structures with high aspect ratio and/or sensitivity to traversal fluid movements, i.e. fluid movements parallel to the substrate's surface.

Inclined positioning in a tall container provides better results for porous structures with good structural strength. It allows for solvent to move throughout the volume of the structure. This immersion method also requires better adherence on the substrate.

Flat positioning in a short/wide container is used for microstructures with reduced height, but large footprint. An example of such microstructures is diffractive optical elements. This method allows for a faster and more homogeneous development of the sample, due to the movement of the dissolved photoresist away from the surface of the sample. Additional gentle moving of the container, in order to produce fluid movements, can speed the process, however, it also increases the risk of exfoliation (see figure 6.29). If the container is held stationary, this method can also be applied to simultaneously develop multiple samples that are usually used with the first immersion method.

If our objective would be to fabricate microstructures with very small features that are highly sensitive to fluid movement during immersion-based sample development, then we can use the multi-solvent approach (see figure 6.27). In other words, we can switch developers without pulling the sample out of the fluid. For example, some acrylic-based photoresists that are optimized for TPP are developed using immersion in a solvent such as IPA or propylene glycol mono-methyl ether acetate (PGMEA) for a specific duration, and then air dried. The problem is that when air-drying a sample with micrometer scale features (or below), the surface tension of the evaporating solvent can affect the sample's geometry. One solution to this issue is to switch the immersion fluid to a more volatile solution that would impart lower forces

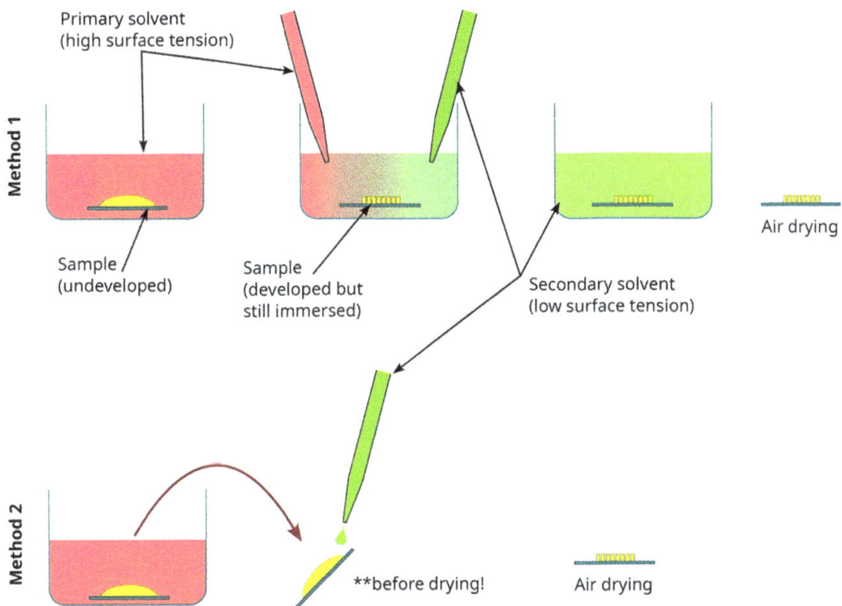

Figure 6.27. Two common methods of solvent switching for high-resolution microstructures, in order to mitigate possible damage induced via surface tension of the evaporating solvent in immersion-based sample development.

through the surface tension. A specific example would be the Novec 7100 engineering fluid.

We can achieve this either by successively subtracting the solvent, while also adding the new more volatile solvent, which results in a change of concentration in favor of the more volatile solvent, or simply washing the sample with the more volatile solvent immediately after finishing the immersion process (before drying).

6.4.3 Possible fabrication issues for TPA additive manufacturing

Experimental results can drift away from the design due to various reasons. If there are impurities in the photoresist, or the solution is not homogeneous, near-field focusing can occur. Localized high energy density can produce microexplosions, which can affect the final result. Laser average power that is higher than required can produce the same effects and does not necessarily appear at the start of the fabrication process. If an explosion occurs, for whatever reason, it results in a polymeric structure of random shape and form. Since the polymer is still transparent, but has a different refractive index than the photoresist, it can produce near-field focusing, similar to impurities. This in turn produces new explosions, continuing as a chain reaction as the beam moves, until the laser turns off. An example of polymer explosion during processing is presented in figure 6.28.

Polymer shrinking is another factor that must be considered when designing a structure. If the designed structure has a large area or volume, the shrinkage process pulls the resulting polymer away from the substrate, which determines exfoliation.

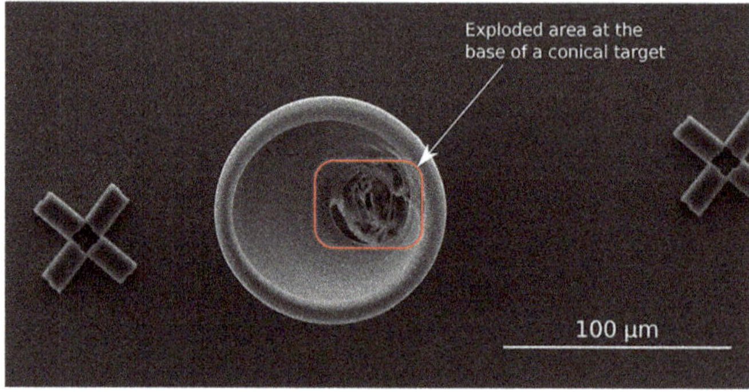

Figure 6.28. Conical microtarget with exploded polymer at the base of the structure.

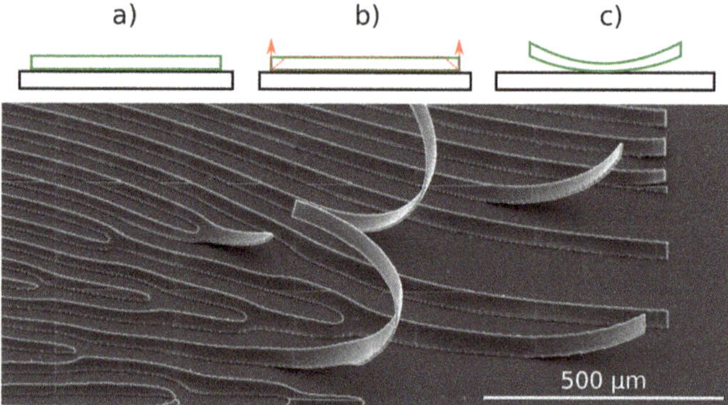

Figure 6.29. Diffractive optical element with exfoliated features determined by polymer shrinkage. (a) polymerized structure; (b) flexural stress due to shrinkage at polymer–air contact surface; (c) flexural strain resulting after polymer structure detaches from the substrate.

The surface tension of the evaporating solvent can also initiate and support exfoliation. The exfoliation process is presented in figure 6.29. The polymer–substrate contact surface cannot be lowered under the influence of shrinkage, because of adherence. However, the polymer–air contact surface does not have this restriction. This results in forces pulling towards the mass center at the polymer–air contact surface, which results in flexural stress and, if the polymer detaches from the substrate, flexural strain.

Exfoliation is mainly influenced by photoresist type and, as described above, surface adherence. Each photoresist type is recommended for a specific application due to its properties, among which shrinkage is often mentioned. We used glass and SIO_2 substrates for our experiments. Substrate adherence can be improved through several methods. The simplest method is cleaning the surface with a solvent such as IPA, acetone or ethanol. Another method is surface treatment with a plasma cleaning step. This step can clean and/or passivate the substrate.

Another issue that is encountered, usually for small and closely packed structures, is polymer fusing, or welding. This effect manifests as adjacent structures changing

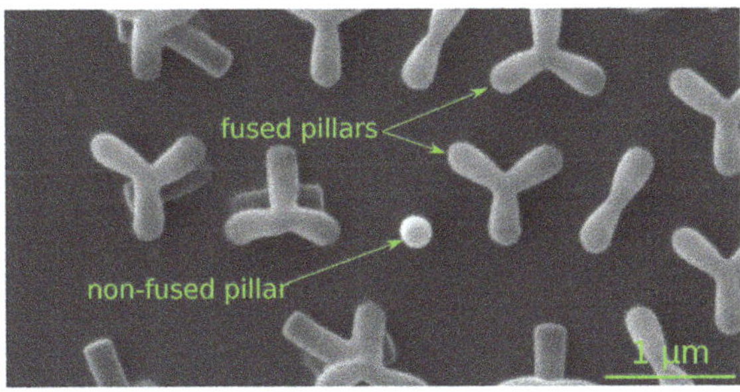

Figure 6.30. Top-view SEM micrograph showing structure welding.

shape or bending towards one another, ultimately fusing together. This process leaves no specific structural signs that could determine whether they were separate structures. Shape bending and clustering is influenced by several factors, such as electrostatic forces (photoresists are dielectric), surface tension of the evaporating solvent during development, elastic forces of the polymeric material, Van der Waals forces, and others. Welding happens primarily due to the mechanism of polymerization. The polymeric chains that form under irradiation do not follow a regular or periodic spatial distribution. Instead, chains are randomly packed together and wrap around each other until the irradiated volume is saturated, i.e. there are not enough monomer and photoinitiator molecules to sustain further polymerization. Before sample development and *drying*, the polymer is flexible and contains randomly distributed molecules of unactivated photoinitiator, monomer and solvent/thinner. When structures come into contact before drying, polymer chains intertwine, resulting in welding. After development and drying, the polymer structures become rigid and welding is eliminated. An example of structure welding is presented in figure 6.30.

In this subsection we have argued that small structures can weld together, large-area structures can exfoliate and complex structures can produce micrometer-level explosions. All these phenomena are possible if some aspects of this technology are not taken into consideration. Even so, additive micro-manufacturing using TPP represents a technology with great application potential in various domains. When all necessary aspects of geometry, material and laser processing are considered, we can obtain microstructures that show high fidelity towards their design. An example can be observed in figure 6.31, which represents the Bran Castle, designed in SketchUp Make 2017 and fabricated with Nanoscribe Photonic Professional [1], by the authors, at the Center for Advanced Laser Technologies - National Institute for Laser, Plasma and Radiation Physics, near Bucharest.

6.5 Subtractive manufacturing through laser-assisted etching

The working principle of micrometer-level 3D printing using laser direct writing is often the same: strong focusing optics generate a voxel inside a transparent material where multiphoton interactions occur. There isn't, however, a universal technology for all 3D printed microstructures. For lab-on-a-chip applications, most

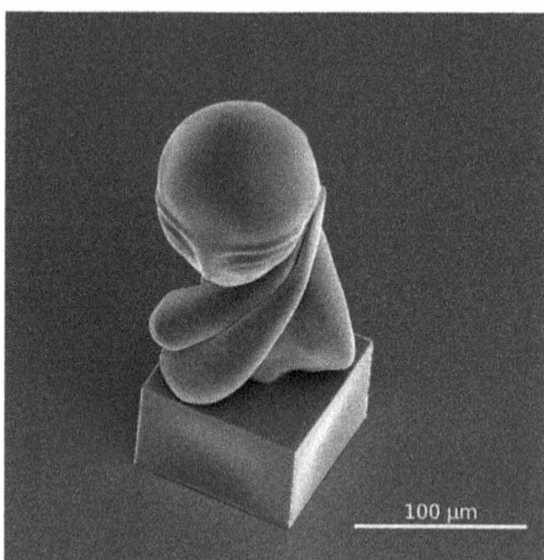

Figure 6.31. SEM micrograph of Bran Castle (left side) and Mademoiselle Pogany by Constantin Brancusi (right side), 30° angle view, designed and fabricated for demonstrative purposes.

encountered materials are polymers. However, in most cases, polymers are not reusable, as they can get contaminated and cannot be cleaned. Moreover, some structures can be fabricated more efficiently through subtractive manufacturing techniques, rather than additive manufacturing. Photosensitive glass is an alternative material for micrometer-level 3D printing, albeit through subtractive manufacturing. It boasts several advantages when compared to polymers, for microfluidic applications. The most important advantages are reusability, the lack of shrinkage and higher laser fluence tolerance.

For microfluidic devices, subtractive manufacturing provides a significantly more time-efficient fabrication method. As technological solutions involving 3D laser direct writing, we can employ TPP using positive photoresists or laser-assisted etching of photosensitive glasses. Positive photoresists are based on cross-linking. This means they require a certain pre- and post-exposure procedure. This usually involves spin coating the pre-polymer, pre-baking, exposure, post-baking and then development. The objective of the baking step is to eliminate the solvent from the photoresist. Solvent evaporation must be homogeneous throughout the whole volume. As such, it must be done on thin films with thickness varying from a couple of micrometers up to tens of micrometers. Therefore, using cross-linking positive photoresists impose limitations regarding the height of the structures. An alternative is to construct a negative mask of the structure and then stamp-print it on a polymer, such as polydimethylsiloxane. While this method is often used, it limits the resolution of the structures, and also limits the geometry to 2.5D.

Photosensitive glass provides an alternative to polymers, for subtractive manufacturing. The glass we used, Foturan, does not require any additional pre-processing before exposure. It does not impose the same limitations regarding the height of the structures and is reusable. It is often processed used femtosecond laser pulses, but we have proven it can be successfully and more efficiently processed using picosecond pulses.

Foturan glass processing involves three main steps: **exposure, annealing** and **etching**. Exposure ionizes a certain element in the glass, and the freed electrons are captured by another element. Annealing provides structural and chemical modifications to the glass. Etching removes the exposed volume.

More precisely, **exposure** with radiation with $\lambda < 350$ nm ionizes Ce^{3+}, resulting in $Ce^{4+} + e^-$. The Ce^{3+} ions in the glass matrix act as a photoreactive agent. Ag^+ ions collect the e^- released during exposure, and create atomic Ag.

Thermal **annealing** is done in two steps. First step represents *nucleation tempering*, where atomic Ag forms clusters/nanoparticles (structural modifications). This is done at a temperature of ≈ 500 °C. The second step of the annealing represents the crystallization step, when the crystalline Li metasilicate forms around the Ag nanoparticles (chemical modifications). This is done at a higher temperature, of > 560 °C (we used 600 °C).

The last step represents selective **etching** of the crystalline metasilicate volume using diluted hydrofluoric acid. The acid etches the metasilicate around 20 times faster than the unexposed glass.

References

[1] Nanoscribe GmbH *https://www.nanoscribe.de* (last accessed 5 January 2019)

[2] Albu C, Dinescu A, Filipescu M, Ulmeanu M and Zamfirescu M 2013 Periodical structures induced by femtosecond laser on metals in air and liquid environments *Appl. Surf. Sci.* **278** 347–51

[3] Ali T M, Fanny L, Jacques L and Jean-Pierre F 2013 Photopolymerization reactions: On the way to a green and sustainable chemistry *Appl. Sci.* **3** 490–514

[4] Anghel I, Jipa F, Andrei A, Simion S, Dabu R, Rizea A and Zamfirescu M 2013 Femtosecond laser ablation of TiO_2 films for two-dimensional photonic crystals *Opt. Laser Technol.* **52** 65–9

[5] Birnbaum M 1965 Semiconductor surface damage produced by ruby lasers *J. Appl. Phys.* **36** 3688–9

[6] Bonse J and Gräf S 2020 Maxwell meets marangoni–a review of theories on laser-induced periodic surface structures *Laser Photon. Rev.* **14** 2000215

[7] Călin B -S, Dobrea C, Tiseanu I and Zamfirescu M 2021 Laser microfabrication of conical microtargets for laser driven particle acceleration *J. Laser Appl.* **33** 012054

[8] Feng W, Wan Y C and Wang X 2020 PMMA-based microsphere mask for sub-wavelength photolithography *Nanomanuf. Metrol.* **3** 199–204

[9] Fouassier J P, Allonas X and Burget D 2003 Photopolymerization reactions under visible lights: principle, mechanisms and examples of applications *Prog. Org. Coat.* **47** 16–36

[10] Gaković B, Zamfirescu M, Panjan P, Luculescu C, Albu C and Petrović S 2024 Laser ablation and LIPSS formation at static and dynamic multi-pulse regime on protective Al_2O_3/ TiAlN coating *Opt. Quantum Electron.* **56** 555

[11] Jipa F, Dinescu A, Filipescu M, Anghel I, Zamfirescu M and Dabu R 2014 Laser parallel nanofabrication by single femtosecond pulse near-field ablation using photoresist masks *Opt. Express* **22** 3356

[12] Jipa F, Ionel L and Zamfirescu M 2024 Advances in design and fabrication of micro-structured solid targets for high-power laser-matter interaction *Photonics* **11** 1008

[13] Jipa F, Zamfirescu M, Velea A, Popescu M and Dabu R 2013 *Femtosecond Laser Lithography in Organic and Non-Organic Materials Updates in Advanced Lithography* (IntechOpen)

[14] Lindenmann N, Balthasar G, Hillerkuss D, Schmogrow R, Jordan M, Leuthold J, Freude W and Koos C 2012 Photonic wire bonding: a novel concept for chip-scale interconnects *Opt. Express* **20** 17667–77

[15] Liu J M 1982 Simple technique for measurements of pulsed Gaussian-beam spot sizes *Opt. Lett.* **7** 196

[16] Odian G 2004 *Principles of Polymerization* (John Wiley & Sons)

[17] Oron M and So/rensen G 1979 New experimental evidence of the periodic surface structure in laser annealing *Appl. Phys. Lett.* **35** 782–4

[18] Paun I A, Zamfirescu M, Luculescu C R, Acasandrei A M, Mustaciosu C C, Mihailescu M and Dinescu M 2017 Electrically responsive microreservoires for controllable delivery of dexamethasone in bone tissue engineering *Appl. Surf. Sci.* **392** 321–31

[19] Pavel N, Salamu G, Jipa F and Zamfirescu M 2014 Diode-laser pumping into the emitting level for efficient lasing of depressed cladding waveguides realized in Nd:YVO$_4$ by the direct femtosecond-laser writing technique *Opt. Express* **22** 23057

[20] Radu C, Simion S, Zamfirescu M, Ulmeanu M, Enculescu M and Radoiu M 2011 Silicon structuring by etching with liquid chlorine and fluorine precursors using femtosecond laser pulses *J. Appl. Phys.* **110** 034901

[21] Reif J, Varlamova O, Uhlig S, Varlamov S and Bestehorn M 2014 On the physics of self-organized nanostructure formation upon femtosecond laser ablation *Appl. Phys.* A **117** 179–84

[22] Sajin G, Bunea A C, Craciunoiu F, Dinescu A, Zamfirescu M and Dabu R 2011 CRLH mm-wave directional coupler on silicon substrate microprocessed by laser ablation *2011 IEEE Int. Conf. on Microwaves, Communications, Antennas and Electronic Systems (COMCAS 2011)* pp 1–4

[23] Salamu G, Jipa F, Zamfirescu M and Pavel N 2014 Cladding waveguides realized in Nd: YAG ceramic by direct femtosecond-laser writing with a helical movement technique *Opt. Mater. Express* **4** 790

[24] Sima F and Sugioka K 2021 Ultrafast laser manufacturing of nanofluidic systems *Nanophotonics* **10** 2389–406

[25] Sipe J E, Young J F, Preston J S and Van Driel H M 1983 Laser-induced periodic surface structure. I. Theory *Phys. Rev.* B **27** 1141–54

[26] Ulmeanu M, Anghel I, Filipescu M, Luculescu C, Enculescu M and Zamfirescu M 2013 Periodic arrays of nanostructures in silicon and gallium arsenide by near-field enhanced laser irradiation in liquid precursors *Colloids Surf.* A **418** 47–51

[27] Ulmeanu M, Zamfirescu M, Rusen L, Luculescu C, Moldovan A, Stratan A and Dabu R 2009 Structuring by field enhancement of glass, Ag, Au, and Co thin films using short pulse laser ablation *J. Appl. Phys.* **106** 114908

[28] Velea A *et al* 2012 Photoexpansion and nano-lenslet formation in amorphous As$_2$S$_3$ thin films by 800 nm femtosecond laser irradiation *J. Appl. Phys.* **112** 033105

[29] Zamfirescu M, Ulmeanu M, Bunea A, Sajin G and Dabu R 2011 Ultrashort pulsed lasers—efficient tools for materials micro-processing *Recent Advances in Nanofabrication Techniques and Applications* ed B Cui (InTech)

High Resolution Laser Microprocessing
Implementation and techniques
Bogdan Ştefăniţă Călin, Marian Zamfirescu and Niculae Puşcaş

Chapter 7

Python scripts

Python scripts have been tested under `Python` ver. 3.8 with libraries `numpy` ver 1.24.4, `scipy` ver1.10.1 and `maltplotlib` ver 3.3.2.

7.1 Probability distribution of an electron in the *YZ* plane

This script graphically represents the probability distribution in the *YZ* plane of a hydrogen atom.

```python
#!/usr/bin/env python3
# -*- coding: utf-8 -*-

#======== Import libraries
import numpy as np
from scipy import misc, special
from math import *
import matplotlib.pyplot as plt

#======== Probability functions

def Rnl(n, l, r):
    '''
    radial part of the wavefunction. r may be an array
    '''
    rho = np.abs(r) * 2. / n
    L = special.eval_genlaguerre(n - l - 1, 2 * l + 1, rho)
    p = np.product(np.arange(n - l, n + l + 1.0))
    return np.sqrt((2./n)**3 / (2*n*p)) * np.exp(-rho/2.) * \
        rho**l * L

def psinlm(n, l, m, x, y, z):
```

doi:10.1088/978-0-7503-3239-2ch7   7-1

```python
        '''
    hydrogen atom wavefunction. x, y, z may be arrays of equal length
    '''
    r = np.sqrt(x**2 + y**2 + z**2)
    phi = np.arctan2(y, x)
    theta = np.arctan2(np.sqrt(x**2 + y**2), z)
    R = Rnl(n, l, r)
    Ylm = special.sph_harm(m, l, phi, theta)
    return R * Ylm

def image_psi(n, l, m, imgsize):
    x = 3 * n**2 * ((np.arange(imgsize) + 0.5) / imgsize * 2 - 1)
    xv, zv = np.meshgrid(x, -x)
    psi = np.real(psinlm(n, l, m, xv, np.zeros_like(xv), zv))
    psi /= np.max(np.abs(psi))
    img_r = np.interp(psi, [-1, -.5, 0, .5, 1], [.5, .92, 1, 0, 0])
    img_g = np.interp(psi, [-1, -.5, 0, .5, 1], [0, 0, 1, 0, 0])
    img_b = np.interp(psi, [-1, -.5, 0, .5, 1], [0, 0, 1, .97, .5])
    img = np.stack((img_r, img_g, img_b), axis=2)
    img = np.clip(img * 256, 0, 255).astype('uint8')
    return img

def image_abspsi(n, l, m, imgsize):
    x = 3 * n**2 * ((np.arange(imgsize) + 0.5) / imgsize * 2 - 1)
    xv, zv = np.meshgrid(x, -x)
    psi = psinlm(n, l, m, xv, np.zeros_like(xv), zv)
    img = np.hypot(psi.real, psi.imag)
    img = np.clip(img * 256 / np.max(img), 0, 255).astype('uint8')
    return img

def image_rpsi2(n, l, m, imgsize):
    x = (2.2+1.1/n) * n**2 * ((np.arange(imgsize) + 0.5) / imgsize * \
        2 - 1)
    xv, zv = np.meshgrid(x, -x)
    psi = psinlm(n, l, m, xv, np.zeros_like(xv), zv)
    img = psi.real**2 + psi.imag**2
    img *= np.hypot(xv, zv)**2
    img = np.clip(img * 256 / np.max(img), 0, 255).astype('uint8')
    return img

def image_xpsi2(n, l, m, imgsize):
    x = (2.2+1.1/n) * n**2 * ((np.arange(imgsize) + 0.5) / imgsize * \
        2 - 1)
    xv, zv = np.meshgrid(x, -x)
    psi = psinlm(n, l, m, xv, np.zeros_like(xv), zv)
```

```python
    img = psi.real**2 + psi.imag**2
    img *= np.abs(xv)
    img = np.clip(img * 256 / np.max(img), 0, 255).astype('uint8')
    return img

# plt.figure()

#========= Script set up to save images rather than show plots
# Script saves images in the same folder

for n in range(1, 2+1):
    for l in range(n):
        for m in range(l + 1):
            fname = 'psi_n{}l{}m{}.png'.format(n, l, m)
            print('drawing', fname) #used to check status
            image = image_psi(n, l, m, imgsize=1600)
            plt.imsave(fname, image)

            fname = 'abs(psi)_n{}l{}m{}.png'.format(n, l, m)
            print('drawing', fname)
            image = image_abspsi(n, l, m, imgsize=1600)
            plt.imsave(fname, image)

            if m == 0:
                fname = 'r*abs(psi)**2_n{}l{}m{}.eps'.format(n, l, m)
                print('drawing', fname)
                image = image_rpsi2(n, l, m, imgsize=1600)
                plt.imsave(fname, image, cmap="viridis", format='eps')
            else:
                fname = 'x*abs(psi)**2_n{}l{}m{}.png'.format(n, l, m)
                print('drawing', fname)
                image = image_xpsi2(n, l, m, imgsize=1600)
                plt.imsave(fname, image)
```

7.2 Transverse modes in a laser cavity

This script graphically represents several possible transverse modes in a laser cavity.
Note that in its current form, shown below, the script can be hardware intensive for
older machines.

```python
#!/usr/bin/env python3
# -*- coding: utf-8 -*-

#========= Import libraries

import numpy as np
```

```python
import matplotlib.pyplot as plt
from scipy import special

#========= intensity profiles for Hermite polynomials
def intprofile(x, y, m, n):
    h_m = special.eval_hermite(m, x)
    h_n = special.eval_hermite(n, y)
    i_0=1
    w_0=3
    w=4
    transv = i_0 * (w_0/w)**2 *(h_m (np.sqrt(2)*x/w)* \
        np.exp(-x**2/w**2))**2 * (h_n (np.sqrt(2)*y/w)* \
        np.exp(-y**2/w**2))**2
    return transv

#========= Data initialization
xmin = -5
xmax = 5
ymin = xmin
ymax = xmax

labeltextx = -0.025
labeltexty = 0.1

x = np.linspace(-5, 5, 2000)
y = np.linspace(-5, 5, 2000)

fig, axs = plt.subplots(3, 3)

i_0=3
w_0=1
w=2
tmpx = len(x)
tmpy = len(y)
intensity = np.zeros((tmpx, tmpy))

var_cmap = "viridis"

m=0
n=0
h_m = special.eval_hermite(m, np.sqrt(2)*x/w)
h_n = special.eval_hermite(n, np.sqrt(2)*y/w)

for i in range(len(x)):
    transvx = i_0 * (w_0/w)**2 *(h_m[i]*np.exp(-x[i]**2/w**2))**2
```

```python
    for j in range(len(y)):
        transvy = (h_n[j]*np.exp(-y[j]**2/w**2))**2
        intensity[i][j] = (transvx*transvy)

axs[m][n].imshow(intensity, cmap=var_cmap,
                 extent = [xmin, xmax, ymin, ymax])

axs[m][n].axes.get_xaxis().set_ticks([])
axs[m][n].axes.get_yaxis().set_ticks([])

props = dict(boxstyle='square', facecolor='gainsboro')
axs[m][n].text(labeltextx, labeltexty,
               '$TEM_{'+str(m) + str(n) + 'q}$',
               transform=axs[m][n].transAxes, fontsize=8,
         verticalalignment='top', bbox = props)

m=1
n=0
h_m = special.eval_hermite(m, np.sqrt(2)*x/w)
h_n = special.eval_hermite(n, np.sqrt(2)*y/w)

for i in range(len(x)):
    transvx = i_0 * (w_0/w)**2 *(h_m[i]*np.exp(-x[i]**2/w**2))**2
    for j in range(len(y)):
        transvy = (h_n[j]*np.exp(-y[j]**2/w**2))**2
        intensity[i][j] = (transvx*transvy)

axs[m][n].imshow(intensity, cmap=var_cmap,
                 extent = [xmin, xmax, ymin, ymax])

axs[m][n].axes.get_xaxis().set_ticks([])
axs[m][n].axes.get_yaxis().set_ticks([])

props = dict(boxstyle='square', facecolor='gainsboro')
axs[m][n].text(labeltextx, labeltexty,
               '$TEM_{'+str(m) + str(n) + 'q}$',
               transform=axs[m][n].transAxes, fontsize=8,
         verticalalignment='top', bbox = props)

m=2
n=0
h_m = special.eval_hermite(m, np.sqrt(2)*x/w)
h_n = special.eval_hermite(n, np.sqrt(2)*y/w)

for i in range(len(x)):
```

```python
        transvx = i_0 * (w_0/w)**2 *(h_m[i]*np.exp(-x[i]**2/w**2))**2
        for j in range(len(y)):
            transvy = (h_n[j]*np.exp(-y[j]**2/w**2))**2
            intensity[i][j] = (transvx*transvy)

axs[m][n].imshow(intensity, cmap=var_cmap,
                extent = [xmin, xmax, ymin, ymax])

axs[m][n].axes.get_xaxis().set_ticks([])
axs[m][n].axes.get_yaxis().set_ticks([])

props = dict(boxstyle='square', facecolor='gainsboro')
axs[m][n].text(labeltextx, labeltexty,
            '$TEM_{'+str(m) + str(n) + 'q}$',
            transform=axs[m][n].transAxes, fontsize=8,
        verticalalignment='top', bbox = props)

m=0
n=1
h_m = special.eval_hermite(m, np.sqrt(2)*x/w)
h_n = special.eval_hermite(n, np.sqrt(2)*y/w)

for i in range(len(x)):
    transvx = i_0 * (w_0/w)**2 *(h_m[i]*np.exp(-x[i]**2/w**2))**2
    for j in range(len(y)):
        transvy = (h_n[j]*np.exp(-y[j]**2/w**2))**2
        intensity[i][j] = (transvx*transvy)

axs[m][n].imshow(intensity, cmap=var_cmap,
                extent = [xmin, xmax, ymin, ymax])

axs[m][n].axes.get_xaxis().set_ticks([])
axs[m][n].axes.get_yaxis().set_ticks([])

props = dict(boxstyle='square', facecolor='gainsboro')
axs[m][n].text(labeltextx, labeltexty,
            '$TEM_{'+str(m) + str(n) + 'q}$',
            transform=axs[m][n].transAxes, fontsize=8,
        verticalalignment='top', bbox = props)

m=1
n=1
h_m = special.eval_hermite(m, np.sqrt(2)*x/w)
h_n = special.eval_hermite(n, np.sqrt(2)*y/w)
```

```python
for i in range(len(x)):
    transvx = i_0 * (w_0/w)**2 *(h_m[i]*np.exp(-x[i]**2/w**2))**2
    for j in range(len(y)):
        transvy = (h_n[j]*np.exp(-y[j]**2/w**2))**2
        intensity[i][j] = (transvx*transvy)

axs[m][n].imshow(intensity, cmap=var_cmap,
                 extent = [xmin, xmax, ymin, ymax])

axs[m][n].axes.get_xaxis().set_ticks([])
axs[m][n].axes.get_yaxis().set_ticks([])

props = dict(boxstyle='square', facecolor='gainsboro')
axs[m][n].text(labeltextx, labeltexty,
               '$TEM_{'+str(m) + str(n) + 'q}$',
               transform=axs[m][n].transAxes, fontsize=8,
           verticalalignment='top', bbox = props)

m=2
n=1
h_m = special.eval_hermite(m, np.sqrt(2)*x/w)
h_n = special.eval_hermite(n, np.sqrt(2)*y/w)

for i in range(len(x)):
    transvx = i_0 * (w_0/w)**2 *(h_m[i]*np.exp(-x[i]**2/w**2))**2
    for j in range(len(y)):
        transvy = (h_n[j]*np.exp(-y[j]**2/w**2))**2
        intensity[i][j] = (transvx*transvy)

axs[m][n].imshow(intensity, cmap=var_cmap,
                 extent = [xmin, xmax, ymin, ymax])

axs[m][n].axes.get_xaxis().set_ticks([])
axs[m][n].axes.get_yaxis().set_ticks([])

props = dict(boxstyle='square', facecolor='gainsboro')
axs[m][n].text(labeltextx, labeltexty,
               '$TEM_{'+str(m) + str(n) + 'q}$',
               transform=axs[m][n].transAxes, fontsize=8,
           verticalalignment='top', bbox = props)

m=0
n=2
h_m = special.eval_hermite(m, np.sqrt(2)*x/w)
h_n = special.eval_hermite(n, np.sqrt(2)*y/w)
```

```python
for i in range(len(x)):
    transvx = i_0 * (w_0/w)**2 *(h_m[i]*np.exp(-x[i]**2/w**2))**2
    for j in range(len(y)):
        transvy = (h_n[j]*np.exp(-y[j]**2/w**2))**2
        intensity[i][j] = (transvx*transvy)

axs[m][n].imshow(intensity, cmap=var_cmap,
            extent = [xmin, xmax, ymin, ymax])

axs[m][n].axes.get_xaxis().set_ticks([])
axs[m][n].axes.get_yaxis().set_ticks([])

props = dict(boxstyle='square', facecolor='gainsboro')
axs[m][n].text(labeltextx, labeltexty,
            '$TEM_{'+str(m) + str(n) + 'q}$',
            transform=axs[m][n].transAxes, fontsize=8,
        verticalalignment='top', bbox = props)

m=1
n=2
h_m = special.eval_hermite(m, np.sqrt(2)*x/w)
h_n = special.eval_hermite(n, np.sqrt(2)*y/w)

for i in range(len(x)):
    transvx = i_0 * (w_0/w)**2 *(h_m[i]*np.exp(-x[i]**2/w**2))**2
    for j in range(len(y)):
        transvy = (h_n[j]*np.exp(-y[j]**2/w**2))**2
        intensity[i][j] = (transvx*transvy)

axs[m][n].imshow(intensity, cmap=var_cmap,
            extent = [xmin, xmax, ymin, ymax])

axs[m][n].axes.get_xaxis().set_ticks([])
axs[m][n].axes.get_yaxis().set_ticks([])

props = dict(boxstyle='square', facecolor='gainsboro')
axs[m][n].text(labeltextx, labeltexty,
            '$TEM_{'+str(m) + str(n) + 'q}$',
            transform=axs[m][n].transAxes, fontsize=8,
        verticalalignment='top', bbox = props)

m=2
n=2
h_m = special.eval_hermite(m, np.sqrt(2)*x/w)
```

```python
h_n = special.eval_hermite(n, np.sqrt(2)*y/w)

for i in range(len(x)):
    transvx = i_0 * (w_0/w)**2 *(h_m[i]*np.exp(-x[i]**2/w**2))**2
    for j in range(len(y)):
        transvy = (h_n[j]*np.exp(-y[j]**2/w**2))**2
        intensity[i][j] = (transvx*transvy)

axs[m][n].imshow(intensity, cmap=var_cmap,
                 extent = [xmin, xmax, ymin, ymax])

axs[m][n].axes.get_xaxis().set_ticks([])
axs[m][n].axes.get_yaxis().set_ticks([])

props = dict(boxstyle='square', facecolor='gainsboro')
axs[m][n].text(labeltextx, labeltexty,
               '$TEM_{'+str(m) + str(n) | 'q}$',
               transform=axs[m][n].transAxes, fontsize=8,
           verticalalignment='top', bbox = props)

fig.tight_layout(h_pad=0, w_pad=-8)
plt.savefig("Transverse_modes.eps", format='eps')

plt.show()
```

7.3 Focus of a Gaussian beam

This script graphically represents the focal area of a Gaussian beam, with some of the most important parameters involved, i.e. beam diameter, depth of focus, wavefront, etc.

```python
#!/usr/bin/env python3
# -*- coding: utf-8 -*-

#======== Import libraries
import numpy as np
import matplotlib.pyplot as plt
import matplotlib.style
import matplotlib as mpl

def beamsize(z, w_0, lm):
    w_z = w_0 * np.sqrt(1 + ((lm*z)/(np.pi * w_0**2))**2)
    return w_z

def curv(z, w_0, lm):
    R_z = z*(1 + ((np.pi * w_0**2)/(lm*z))**2)
```

```python
        return R_z

def u(z, w_0, k, r, R_z, w_z):
    u = w_0 / w_z * np.exp(-r**2 / w_z**2) * np.exp(-1j * \\
    (k*z - 1/(np.tan(z/z_0)))) * np.exp(-1j*(k*r**2)/(2*R_z))
    return u

zmin = -5E-06
zmax = 5E-06
rmin = 0
rmax = 5E-06

z = np.linspace(zmin, zmax, 1000)
r = np.linspace(rmin, rmax, 1000)

w_0 = 632E-09
lm = 632E-09
k = 2*np.pi / lm
z_0 = k*w_0**2/2
theta = np.rad2deg(lm/(np.pi*w_0))

fig, ax = plt.subplots()

R_z = curv(z, w_0, lm)
w_z = beamsize(z, w_0, lm)

gaussint = []
for i in range(len(z)):
    tmp_vec = np.abs(u(z[i], w_0, k, r, curv(z[i], w_0, lm),
                beamsize(z[i], w_0, lm)))
    gaussint.append(tmp_vec)

norad = 20
z_front = 0
for i in range(norad-1):
    z_front = z_front + zmax/norad
    rad1 = curv(z_front, w_0, lm)
    w_1up = beamsize(z_front, w_0, lm)
    ang1 = np.rad2deg(np.arctan(w_1up/rad1))
    ax.add_patch(mpl.patches.Arc((0, z_front-rad1),
                2*rad1, 2*rad1, angle=0, linewidth=0.5,
                theta1 = 90-ang1, theta2=90+ang1,
                color='blue'))

z_front = z_front + zmax/norad
```

```python
rad1 = curv(z_front, w_0, lm)
w_1up = beamsize(z_front, w_0, lm)
ang1 = np.rad2deg(np.arctan(w_1up/rad1))
ax.add_patch(mpl.patches.Arc((0, z_front-rad1),
                2*rad1, 2*rad1, angle=0, linewidth=0.5,
                theta1 = 90-ang1, theta2=90+ang1,
                color='blue',
                label='Wavefront■\ncurvature,■$R(z)$'))

ax.plot([0, w_1up], [0, z_front], 'r--',
                linewidth=0.5,
                label="Asymptotes■\ndefining■$■\\theta_G$")
ax.plot([0, -w_1up], [0, z_front], 'r--', linewidth=0.5)


colbar = ax.imshow(np.flipud(gaussint),
                extent=[rmin, rmax, zmin, zmax],
                aspect='auto',
                cmap="viridis")
ax.imshow(np.fliplr(np.flipud(gaussint)),
                extent=[-rmax, rmin, zmin, zmax],
                aspect='auto',
                cmap="viridis")

cbar = fig.colorbar(colbar, ax=ax)
cbar.set_label('Intensity', size=10)
cbar.ax.tick_params(labelsize=6)

ax.plot(beamsize(z, w_0, lm), z, 'w-',
                label='Beam■size,■$w(z)$',
                linewidth=0.5)

ax.plot((-beamsize(z, w_0, lm)), z, 'w-', linewidth=0.5)

ax.plot(z*0, z, 'k-.', linewidth=0.5)

for i in range(1):
    z_cutpos = 300*(i+1)
    ax.plot(r, gaussint[z_cutpos][:]*1E-06+z[z_cutpos],
                'k-', linewidth=0.5,
                label='Transverse■intensity■\n' +
                    'profile■at■$z=-2■\mu■m$')
    ax.plot(-r, gaussint[z_cutpos][:]*1E-06+z[z_cutpos],
                'k-',
                linewidth=0.5)
```

```python
y_formatter = matplotlib.ticker.ScalarFormatter(useOffset=False)
ax.yaxis.set_major_formatter(y_formatter)

x_formatter = matplotlib.ticker.ScalarFormatter(useOffset=False)
ax.xaxis.set_major_formatter(x_formatter)

ax.get_xaxis().set_major_formatter(
    matplotlib.ticker.FuncFormatter(lambda x, p: format(int(x*1E+06))))

ax.get_yaxis().set_major_formatter(
    matplotlib.ticker.FuncFormatter(lambda y, p: format(int(y*1E+06))))

plt.xlabel("Radius ($\cdot 10^{-6} m$)")
plt.ylabel("Propagation direction, z ($\cdot 10^{-6} m$)")

ax.legend(facecolor="lightsteelblue",
          loc='upper center',
          bbox_to_anchor=(0.9, 1),
          shadow=True,
          fontsize=6)

plt.axis([-rmax, rmax, zmin, zmax])

plt.savefig("tester_gauss.eps", format='eps')
plt.show()
```

7.4 Electron wavefunctions in the Hydrogen atom.

This script is used to calculate and graphically represent radial wavefunctions/ probability distribution of an electron in the hydrogen atom. Note, however, that proper values of the principal and orbital quantum numbers must be introduced when prompted, in order to receive results.

```python
#!/usr/bin/env python3
# -*- coding: utf-8 -*-

#======== Import libraries
import numpy as np
import matplotlib.pyplot as plt
from scipy.special import genlaguerre
from scipy.integrate import simps
import matplotlib.style
import matplotlib as mpl
```

```python
plt.rcParams.update({'font.size': 10})

a = 5.29*10**-11
numValues = 20000

#======== Radial
def R(n,l,r):
    '''
    Returns the radial wavefunction for each value of r
    where n is the principle quantum number and l is
    the angular momentum quantum number.
    '''
    y=np.zeros(numValues)
    Lag = genlaguerre(n-l-1, 2*l+1)
    for i in range(len(Lag)+1):
        i = float(i)
        y = y + (((r/(n*a))**(l))*(np.exp(-r/(a*n)))* \
        Lag[int(i)]*(((2*r)/(a*n))**i))
    return y

# Normalisation function for better plotting of multiple functions
def normalise(x,y):
    integral = simps(np.absolute(y),x)
    print(integral, "simps")
    print(type(integral))
    y = y/np.absolute(integral)
    return y

def plotPsi(r,psi, n, l):
    plt.subplot(311)
    plt.plot(r, psi, label="$n=$"+str(int(n)) + ",l=" +
                str(int(l)) +
                '$')
    plt.legend(loc='upper center',
        bbox_to_anchor=(1, 1),
        shadow=True,
        fontsize=10)
    plt.grid(False)
    plt.xlabel("$r/a_0$")
    plt.ylabel("Radial wave\nfunction $\Psi$", fontsize=10)

def plotPsiSquared(r,psi, n, l):
    psiSquared = psi**2
    psiSquared = normalise(r,psiSquared)
```

```python
        plt.plot(r, psiSquared ,
                    label="$n■=■"+str(int(n)) +
                    ",■l■=■" +
                    str(int(l)) +'$')
        plt.legend(loc='upper■center',
                    bbox_to_anchor=(1, 1),
                    shadow=True,
                    fontsize=10)
        plt.grid(False)
        plt.xlabel("$r/a_0$")
        plt.ylabel("Radial■probability■\ndensity■$\Psi^2$", fontsize=10)

def plotRadialDistribution(r,psi, n, l):
        radialDistribution = 4*np.pi*(r**2)*psi**2
        radialDistribution = normalise(r, radialDistribution)

        plt.plot(r, radialDistribution ,
                    label="$n■=■"+str(int(n)) + ",■l■=■" +
                    str(int(l)) +'$')
        plt.legend(loc='upper■center',
                    bbox_to_anchor=(1, 1),
                    shadow=True,
                    fontsize=10)
        plt.grid(False)
        plt.xlabel("$r/a_0$")
        plt.ylabel("Radial■probability■\ndistribution■$4\pi■r^2■\Psi$",
                    fontsize=10)

def plotAll(r,psi, n, l):
    plt.figure(1)
    plt.subplot(311)
    plotPsi(r,psi, n, l)
    plt.subplot(312)
    plotPsiSquared(r,psi, n, l)
    plt.subplot(313)
    plt.subplots_adjust(bottom=-0.5)
    plotRadialDistribution(r,psi, n, l)
    plt.savefig('tester.eps', format='eps', bbox_inches='tight')

def graphs(r, psi,choice, n, l):
    if choice == 1:
        plotPsi(r,psi)
    elif choice == 2:
        plotPsiSquared(r,psi)
    elif choice == 3:
```

```python
        plotRadialDistribution(r,psi)
    else:
        plotAll(r,psi, n, l)

def main():
    from cycler import cycler
    mpl.rcParams['axes.prop_cycle'] = cycler(color='bgrcmyk')
    numPsi = int(input("How many wavefunctions do you want to draw?"))
    width = float(input("Plotting radius in units of a?")) * a
    print("1 - Plot the wavefunction")
    print("2 - Plot the probability density")
    print("3 - Plot the radial distribution function")
    print("4 - Plot all of the above")
    choice = int(input("Options:"))
    for i in range(numPsi):
        print("Wavefunction " + str(i+1))
        n = float(input("Enter principal quantum number, n:"))
        l = float(input("Enter orbital quantum number, l:"))
        r = np.linspace(0,width, numValues)
        psi = R(n,l,r)
        r = r/a
        psi = normalise(r, psi)
        graphs(r, psi,choice, n, l)

    plt.show()

if __name__ == '__main__':
    main()
```

7.5 Modelocking in a laser cavity.

This script is used generate longitudinal modes in a laser cavity with and without a fixed phase relationship. Some data is randomly generated and therefore the script will produce different results each run.

```python
#!/usr/bin/env python3
# -*- coding: utf-8 -*-

#======== Import libraries

import numpy as np
import matplotlib.pyplot as plt

#======== Data initialization
N = 7
```

```python
omg = np.linspace(0, N, N)
tmin=0
tmax=20
t = np.linspace(tmin, tmax, 1000)

#======== Plot
fig, ax = plt.subplots(N, 2)

inten = 0
inten_sum = 0

for i in range(N-1):
    ranno = 10*np.random.random(1)[0]
    #ranno = 1 #uncomment this to remove randomness
    inten = (np.sin( 2*omg[i+1]*(t+ranno)/ 2) /
            np.sin(omg[i+1]*(t+ranno) / 2))**2
    inten_sum = inten_sum + inten
    ax[i][0].plot(t, inten, 'b-')
    ax[i][0].axes.get_xaxis().set_ticks([])
    ax[i][0].tick_params(axis='y', labelsize=8)

ax[N-1][0].plot(t, inten_sum, 'b-')
ax[N-1][0].tick_params(axis='y', labelsize=8)
ax[N-1][0].tick_params(axis='x', labelsize=8)

ax[N-1][0].set_ylabel('Sum of\n all modes', fontsize=8)

inten = 0
inten_sum = 0

for i in range(N-1):
    #ranno = N* np.random.random(1)[0]/2
    ranno = 1 #uncomment this to remove randomness
    inten = (np.sin( 2*omg[i+1]*(t)/ 2) / np.sin(omg[i+1]*(t) / 2))**2
    inten_sum = inten_sum + inten
    ax[i][1].plot(t, inten, 'b-')

    ax[i][1].arrow(0, 4, 0, -10,
            linewidth = 1,
            linestyle=":",
            color="red")
    ax[i][1].arrow(5.366, 4, 0, -10,
            linewidth = 1,
            linestyle=":",
            color="red")
```

```python
    ax[i][1].arrow(10.756, 4, 0, -10,
              linewidth = 1,
              linestyle=":",
              color="red")
    ax[i][1].arrow(16.15, 4, 0, -10,
              linewidth = 1,
              linestyle=":",
              color="red")

    ax[i][1].axes.get_xaxis().set_ticks([])
    ax[i][1].axes.get_yaxis().set_ticks([])

ax[N-1][1].plot(t, inten_sum, 'b-')

ax[N-1][1].arrow(0, 25, 0, -50,
              linewidth = 1,
              linestyle=":",
              color="red")
ax[N-1][1].arrow(5.366, 25, 0, -50,
              linewidth = 1,
              linestyle=":",
              color="red")
ax[N-1][1].arrow(10.756, 25, 0, -50,
              linewidth = 1,
              linestyle=":",
              color="red")
ax[N-1][1].arrow(16.15, 25, 0, -50,
              linewidth = 1,
              linestyle=":",
              color="red")

ax[N-1][1].tick_params(axis='y', labelsize=8)
ax[N-1][1].tick_params(axis='x', labelsize=8)
ax[N-1][1].axes.get_yaxis().set_ticks([])

ax[N-1][1].set_xlabel("$t/T$", x=0, fontsize=12)

fig.text(0.01, 0.5, '$Intensity \;(a.u.)$',
              va='center',
              rotation='vertical',
              fontsize=12)

ax[0][0].set_title("Random phase difference \n(not mode-locked)",
              fontsize=10)
ax[0][1].set_title("Fixed phase difference \n(mode-locked)",
              fontsize=10)
```

```python
fig.tight_layout(h_pad=0, w_pad=0)

#ax[1].legend(facecolor="lightsteelblue",
             loc='upper center',
             bbox_to_anchor=(0.8, 1.05),
             shadow=True,
             fontsize=8)

plt.savefig("tester_modelocking.eps", format='eps')
plt.show()
```

7.6 Electric susceptibility

This script graphically represents the electric susceptibility as a function of frequency, relativeto the phase.

```python
#!/usr/bin/env python3
# -*- coding: utf-8 -*-

# Plot of electric susceptibility as a function of phase

# ======== Import Libraries
import matplotlib.pyplot as plt
import numpy as np

# ======== Data initialization
N = 0.02#1E+23
q = 1.6E-19
eps0 = 8.85E-12
m = 9.1E-31
wp = np.sqrt((N*q**2)/(eps0*m)) # see eq. (2.59)
gamma = 0.5
w0 = 3

w = np.linspace(0, 10, 100)

chi = (wp**2) / (w0**2 - w**2 - 1j*w*gamma) #see eq. (2.60)
phase = np.angle(chi, deg=True)
maxval = np.max(np.absolute(chi))

# ======== Plot
```

```python
fig, ax1 = plt.subplots()

color = 'tab:red'
ax1.set_xlabel('Frequency,■$\omega$')
ax1.set_ylabel('Amplitude', color=color)
ax1.plot(w, np.absolute(chi), color=color)
ax1.tick_params(axis='y', labelcolor=color)

ax2 = ax1.twinx()   # shared X axis

color = 'tab:blue'
ax2.set_ylabel('Phase■(degrees)', color=color)
ax2.plot(w, phase, color=color)
ax2.tick_params(axis='y', labelcolor=color)

fig.tight_layout()
#plt.savefig("chi.eps", format='eps') #optional
plt.show()
```

7.7 Electric permittivity

This script graphically represents the electric susceptibility as a function of frequency, relative to the phase.

```python
#!/usr/bin/env python3
# -*- coding: utf-8 -*-

# Plot of electric susceptibility as a function of phase

# ======== Import Libraries
import matplotlib.pyplot as plt
import numpy as np

# ======== Data initialization
N = 0.02#1E+23
q = 1.6E-19
eps0 = 8.85E-12
m = 9.1E-31
wp = np.sqrt((N*q**2)/(eps0*m)) # see eq. (2.59)
gamma = 0.5
w0 = 3

w = np.linspace(0, 10, 100)
```

```python
chi = (wp**2) / (w0**2 - w**2 - 1j*w*gamma) #see eq. (2.60)
phase = np.angle(chi, deg=True)
maxval = np.max(np.absolute(chi))

# ======== Plot

fig, ax1 = plt.subplots()

color = 'tab:red'
ax1.set_xlabel('Frequency,■$\omega$')
ax1.set_ylabel('Amplitude', color=color)
ax1.plot(w, np.absolute(chi), color=color)
ax1.tick_params(axis='y', labelcolor=color)

ax2 = ax1.twinx()   # shared X axis

color = 'tab:blue'
ax2.set_ylabel('Phase■(degrees)', color=color)
ax2.plot(w, phase, color=color)
ax2.tick_params(axis='y', labelcolor=color)

fig.tight_layout()
#plt.savefig("chi.eps", format='eps') #optional
plt.show()
```

7.8 Refractive index and attenuation

This script graphically represents the ordinary refractive index, attenuation constant and extinction coefficient in the context of Lorentz dielectrics.

```python
#!/usr/bin/env python3
# -*- coding: utf-8 -*-

# ======== Import Libraries
import matplotlib.pyplot as plt
import numpy as np

# ======== Data initialization
N = 0.02#1E+23
q = 1.6E-19
eps0 = 8.85E-12
m = 9.1E-31
wp = np.sqrt((N*q**2)/(eps0*m))
gamma = 0.5
w0 = 3
```

```python
w = np.linspace(0, 10, 10000)

eps_real = 1 + wp**2* (w0**2 - w**2) / ( (w0**2 - w**2)**2 +
                w**2*gamma**2 )
eps_imag = wp**2 * (w*gamma) / ( (w0**2 - w**2)**2 +
                w**2*gamma**2 )

eps = eps_real + 1j*eps_imag

n0 = np.sqrt( (np.abs(eps) + eps_real)/2 )
kappa = np.sqrt( (np.abs(eps) - eps_real)/2 )

alpha = 2*kappa*w/10 #3E+08

#======== Plot

fig, ax1 = plt.subplots()
fig.set_figheight(5)
fig.set_figwidth(9)

ax1.set_xlabel('Frequency,■$\omega$')
ax1.set_ylabel("Complex■refractive■index,■$n(\omega)$", color='k')
ax1.plot(w, n0, '-r', label='Refractive■index,■n')

#horizontal line to show refractive index reaching values <1
ax1.plot(w, w/w, '--r')


ax2 = ax1.twinx()   # shared X axis

ax2.plot(w, kappa, '-b', label='Extinction■coefficient,■$\kappa$')

ax2.plot(w, alpha, '-g', label='Attenuation■constant,■$\\alpha$')
ax1.legend(loc=(0.7, 0.92))
ax2.legend(loc=(0.7, 0.8))

fig.tight_layout()
plt.savefig("index.eps", format='eps')
plt.show()
```

7.9 Laser amplification. Gain and extraction efficiency

This script computes the gain and extraction efficiency of a laser amplifier.

```python
# ========= Import libraries
from pylab import figure, show, savefig
from numpy import arange, exp, log

# ========= Data for output file
# File name to save PNG
filename = '4_gain_vs_eff.png'
savePlot = False
# ---> Switch to True to save png file

# ========= Simulation grid - F_in / F_sat
F_in_F_sat = arange(0.01, 5, 0.1)

# ===================== COMPUTATIONS ==========================
G0 = 3
G       = 1/F_in_F_sat *  log(1 + (exp(F_in_F_sat) - 1) * G0)
eff_extr = 1/    log(G0) * (log(1 + (exp(F_in_F_sat) - 1) * G0) \
                            - F_in_F_sat)

# ==================== PLOTS  ===============================
fig_size =(8, 5)
fig = figure(1, figsize=fig_size)

ax1 = fig.add_subplot(111)
# Second axes shares the same x-axis
ax2 = ax1.twinx()

ax1.plot(F_in_F_sat, G, 'b')          # Gain - blue
ax2.plot(F_in_F_sat, eff_extr, 'r')   # Extraction eff. - red


# ===== Axis 1 config
font = {'family': 'arial','weight': 'normal', 'size': 20 }

ax1.set_xlim((0, 5))
ax1.set_ylim((1, 3))
ax1.set_xlabel(r'$F_{in}/F_{sat}$', fontsize=20)
ax1.set_ylabel(r'Gain ($F_{out}/F_{in})$',
               fontdict=font, color = 'b')
ax1.tick_params(axis='y', labelcolor = 'b')

# ===== Axis 2 config
ax2.set_xlim((0, 5))
ax2.set_ylim((0, 1))
ax2.set_ylabel(r'Extaction efficiency ($\eta_{ext}$)',
```

```
              fontdict=font, color = 'r',
              labelpad=30, rotation=270)
ax2.tick_params(axis='y', labelcolor= 'r')


ax1.xaxis.set_tick_params(labelsize=18)
ax1.yaxis.set_tick_params(labelsize=18)
ax2.xaxis.set_tick_params(labelsize=18)
ax2.yaxis.set_tick_params(labelsize=18)

fig.tight_layout()   # avoid clipped labels

# ========= Save plot
if savePlot:
    savefig(filenamePNG , dpi=300)

show()
# ================================================================
#                            END
# ================================================================
```

7.10 Multipass laser amplifier

This script computes the number of roundtrips in a multipass laser amplifier working in *high-efficiency, low-gain* regime.

```
# ========= Import libraries
from pylab import figure , show, append , array , savefig
from numpy import exp, log, pi

# ========= Data for output file
# File name to save PNG
filenamePNG = '4_multipass_output.png'
savePlot = False
# ---> Switch to True to save png file

#========= Pump parameters
lambda_Pump = 532           # nm  - Laser Pump wavelength
D_pump   = 0.5              # cm  - Pump beam diameter (532nm)
E_pump   = 0.5              # J   - Pump pulse energy

# ========= Seed parameters
lambda_seed = 800           # nm  - Seed Laser wavelength
D_seed   = D_pump          # mm  - Seed beam diameter
E_seed   = 0.006           # J   - Seed pulse energy
```

```python
# ========= Active  medium
lambda_L = lambda_seed
F_sat     = 0.9                    # J/cm2 - Saturation Fluence
F_LIDT    = 4.0                    # J/cm2 - Damage Threshold Fluence
                                   # (LIDT - laser induced damage threshold)
n_P       = 0.92                   # Pump quantum efficiency
n_F       = 0.8                    # Fluorescence quantum efficiency
n_Q       = lambda_Pump/lambda_L


# ===================== COMPUTATIONS ==========================
F_pump = E_pump/(pi/4*D_pump**2)    # Pump Fluence
F_ac = n_P*n_Q*n_F*F_pump           # Accumulated Fluence
F_seed = E_seed/(pi/4*D_pump**2)    # Seed Fluence (F_in = F_seed)

if (F_pump < F_LIDT):
    text = "Fluence■{0:5.3f}■<■Threshold■{1:5.3f}■\n"
    print(text.format(F_pump, F_LIDT))
    # see https://www.python-course.eu/python3_formatted_output.php
elif (F_pump >= F_LIDT):
    text = "Fluence■{0:5.3f}■>■Threshold■{1:5.3f}"
    print(text.format(F_pump, F_LIDT))
    print("Fluence■grater■than■LIDT■!!!!■\n")

# ===== Initialization  at  first  round  trip  (seed)
F_in  = F_seed
F_av  = F_ac
F_out = 0

# Arrays  initialization
E_out    = array([])
G_0      = array([])
N_range = range(1,17)

print( "■F_av■|■■G_0■■|■E_out")
print( "------|--------|------")


for i in N_range:

    # Gain,  Fluence and  Energy  at loop i
    G_0_i = exp (F_av/F_sat)
    F_out = F_sat * log(1+(exp(F_in/F_sat)-1)*G_0_i)
    E_out_i = F_out*pi/4*D_seed **2
```

```python
    text = "{0:5.3f}■|■{1:5.3f}■|■{2:5.3f}"
    print(text.format(F_av , G_0_i , E_out_i))

    F_av=F_av-(F_out-F_in)

    # Next loop OUT becomes IN
    F_in = F_out*0.98**2
    G_0 = append(G_0,G_0_i)          # store Gain data as array
    E_out = append(E_out,E_out_i)    # store Energy_out data as array

# - end loop -

N_range_plus = append(0,N_range)
E_out = append(E_seed,E_out)


# ===================== PLOTS  ==============================
fig_size =(8, 5)
fig = figure(1, figsize=fig_size)
ax1 = fig.add_subplot(111)
ax2 = ax1.twinx()

ax1.plot(N_range,  G_0, 'b-o')            # Plot Energy in blue
ax2.plot(N_range_plus,  E_out,'r-o')      # Plot Gain in red

# ===== Axis 1 config
font = {'family': 'arial', 'weight': 'normal', 'size': 20}
ax1.set_xlim((min( N_range_plus)*0.9, max(N_range_plus)*1.01))
ax1.sct_xlabel('N■(number■of■passes)', fontdict=font)
ax1.set_ylabel('Gain', fontdict=font, color='b')
ax1.tick_params(axis='y', labelcolor='b')
ax1.xaxis.set_tick_params(labelsize=18)
ax1.yaxis.set_tick_params(labelsize=18)

# ===== Axis 2 config
ax2.set_ylabel('Output■Energy■(J)', fontdict=font, color='r',
               rotation=270, labelpad=30)
ax2.tick_params(axis='y', labelcolor='r')
ax2.xaxis.set_tick_params(labelsize=18)
ax2.yaxis.set_tick_params(labelsize=18)

#=== Set axis limits
ax1.set_ylim((1, 4.1))
ax2.set_xlim((0, 16.1))
ax2.set_ylim((0, 0.25))
```

```
fig.tight_layout()   # avoid clipped labels

# ======== Save plot
if savePlot:
    savefig(filenamePNG , dpi=300)

show()
# ================================================================
#                              END
# ================================================================
```

7.11 Dispersion of ultrafast laser pulses

This script computes the second-order and third-order dispersion for a slab of transparent material. Refractive index as function of wavelength is reconstructed from Sellmeier coefficients.

```
#========= Import libraries
import numpy as np
from pylab import figure , show, savefig
import matplotlib.ticker as ticker
from matplotlib import cm, colors

#========= Data for output file
# File name to save PNG
filenamePNG = '4_GVD_SiO2.png'
label = r'SiO$_2$ refractive index'
savePlot = False
# ---> Switch to True to save png file

#========= Parameters
wb = 0.250            # blue part of the spectrum interval - in um
wr = 1.650            # red part of the spectrum interval - in um
c = 0.299792458       # in micrometers/femtosec. (3e8 *1e6/1e15)
lambda_0 = 0.8        # laser wavelength in micrometers - in um
tau0_pulse = 10       # pulse duration in fsec.
L = 10                # thickness of material in mm
# ================================================================
# Sellmeier equation parameters for SiO2
# Ref: J. Opt. Soc. Am. 55, 1205-1208 (1965).
B1, B2, B3 = 0.6961663, 0.4079426, 0.8974794
C1, C2, C3 = 0.0684043, 0.1162414, 9.896161   # um

# ================================================================
```

```python
# ================================================================
def dn_dl(lambda_um, n_vals):
    d_lambda = lambda_um[1] - lambda_um[0]
    return np.gradient(n_vals, d_lambda)

def d2n_dl2(lambda_um, n_vals):
    d_lambda = lambda_um[1] - lambda_um[0]
    dndl = np.gradient(n_vals, d_lambda)
    d2n_dl2 = np.gradient(dndl, d_lambda)
    return d2n_dl2

def d3n_dl3(lambda_um, n_vals):
    d_lambda = lambda_um[1] - lambda_um[0]
    dndl = np.gradient(n_vals, d_lambda)
    d2n_dl2 = np.gradient(dndl, d_lambda)
    d3n_dl3 = np.gradient(d2n_dl2, d_lambda)
    return d3n_dl3
# ================================================================

# ================================================================
def vG_func(lambda_um, n_index_lambda):
    c = 0.299792458 # microm/fsec
    dn_gradient = dn_dl(lambda_um, n_index_lambda)
    vG = c / (n_index_lambda - lambda_um * dn_gradient)
    return vG
# ================================================================

# ----------------------------------------------------------------
def vP_func(lambda_um, n_index_lambda):
    c = 0.299792458  # microm/fsec
    vF = c / np.array(n_index_lambda)
    return vF  # microm/fsec

# ================================================================
def tod_func(lambda_um, n_index_lambda):
    c_mps = 299792458  # m/s
    lambda_m = lambda_um * 1e-6
    d2n_gradient = d2n_dl2(lambda_m, n_index_lambda)
    d3n_gradient = d3n_dl3(lambda_m, n_index_lambda)

    TOD_mps = - lambda_m ** 4 / (4 * np.pi ** 2 * c_mps ** 3) * \
              (3 * d2n_gradient + lambda_m * d3n_gradient)
    TOD = TOD_mps * 1e45 / 1e3  # fs^3/mm
    return TOD
```

```python
# ================================================================

# ================================================================
def gvd_func(lambda_um, n_index_lambda):
    c_mps = 299792458 # m/s
    lambda_m = lambda_um * 1e-6
    d2n_gradient = d2n_dl2(lambda_m, n_index_lambda)
    GVD_mps = lambda_m ** 3 / (2 * np.pi * c_mps ** 2) * \
                            d2n_gradient
    GVD = GVD_mps * 1e30 / 1e3   # fs^2/mm
    return GVD
# ================================================================

# ================================================================
def n_Sellmeier(lambda_um, b1, b2, b3, c1, c2, c3):
    # Sellmeier equation
    lambda2 = lambda_um ** 2
    n_index = np.sqrt(1 \
                    + b1 * lambda2 / (lambda2 - c1 ** 2) \
                    + b2 * lambda2 / (lambda2 - c2 ** 2) \
                    + b3 * lambda2 / (lambda2 - c3 ** 2))
    return n_index
# ================================================================

# ================================================================
# ===================== Load data from file ====================
# path = ""
# fname = "fusedSilica.txt"
# filename = path + fname
# data = np.loadtxt(filename, skiprows=1)
# lambda_um       = data[:,0]
# n_index_lambda = data[:,1]
# ================================================================

# ================================================================

step_lambda = 10000
lambda_um = np.linspace(wb, wr, step_lambda)

# ==================== COMPUTATIONS ========================
# Spectral bandwidth
Delta_lambda = lambda_0**2 / c * (0.441/tau0_pulse)
n_index_lambda = [n_Sellmeier(l_n, B1, B2, B3, C1, C2, C3)
                                for l_n in lambda_um]
VG = vG_func(lambda_um, n_index_lambda)
```

```python
VP  = vP_func(lambda_um, n_index_lambda)
GVD = gvd_func(lambda_um, n_index_lambda)
TOD = tod_func(lambda_um, n_index_lambda)


# ===================== PLOTS =============================
fig_size = (8, 9)
fig = figure(1, figsize=fig_size)


font = {'family': 'arial', 'color': 'black',
        'weight': 'normal', 'size': 22, }
fontText = {'family': 'arial', 'color': 'black',
        'weight': 'normal','size': 16}


ax1  = fig.add_subplot(311)
axVg = fig.add_subplot(312)
axVp = axVg.twinx()   # shares the same y-axis
ax2  = fig.add_subplot(313)
ax3  = ax2.twinx()   # shares the same y-axis


ax1.plot(lambda_um * 1000, n_index_lambda, 'k', label=label)
axVg.plot(lambda_um * 1000, VG, 'b', label=r'GV')
axVp.plot(lambda_um * 1000, VP, 'r', label=r'GP')
ax2.plot(lambda_um * 1000, GVD, 'b', label=r'GVD')
ax3.plot(lambda_um * 1000, TOD, 'r', label=r'TOD')


ax1.set_xlim((wb * 1000 * 1.01, wr * 1000 * 0.99))
ax1.set_ylim((np.min(n_index_lambda), np.max(n_index_lambda)))
axVg.set_xlim((wb * 1000 * 1.01, wr * 1000 * 0.99))
axVp.set_xlim((wb * 1000 * 1.01, wr * 1000 * 0.99))
ax2.set_xlim((wb * 1000 * 1.01, wr * 1000 * 0.99))


ax2.set_ylim((-40, 200))
ax3.set_ylim((-40, 200))
axVg.set_ylim( ( np.min(VG) *0.995 , np.max(VP)*1.005 ) )
axVp.set_ylim( ( np.min(VG) *0.995 , np.max(VP)*1.005 ) )


ax1.set_ylabel('n■index', fontdict=font)
axVg.set_ylabel(r'$v_G\■■(\mu■/fs)$', fontdict=font)
axVp.set_ylabel(r'$v_P\■■(\mu■/fs)$', rotation=270,
                labelpad=30, fontdict=font)
ax2.set_ylabel(r'$GVD\■(fs^2/mm)$', fontdict=font)
ax2.set_xlabel(r'$Wavelength\■(nm)$', fontdict=font)
ax3.set_ylabel(r'$TOD\■(fs^3/mm)$', rotation=270,
                labelpad=30, fontdict=font)
```

```python
ax1.xaxis.set_major_locator(ticker.NullLocator())
axVg.xaxis.set_major_locator(ticker.NullLocator())
axVp.xaxis.set_major_locator(ticker.NullLocator())

axVg.tick_params(axis='y', labelcolor='blue')
axVp.tick_params(axis='y', labelcolor='red')
ax2.tick_params(axis='y', labelcolor='blue')
ax3.tick_params(axis='y', labelcolor='red')

# set legends
ax1.legend(loc='upper right', fontsize=16)

# set ticks parameters
ax1.xaxis.set_tick_params(labelsize=16)
ax1.yaxis.set_tick_params(labelsize=16)
axVg.yaxis.set_tick_params(labelsize=16)
axVp.yaxis.set_tick_params(labelsize=16)
ax2.xaxis.set_tick_params(labelsize=16)
ax2.yaxis.set_tick_params(labelsize=16)
ax3.yaxis.set_tick_params(labelsize=16)

# Find value at index
idx = np.abs(lambda_um - lambda_0).argmin()
n_at_lambda0 = n_index_lambda[idx]
VG_at_lambda0 = VG[idx]
VP_at_lambda0 = VP[idx]
GVD_at_lambda0 = GVD[idx]
GVD_at_lambda0 = GVD[idx]
TOD_at_lambda0 = TOD[idx]

# ========== plot a reference line at lambda_0
axVg.plot([lambda_0 * 1000, lambda_0 * 1000], [
                np.min(VG)*0.995, VP_at_lambda0], 'k:')
ax1.plot([lambda_0 * 1000, lambda_0 * 1000],
                [np.min(n_index_lambda), n_at_lambda0], 'k:')
ax2.plot([lambda_0 * 1000, lambda_0 * 1000],
                [-40, GVD_at_lambda0], 'k:')

# ========== plot a gaussian spectrum at the bottom of graphs
ax4 = axVg.twinx()  # shares the same y-axis
I0 = VP_at_lambda0 * 0.8
I_gauss = I0 * np.exp(-(lambda_um - lambda_0) ** 2 \
                        / Delta_lambda ** 2)
```

```python
colourmap = cm.get_cmap('jet')
normalize = colors.Normalize(vmin=(lambda_0 - 0.07),
                             vmax=(lambda_0 + 0.07))
npts = 150
for i in range(2 * npts - 1):
    ax4.fill_between([lambda_um[i * 10 + idx - npts * 10] * 1000,
    lambda_um[i * 10 + 1 + idx - npts * 10] * 1000],
    [I_gauss[i * 10 + idx - npts * 10],
    I_gauss[i * 10 + 10 + idx - npts * 10]],
    color=colourmap(normalize(lambda_um[i * 10 + idx - npts * 10])),
    alpha=0.2)
ax4.yaxis.set_major_locator(ticker.NullLocator())
ax4.set_ylim((0, np.max(VP)))


# ================= text on plots
string =     r'$\lambda_0■=■{:5.1f}\■um$■'  + '\n' + \
             r'$n(\lambda_0)■=■{:7.4}■$'
ax1.text(lambda_0*1000*0.95, n_at_lambda0*1.005,
         string.format(lambda_0*1000, n_at_lambda0),
         fontdict=fontText )

string = r'$GVD(\lambda_0)■=■{:5.3}\■fs^2/mm$■'        + '\n' + \
         r'$TOD(\lambda_0)■=■{:5.3}\■fs^3/mm$'

ax2.text(lambda_0*1000*0.95, GVD_at_lambda0*3,
         string.format(GVD_at_lambda0, TOD_at_lambda0),
         fontdict=fontText )

string = r'$v_P(\lambda_0)■=■■{:5.3}\■fs/\mu■m$■'        + '\n' + \
         r'$v_G(\lambda_0)■=■{:5.3}\■fs/\mu■m$'

axVg.text(lambda_0*1000*1.2, VG_at_lambda0*0.95,
          string.format( VP_at_lambda0, VG_at_lambda0),
          fontdict=fontText )


fig.tight_layout()  # avoid clipped labels

# ======== Save plot
if savePlot:
    savefig(filenamePNG , dpi=300)

show()


    # ================================================================
    #                            END
    # ================================================================
```

7.12 Fourier synthesis of temporal pulse shapes

This script computes temporal pulse shape from a phase shift function. The phase includes dispersion with group delay dispersion and third-order dispersion, V-shape function, sine etc.

```python
# ================== Import Libraries
from pylab import figure, show, savefig, legend, tight_layout
import matplotlib.ticker as ticker
import numpy as np
from scipy.fftpack import fft, ifft, fftshift, ifftshift, fftfreq

# ================== Data for output file

# File name to save PNG
filenamePNG = '4_fourier_synthesis_V-shape200.png'
savePlot = False
# ---> Switch to True to save png file

textLabel = ''
textLabel += r'V-Shape'
textLabel += '\n'
textLabel += r'with■200■fs■delay'

ylim = 0.5                              # y limit scale in plot

c = 0.299792458                         # um/fs
# ================== Generate time and frequency space grid
pow2 = 2**15
t = np.linspace(-1000, 1000, pow2) # time grid in fs
f = fftfreq(len(t), t[1] - t[0])   # frequency grid in fs^-1
w = 2 * np.pi * f                  # angular frequencies

# ================== Pulse parameters
lambda_0 = 0.8                     # central wavelength (um)
tau0 = 10                          # pulse duration (fs)
w0 = 2 * np.pi * c/lambda_0        # angular frequency (rad/fs)
w += w0

Et = np.exp(-t**2 / (2 * tau0**2)) # Initial pulse shape
```

```python
Ew = fft(Et)                             # Initial spectrum
# ================== GDD and TOD
GDD = 0                                  # fs^2
TOD = 0                                  # fs^3
label = textLabel.format(GDD,TOD)


# ============== Compute Phase phi and Filtering factor H
phi  = 0                                 # cumulate various functions

# Dispersion of the system with GDD and TOD
phi += 0.5 * GDD * (w-w0)**2 + (1/6) * TOD * (w-w0)**3

# ================== V-Shape
a1, a2, a3  = -100, -100, 51.2
phi += [(((( np.abs(omega-w0)) - a3) * a1)) * (omega<w0)\
                           for omega in fftshift(w) ]
phi += [(((( np.abs(omega-w0)) - a3) * a2)) * (omega>w0) \
                           for omega in fftshift(w) ]

# ================== Sin function
# k, q = 2, 2
# phi += np.pi * 2* np.sin(2*np.pi * w / np.max(w) *pow2/q)

phi = np.mod(phi,2*np.pi)                 # wrapping the phase
spectral_phase = fftshift(phi)

# phi = np.cumsum(phi)

# Intensity shapping is possible, ex. by clipping
I_shapping = 1

# Phase mask (or filtering function)
H = np.exp(1j * phi) * I_shapping

# ================== After dispersion
Ew_disp = Ew * H                         # spectral domain
Et_disp = np.fft.ifft(Ew_disp)           # temporal domain

It = np.abs(Et_disp)**2                   # temporal pulse shape
Iw = np.abs(ifftshift(Ew_disp))**2  # spectral intesity

# ================== Transform angular frequency to wavelengths
lambda_nm = 2 * np.pi * c  / ifftshift(w) * 1000

# ===================== PLOTS ==============================
```

```python
fig_size = (5, 7)
fig = figure(1, figsize=fig_size)
ax11 = fig.add_subplot(211)
ax21 = fig.add_subplot(212)
ax321 = ax21.twinx()

ax11.plot(t, It, label=label, linewidth=2)
idx = np.abs((lambda_nm-800)).argmin()

# crop data around I_max
xcropA = idx-200
xcropB = idx+200
ax21.plot( lambda_nm[xcropA : xcropB ],
          Iw[xcropA : xcropB ]/ np.max(Iw[xcropA : xcropB ]),
          'r', linewidth=2)
ax321.plot(lambda_nm[xcropA : xcropB ],
          spectral_phase[xcropA : xcropB ],
          'tab:green', linewidth=2)

font = {'family': 'arial', 'weight': 'normal', 'size': 22}
ax11.set_xlabel(r'Time■(fs)',           fontdict=font, color = 'black')
ax11.set_ylabel(r'Intensity■(a.u.)', fontdict=font, labelpad=30)
ax21.set_xlabel(r'Wavelength■(nm)',   fontdict=font)
ax21.set_ylabel(r'Intensity■(a.u.)', fontdict=font,
                                      labelpad=30, color = 'r')
ax321.set_ylabel(r'Phase',fontdict=font, labelpad=30,
                         color = 'tab:green', rotation=270)
ax11.xaxis.set_tick_params(labelsize=14)
ax11.yaxis.set_tick_params(labelsize=14)
ax21.xaxis.set_tick_params(labelsize=14)
ax21.yaxis.set_tick_params(labelsize=14)
ax321.yaxis.set_tick_params(labelsize=14)

ax11.yaxis.set_major_locator(ticker.NullLocator())
ax21.yaxis.set_major_locator(ticker.NullLocator())
ax321.yaxis.set_major_locator(ticker.NullLocator())

ax11.set_xlim((-250, 250))
ax11.set_ylim((-0.01, ylim))
ax21.set_xlim((700, 900))
ax321.set_ylim((-20*np.pi, 20*np.pi))
text_kwargs = dict(ha='center', va='center',
                              fontweight= 'normal',fontsize=14,
                              color='black', fontfamily ='sans-serif')
ax11.text(0, ylim *0.9, label, **text_kwargs)
```

```python
ax11.tick_params(axis='y', labelcolor='blue')
ax21.tick_params(axis='y', labelcolor='red')
ax321.tick_params(axis='y', labelcolor='green')

tight_layout()  # avoid clipped labels

# ======== Save plot
if savePlot:
    savefig(filenamePNG , dpi=300)

show()


# ================================================================
#                             END
# ================================================================
```

7.13 Beam shaping from phase information

This script computes image reconstructed from phase to be applied on a spatial light modulator.

```python
# ================== Import Libraries
import numpy as np
from scipy.fft import fftshift , fft2
from pylab import figure , show, savefig , tight_layout , title
from PIL import Image

# =============== Data for output file ======================
# File name to save PNG
filenamePNG = '4_pulse_shaping_Rec_Image_Rect.png'
savePlot = False
# ---> Switch to True to save png file

# ================================================================
def import_img(file , size =(2**10, 2**10)):
    img = Image.open(file).convert('L')        # as grey level
    img = img.resize(size)                      # resize
    img_array = np.array(img).astype(np.float32)# int to float32
    img_array /= img_array.max() * 2 * np.pi    # norm. to 2pi
    return img_array
# ================================================================


# ================================================================
def img_rec(Phase):
```

```python
    Ampl = 1*np.exp(-20 * (X ** 2 + Y ** 2)) # gaussian beam
    Field = Ampl * np.exp(1j * Phase)         # Field
    IMG_rec = np.abs(fftshift(fft2(Field)))   # reconstruction
    return IMG_rec
# ================================================================


# =============== Compute grid ===============================
N = 2**8
x = np.linspace(-1, 1, N)
X, Y = np.meshgrid(x, x)
# =============== Compute Phase ===============================
# Phase = import_img("mask1.png", size = (N,N) ) # import file
# Phase = np.mod(np.arctan2(Y, X) , 2*np.pi)         # OAM
# Phase = np.mod(200*X+100*Y, 2*np.pi)               # Tilt
# Phase = np.mod(100 * np.exp(-5 * (X ** 2)), 2*np.pi) # line
# Phase = np.mod(100* np.exp(-5 * (X ** 2 + Y ** 2)), 2*np.pi)

# Rectangle shape
Phase  = np.mod(100 * np.exp(-5 * (X ** 2)), 2*np.pi)
Phase += np.mod(100 * np.exp(-5 * (Y ** 2)), 2*np.pi)

IMG_rec = img_rec(Phase)

# ==================== PLOTS ================================
fig_size = (6, 10)
fig = figure(1, figsize=fig_size)
ax11 = fig.add_subplot(211)
title("Phase", fontsize = 20)
ax21 = fig.add_subplot(212)
title("Image", fontsize = 20)
ax11.axis('off')
ax21.axis('off')

ax11.imshow(Phase   , cmap='gray')
ax21.imshow(IMG_rec , cmap='gray')

fig.tight_layout() # avoid clipped labels

# =============== Save plot ================================
if savePlot:
    savefig(filenamePNG , dpi=300)

show()
# ================================================================


#                          END
# ================================================================
```